A Road Leading to The Centre of The Universe

ISBN-13:
978-1537367439

ISBN-10:
1537367439

All rights are reserved.
No part, parts or the entirety of this book may be reproduced by publishing, electronically copied, duplicated by whatever means that form reproduction or duplication, without the prior written consent of the copy rite owner.

Author Peet (P.S.J.) Schutte

TO WHOM IT MAY CONCERN,

Dear Professor,
I am Petrus Stephanus Jacobus Schutte going by the name of Peet, the author of the above-mentioned book(s). I hope you find your reading of this book presented as an open letter a most fruitful experience. It is thought provoking because it strays widely from mainstream science but for that there is a very good reason. The proof is real and carries the arguments in a manner, which is by nature childlike simple to follow. Tracing the argument in detail helps to uncover incorrect principles. The accepting of such principles as incorrect depends on your inquisitive nature and your ability to consider dismissing your personal historian bias. Inspired by my belief that I have uncovered a mistake in mathematics led me to a new approach to cosmology. Admittedly the content of this letter is controversial. That is what my views have been blamed for in the past by those who are accepted and self proclaimed members of Mainstream Physics, and that I accept, but by such acceptance I admit that my line of thought is and extraordinary. Since the concepts I follow starts with the start, which, was named as the Big Bang, I start by tracing a new approach and add to the already existing theories. Every one accepts the Big Bang Beginning. In admitting that we then can proceed from that point without further arguing or debating about that part. In admitting to such a start we can conclude that with that comes an inconsistency not yet discovered. The official father of the Big Bang being accepted is a person by the name of Father LE MAÎTRE.

Father LE MAÎTRE, GEORGE ÉDOUARD (1894-1966) was a Belgian priest and cosmologist who was the first person to embrace the fact that the universe expanded from an infant stage. His model of an expanding Universe (1927) was superior to that of W. de Sitter in that it took into account mass, gravitation and the curvature of space. Similar models were proposed in the early 1920s by the Russian mathematician Alexander Alexandrovich Friedmann (1888-1925) but Friedman compiled various such possibilities. Lemaître argued further (1931) that the quantum theory supported an origin in the explosion of a 'primeval atom' or 'cosmic egg' into which was originally concentrated all mass and energy. As modified by A.S. Eddington, Lemaître's model provided the springboard for G. Gamow's Big Bang theory. In the wider picture of science in general a lot changed to just allow such turnabout in thought since the day of Isaac Newton. From Newton's attraction and contraction many things came into place that allowed changed in the most hardened minds. Accepting facts about the Big Bang concept is quite radical.

LE MAÎTRE'S UNIVERSE
A model of the universe containing a cosmological constant term, named after G.E. Lemaître. In this model, space has a positive curvature but expands forever. The Lemaître universe is both homogeneous and isotropic. The most interesting aspect of such a universe is that it undergoes a

so-called coasting phase in which the cosmic scale factor is roughly constant with time. This theory was an evolutionary process to act as an alternative term for the standard Big Bang theory. The word 'hot' was initially used to distinguish it from a rival theory, which had a cold initial phase. The existence of the cosmic background radiation requires that the Universe must have been hot in the past if the Big Bang picture is correct.

The Big Bang theory accounts for the expansion of the Universe; the existence of the cosmic background radiation; and the abundance of light nuclei such as helium, helium-3, deuterium, and lithium-7, which are predicted to have been formed about 1 second after the Big Bang when the temperature was 10^{10} K. The cosmic background radiation provides the most direct evidence that the Universe went through a hot, dense phase. In the Big Bang theory, the background radiation is accounted for by the fact that, for the firs million years or so (i.e. before the decoupling of matter and radiation), the Universe was filled with plasma that was opaque to radiation and therefore in thermal equilibrium with it. This phase is usually called the primordial fireball. When the Universe expanded and cooled to about 3 00 K it became transparent to radiation. The discovery of the microwave background in 1965 resolved a long-standing battle between the Big Bang and its then rival, the steady-state theory, which cannot explain the blackbody form of the microwave background. Ironically, the term Big Bang was initially intended, to be derogatory and was coined by F. Hoyle, one of the strongest advocates of the steady state.

BIG BANG CHRONOLOGY

Era	Time after Big Bang	Temperature
Planck era	0 to 10^{43} s	7 to 10^{34} K
radiation era[a]	10^{-43} s to 30 000 years	10^{34} to 10^4 K
matter era[b]	30 000 years to present	10^4 to 3 K

[a]The time from about 10^{-6} or 10^{-5} s to about 1 s or so is subdivided into the hydron and lepton eras.
[b]Includes the recombination epoch, which took place about 300 000 years after the Big Bang, at a temperature of about 3 000 K.

At such a point where the Universe came about in such a small confined space present during period the gravity must have been supreme by having the ability to confine the universe into a space the size of less than one atom. The preference choice of gravity is to reduce space. This statement too may see shocking at first to those pure at heart but a statement I do explore later nonetheless but ask the reader at this stage to accept. Gravity holds the sphere as the preference choice with out a doubt. It means that to find gravity we have to investigate the sphere. Gravity holds the strongest locality and holds the sphere as the strongest form of all forms available. This will become better understood and might then be less reluctant accepted when I indicate the value as well as the Location of singularity. In the sphere where space is the least and that will be in the centre of the circle gravity bonds all atoms together in a unit as well as distribute a specific alliance in shape and form. Gravity is the strongest in all cosmic structures holding the form of the sphere and is in the very centre where space is the least therefore the more any star produces gravity, the smaller the star gets as far as volumetric occupation goes. Take the Neutron star and the Black Hole as an example and compare that with the sun and the answer are self-proving. I claim that gravity is all about reducing space and not attracting matter but that I explain on another occasion. Therefore gravity must be where space is the least. Looking at a sphere we find that which holds the sphere true to form is in the centre of the sphere, that then has to be the most intense point of gravity. By confirming the round shape without favouring any specific point, one can see that the sphere as a form is dominated or controlled from one specific location in the centre. From the centre every part of the circle structure and all structural positions of the circle in all circles refer to the centre in perfect aligning. In the middle the sphere bonds all sides there is possible equal and that then has to be where the strongest gravity can be located. The Big Bang was where gravity held the Universe in the least space there ever was. To find the original gravity we therefore have to reduce the sphere to the circle and reduce the circle from there as far as one can go. In our reducing of the Universe we must first acknowledge that the universe constitutes many spheres, which is giving the Universe gravity as a combining unifying part of the sphere giving the sphere form (or gravity) the sphere being a circle in many positions from where gravity secures form. If we wish to go back in time and at the same time maintain some coherency we must concentrate on a single circle because a sphere is a circle by millions of possibilities linked together by a name that changes the concept. Going into our past we must take a circle that is representing the cosmos and reduce the circle to see what comes from there. That is what I did and in that I uncovered a hornets nest. The hornet sting depends on the individual persons view of what a sting is. The sphere is any circle and all circles have two parts. On part is a line that indicates the distance between the two opposing sides being 180^0 apart. The second factor is the pi indicating the result of the square value of the other factor being the straight line which by the square of the shape it indicates. But where as in the normal square the line matches angle touching lines that connects and in the square on will find the surface in the lines. With the circle the square falls inside the square but the lines cross at an angle and the edges hold the form value of the circle being Π. The circle holds a cross inside and a square never crosses the lines but the touch at angles. With the square we refer to the lines indicating the borders of the square by side we face as in example length and breadth. In the case of the circle we use the lines crossing to the inside on the length they represent being a full line or half a line. In the case of the full line the name used for such a line is a diameter and in the case where only half the inside line of the circle apply we gave it the name of a radius. Being fully aware of the various names I prefer to use r to indicate any and all lines that forms a combination to indicate either surface or volume.

To go back into the past one have to reduce the radius because the radius of the circle indicate the size while pi indicate form. By reducing the circle through the radius it becomes another matter of eliminating form because in such reducing of the radius it then becomes a matter of reducing a straight line. When reducing the circle in size one have to reduce the radius or the diameter because the pi factor is the indicator of the form as being a circle. It then becomes a process where it is just dividing the radius defined by the symbol r by halving the answer every time until there can be no dividing any further and such a reducing cannot end by becoming zero. One may divide by two, halving the result every time, which is a normal mathematical expression and by reducing by half can never reach zero. Allowing zero to be accomplished through any legal mathematical equation is not a mathematical fact and I challenge any person to prove such a feat. The numerical procedure may become tiresome or to small for any human to make sense of the outcome but never can such a dividing bring about zero in the ultimate answer. Zero cannot divide nor can zero multiply should I wish to retrace my steps to where I started. In using the method of the dividing by reducing the answer to half the size it will forever allow such a process to continue without ending in zero because no matter how small, the next value will be dividable by two in that will forever be a value in place. This stands as a mathematical fact and I do not have to prove my statement but those persons in academic positions, which portray me as a person trying to per sway facts to fit my convenience, must explain how one can multiply distances filled with zero as a concept. This alone is the biggest obstacle why there is so little exploring going about the space expansion present overall. With that remark I am just referring to the work of Hubble but trying to make all out sense of the concept in general.

The man that (to my humble opinion) took cosmology into a new dimension was HUBBLE, EDWIN POWELL (1899-1953), the American astronomer. He first studied nebulae, concluding in 1917 that the spiral-shaped ones (which we know as galaxies) were different in nature from diffuse nebulae, which he found to be gas clouds illuminated by stars. From 1923, using the 100-inch (2.5-m) telescope at Mount Wilson Observatory, he resolved the outer regions of the spiral nebulae M31 and M33 into star, identifying over 30 Cepheid variables in them. This proved that such 'nebulae' were truly independent star systems like our own – other galaxies. In 1925, he devised the so-called tuning-fork diagram of galaxies, dividing them into ellipticals, spirals, and barred spirals, which he believed to indicate an evolutionary sequence. By 1929, Hubble had good distance measurements for over twenty galaxies, including members of the Virgo Cluster. By comparing distances with their velocities, as revealed by the redshifts in their spectra, he concluded that galaxies were receding with speeds that increased with their distance, a relationship known as the Hubble law. This was powerful evidence that the Universe is expanding. The dynamics of his work was so far reaching everybody (including Einstein had to revise their theories to accommodate his findings. His findings are the most disputed, undisputed observations in all of history. The HUBBLE CLASSIFICATION is a widely used system for classifying galaxies according to their visual appearance, illustrated on the tuning-fork diagram. The sequence is based on three criteria: the relative sizes of the central bulge of stars and the flattened disk; the existence and character of spiral arms; and the resolution of the spiral arms and / or disk into stars and H II regions. The system was originated by E.P. Hubble.

The sequence starts with round elliptical galaxies (EO) showing no disks. Increasing flattening of a galaxy is indicated by a number which is calculated from 10 (a − b)/a, where, a, and b, are the major a minor axes as measured on the sky. No elliptical is known that is flatter than E7. Beyond this, a clear disk is apparent in the ventricular or SO galaxies. The classification then splits into two parallel sequences of disk galaxies showing spiral structure: ordinary spirals, S, and barred spirals, SB. The spiral types are subdivided into Sa, Sb, Sc, Sd (Sba, SBb, SBc, SBd for barred spirals). With each successive subdivision, the arms become less tightly wound (but more easily resolvable into stars and H II regions), and the central bulge becomes less dominant. Two types of irregular galaxy are defined. Irr I galaxies show rather amorphous, irregular structure with perhaps a hint of a spiral arm or bar, and can be placed at the far end of the spiral sequence. Irr II galaxies are sufficiently unusual to defy assignment to any of the other types, although this category encompasses only about 2% of bright or moderately bright galaxies in the nearby Universe. The original, erroneous idea that the sequence might be an evolutionary one led to the ellipticals refers to, as early-type galaxies, and the spirals and Irr I irregulars as late-type galaxies. Colour and amount of interstellar material vary systematically along the Hubble sequence: ellipticals are red and contain little interstellar gas or dust, whereas late spirals and Irr I galaxies are blue, with significant amounts of interstellar material. The relatively faint dwarf spheroidal galaxies were not recognized as a separate type in the Hubble classification. Some variants of the Hubble classification use plus and minus signs to subdivide classes, so that Sa^+ is later than Sa, but earlier than Sb^-. The importance of the HUBBLE CONSTANT is still to this day, underestimated. This "constant" is well explained, for the first time, I might add, in this book. The Symbol H_o is the figure that relates the speed of an object's recession in the expanding Universe to its distance in the Hubble law. It represents the current rate of expansion of the Universe. This important cosmological parameter is usually measured in units of kilometres per second per megaparsec. In the Big Bang theory, H_o varies with time and it is therefore more properly known as the Hubble parameter. Its value is not accurately known but is thought to lie between 50 and 100 km/s/Mpc, recent research tending to favour values towards the lower end of this range. In the HUBBLE DIAGRAM, a graph plots either the redshift, or velocity of recession of galaxies against their apparent magnitude or distance from us. The Hubble law appears in the form of a straight line on such a plot. The original diagram, presented by E.P. Hubble in 1929, was the first indication that the Universe is expanding. The Hubble diagram is mainly used to test the geometry of the Universe, since at large distances any departures from the simple linear form of the Hubble law should show up as a curve. The HUBBLE FLOW is the general outward motion of galaxies resulting from the uniform expansion of the Universe. All motions lie in a radial direction from the observer, and the velocities are proportional to the distance of the galaxies. The real pattern of galaxy motions is not exactly of this form, particularly close to us, because of the mutual gravitational interaction between galaxies; some nearby galaxies are even moving towards the Milky Way. At large distances, however, the discrepancies are small compared with the Hubble flow. All these findings are incorporated in the HUBBLE LAW, which is the mathematical equation of the principle law that governs the expansion of the Universe. According to the law, the apparent recession velocity of galaxies is proportional to their distance from the observer. In mathematical terms, $v = H_o r$, where v is the velocity, r the distance, and H_o the Hubble constant. The law was put forward in 1929 by E. P. Hubble.

The HUBBLE RADIUS is a distance defined as the ratio of the velocity of light, c, to the value of the Hubble constant, H_o, This gives the distance from the observer at which the recession velocity of a galaxy would equal the speed of light. Roughly speaking, the Hubble radius is the radius of the observable Universe. Depending on the precise value of the Hubble constant, the Hubble radius lies between 9 and 18 billion l.y. This data is the basis on which the age of the universe depends and is the HUBBLE TIME. The time required for the Universe to expand to its present size, assuming that the Hubble constant has remained unchanged since the Big Bang. It is defined as the reciprocal of the Hubble constant, $1/H_0$. Depending on the precise value of the Hubble constant, the Hubble time is between 9 and 18 billion years. In the standard Big Bang theory, the actual age of the Universe is always less than the Hubble time, because the expansion was faster in the past.

The Universe is a combination of many material formations holding positions in space. Some of such material was covered in the blanket of heat, distributing into more spacious surroundings as the material expanded from the centre flowing outwards. Hubble's constant is proof that the space between cosmic structures are departing from many centre positions between such objects and this is a trend being located between all the objects through out space but also indicating a definite growth in the radius and such radius growth follow a patter where the growth seems to flow from any such a centre point away from the centre. With out the absolute and undeniable proof coming from the Hubble constant bringing proof beyond any possible doubt in any one's mind that expanding is very much and a very big part of all Cosmic activity the accepting of the Big Bang would not be in place. $F = G (M_1 \times m_2)/r^2$ is in essence a big issue about contraction while Hubble showed the space was not dividing. The space was multiplying. The stars are growing apart and so is the galactica. This then brings in the question of space available.

With me not whishing to go into the formation of structures at this point in the book I would like to point to the fact that my following referring to the solar system is actually referring to a similar solar system that is somewhere and is now a part of a galactica we do not know about. I bring this in to disqualify any academic loophole that may come about from an argument about the solar system coming into place at a later stage of the cosmic developing and therefore the argument I am about to present that such an argument does not apply. To avoid such a loophole we now use a hypothetical but real solar system in space, which formed as the Big Bang took place. There are those who avoid admitting to inconsistencies by arguing that my argument about growth is invalid because the solar system was not in place at the Big Bang. To them we now present a solar system that is identical to the one we know in precise duplication thereof. But it represents a precise duplication of our solar system and was in place ever since the Big Bang. That means with the solar systems being apart in millions of kilometres there was a time the planets and the sun were apart by the measure of kilometres. The Big Bang shows a growth in space. Then there must have been a time when the planets were between fifty-nine and fife hundred and ninety kilometres away from the sun. How did the planets being the size they are at present fit into such a space and still be apart? Material too must be part of the growing. This line of arguing I suppose is much below the Academics pursuit of matters but since I am much lesser in mental standards of developing than they are, such reasoning prompted me to go on some investigating journey. Light journeys through out the cosmos and it will be sensible to follow lights travel.

With objects being apart at some distances and light flowing in straight lines between them it must take light a straight line to travel between cosmic objects. The distance the light has to cover depends on the radius there is between the objects and as such the Universe is then about structures claiming space and space setting objects apart being the radius standing between those objects. The objects are circles by dimensions and the space is also dimensions that are crossed by lines travelling through the dimension. With light being a line and the Big Bang coming from a situation that was a lot more cramped for space than at present, the correct path to follow if I wish to trace the steps of the Cosmos back to the Big Bang is to reduce the straight line between the structures and find where such a line will no longer be a line. The same procedure will apply to the material structures all being in a sphere form. A sphere is a lot of circles forming a unit but not repeating the space claimed the one another Such a circle also apply a straight line only known by another name but still serves the same purpose. Reducing the line will lead us to the beginning of time.

r /2 By dividing the radius r by the half of the value that then reduces r to a point where the left edge of the line reducing will be at the very same place the right hand edge of the line that is reducing will be. At one point the spots that formed the two ends of the line will be at the same spot. Any further dividing will land the left hand spot past the right hand spot in the opposing half where it then will grow once again but in the opposing direction. All possible dividing then ends on one spot where such a one spot shares a location with all other possible sides. The centre then physically is in the single dimension applying as one spot to share a location for all sides. At such a point there is no further dividing possible. On several occasions in the past I have been accused of manipulating the argument to produce none-existing or overrate facts. That is not the case. I am not manipulating facts to create an argument as so many accuse me of. What I am talking about is a mathematical fact that any one can prove by calculating following a very simple procedure. A child is capable of using the two times table and dividing by two every time is the most simple form that mathematics may be used. It is a mathematical fact that a line will reach a point where all sides are at one spot and as such cannot divide any more. At that point all sides share but all sides prevent zero becoming as factor since the sides share on spot. While the different sides are in one place the factor and value is one to all.

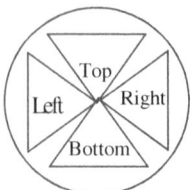  Reducing the radius r from all angles possible throughout the circle will bring about that all possible direction will eventually land on the very same spot with no more dividing possible. Yet zero cannot be a factor since the sides still hold value. A point arrive where more reducing will land the one side on the opposite side of the line but it will not bring about zero in the equation

What this argument further proves is that the circle reducing must then come from all points because the radius might be a line but that line represents a circle through 360^0. Taking that into account it is important to recognise that notwithstanding the size of a line, which any radius of any size is there is another line (or dot) eternally bigger as well as eternally smaller than the line in question. While we are in the third dimension being part of the third dimension then allows that all parts of the third dimension forever can be divided once more until the line in the third dimension is no longer part of the third dimension. When such a line leaves the third dimension it is still dividable because it might not be part of our dimension any more but it can still reduce further as part of the second dimension. By that time it has left our scope by miles does not mean that it end there because from our perspective that is where it ends. Yet it can still reduce infinitely more until it has left the second dimension and then at last forms part of the first dimension. Only then when the line reaches the firs t dimension no further dividing of that line is longer possible. We can never grasp the size of a line that forms the utmost or the least of possibilities and therefore size belongs to the human mind forming conceptions of big and small, but it has no place in the cosmos at large. This concept not only applies to size, but to all limits and divides we wish to create forming borders that we can appreciate. When looking at the circle in the conventional manner, we persist with errors brought about in culture and not by applying some significant modern logic. Take a circle and reduce such a circle constantly to where it no longer can reduce. Reduce it to a point where only form remains part of the circle because the radius has gone beyond human measure and becomes so small it is not noticeable with what ever tools man may use, then what remains is pi since pi does not indicate size but indicate form, and form is all that then will remain. In any circle or sphere the size only depend on the fluctuation of r, as a component to the circle or sphere but that does not affect the form by indication of Π in any way there may be. The conclusion I drew from following this process is that from this line can start at zero because that will be a mathematical impossibility since no line can ever reduce to zero. A line will forever be able to reduce further becoming smaller but it can never reach zero because zero is not on the scale of lines. If a line cannot reduce to zero it then cannot start at zero. A line or spot starting at zero would therefore be shorter than the shortest line possible. For obvious reasons can no line, or any line grow or extend from zero because such a line must then quit zero and become something, thus abandon its original value. That would mean the start of the line has a different value to the end and a line holds conformity through out. When any line is starting from point zero it can never leave zero because of the influence of being zero disqualifies any possibility of growth. If the line then had to grow in all directions at the same pace the line must then become a circle or being three-dimensional, then form a multi circle we named a sphere. Since the Universe is about circles and lines connecting circles I came to conclude that flowing from this fact is that in the universe there can be no zero improvising as a filling ingredient for the space of a point or be unfilled space. In the case of the growing sphere the value of the circle is Π, and that is where creation must have started. That gave me the clue where to start looking for singularity. One would find singularity in the value Π and the value Π will be in all things rotating in a circle. As usual I am again shooting the gun before the hunt started. Lines in mathematics do not start from zero and that is no discovery on my part that was a realisation I came to.

UNIVERSE

Everything that exists, including space, time, and matter. The study of the Universe is known as cosmology. Cosmologists distinguish between the Universe with a capital 'U', meaning the cosmos and all its contents, and universe with a small 'u' which is usually a mathematical model derived from some physical theory. The real Universe consists mostly of apparently empty space, with matter concentrated into galaxies consisting of stars and gas. The Universe is expanding, so the space between galaxies is gradually stretching, causing a cosmological redshift in the light from distant objects. There is growing evidence that space may be filled with unseen dark matter that may have many times the total mass of the visible galaxies. The most favoured concept of the origin of the Universe is the Big Bang theory, according to which the Universe came into being in a hot, dense fireball about 10-20 billion years ago.

UNIVERSAL TIME (UT)

A worldwide standard time-scale, the same as Greenwich Mean Time. Universal Time is the mean solar time on the meridian of Greenwich. It is defined as the Greenwich hour angle of the mean sun plus 12 hours, so that the day begins at midnight rather than noon. It is closely linked to Greenwich Mean Sidereal Time (GMST), since the mean sidereal day is a precisely known fraction of the mean solar day. In practice, UT is determined by a formula from GMST, which in turn is derived directly from such observations of the meridian transits of stars. The version of UT derived directly form such observations is designated UTO, which is slightly dependent on the observing site. When UTO is corrected for the variation in longitude due to the Chandler wobble, a version of Universal Time, UT1, is derived which has genuine worldwide application. When UT1 is compared with International Atomic Time (TAI), it is found to be losing approximately a second a year against TAI. Broadcast time signals use the time-scale known as Coordinated Universal time (UTC). This is TAI with an offset of a whole number of seconds. The offset is adjusted when necessary by the introduction of a leap second, and UTC is always kept within 0.9 s of UT1. On this issue there is much more to explore than the meagrely mentioned. Time stands related to the position an object holds to a centre such an object refers too while in rotation. Kepler found for instance that T^2, which holds the orbit to a rotation specific, is directly dependent on **k** to value the space a^3.

Einstein proved that in the presence of a strong gravity time slows down. Surprisingly with that evidence being around this long nobody since then in science took those statements and made any further progress from there. It was left in some drawer to dry. Science still sticks to the opinion that time did not change slightly since the beginning of the time and holds the same pace ever since. With the entire Universe including all the gravity now present and not excluding one Black hole or dust speck pressed in an area possibly the size of a lepton the gravity extending from that must have been beyond what words can ever describe. If the gravity was that high and Einstein already proved gravity slows time down, then there is one logical conclusion and that is that time was n fact standing still. Mathematically it is incorrect to allow gravity to compress the Universe into a spot smaller that an atom and exclude any other factors and relevancies to change. But before coming to the mathematics I would first like to bring your attention to the practical side. I am promoting a theory in which I am able to prove there is as much contraction (moving in the direction of the Big Crunch) taking the cosmic universe back to the size it had during the Big Bang as there is expansion (moving apart by Hubble's Constant) and the contraction is as much part of the expansion. By contracting the Universe is expanding and everything is based on gravity providing both actions. The universe rides on a balance and we have to locate such a balance. To prove my theory I firstly had to locate the centre of the universe. Even admitting to such a notion sounds like madness or in the least a tasteless joke, but please give me a chance to explain in more detail. I realised that my effort to locate the point holding singularity only stood any chance of success if the reducing of the line enabled me to backtrack the exploding universe to its origins. By applying some basic effort I have located the position from where all movement came and the direction it took moving forward in time…and yes, while I were also doing the finding the centre of the Universe I even located time as such. There are two standard formulas used to calculate a circle. The one use an r to indicate the radius and the other use a D to indicate the diameter, which is double the radius and therefore needs to be divided by a four to eliminate the Newtonian inverse square law amounting to the difference there will be between the two. The one using the radius is Πr^2 and the other formula using the diameter is $\Pi D^2 / 4$.

Thee beginning of my involvement with cosmology were brought about by my personal arriving at my understanding of my future theory started by trying to understand Einstein's view on light in motion. When light depart in opposing directions from one joint point and the light departed will travel in a straight line 180^0 in direction to each other they are all still relevant.

After travelling for one year in opposite directions the light will be one year from the point of origin. Under the normal circumstances the light will be two light years apart. That means if one could stop the light travelling to the left and have that light standing dead still, while the light flowing to the right can make a complete turnabout, it will take the light coming back one year to reach the point of origin once more. It will take the light one more year to reach the other point that was then standing still for one year.

Einstein proved that the normal is not the case with light. Light travelling in opposing directions for one year will be one year from the source it came and the light will be one year apart. But Einstein's claim that this comes about because the light is equal to time did not make sense to me. If the light was time, then time was no factor to light. In that case light will travel through space in a ratio of one meaning the moment it releases from the source it is on the other side of space notwithstanding the distance the space has. That is not the case because light is just a simple speed ratio like any cart or aircraft or spaceship. Light was distance during time duration and that comes down to being pretty fast, but it still remains speed and speed has any other relation to time than placing time as one to light.

After travelling for one year light had a distance of C multiplied by the seconds in one year to each side of the source but the points of light travelling independent as light had that same distance being apart of the two points the light occupies at that moment. The two markers are just as far apart from each other as they are apart from the original starting point. With this in mind the use of C^2 by science might prove convenient, but it also proves with this mockery how big farce such innovative calculation is. There is no chance of anything going at C2 because there is no exceeding of C by light as such. If that were the case light would not be present in the explosion or antigravity. The speed of light is not a force it is a speed. It is a ratio putting space in relation to the time density the speed will establish. This whole argument pointed to Kepler holding 6tthe straight line in relevance to space and time.

From that point I concluded that the link must be the value of a straight line sharing a dimensional value with a half circle and the triangle. If we look at the line supposedly travelling straight we find that the straight flowing is equal to the square relating the triangle. This is completely Kepler indicating gravity being $a^3 = T^2 k$ Look at the dimension and not the number implicated. It is $^{1 + 2 + 3}$ and transfers that to the line being the ^{1} and the ^{2} being the square being equal to the triangle as ^{3}. But it diverts very much from normal mathematics and that is precisely what Kepler's formula also does. With Kepler $a^3 = T^2 k$ and with mathematics the volumetric size of space must either be according to the measure of normal mathematics if it is a cube then $a^3 = L \times B \times H$ and in the case of a sphere the measure will be $a^3 \ 4/3\Pi r^3$. This was a triangle in relation to the square we find in the half circle standing again related to the half circle. It is not standard mathematics and anyone drawing links between mathematics and the speed of light has no idea about what is involves. With that I have again antagonised millions of the most important people with which I have to share a view. I do share their view on the Cosmos but not with their view holding mathematics as a standard fit all and apply anywhere in the cosmos. It is about lines carrying dimensional properties and with that we have to consider the line once again.

Let us find the smallest possible line first. Reducing the line will eventually leave all sides on the same spot. Such a spot must be round in form. The line being the smallest line will start off as a dot. A line so small it has reached a point not dividable any more will have all sides literally on the preside same spot, and I have located singularity in just such a spot. I came to the conclusion that the spot I found had to be singularity purely on the grounds that that spot holds only one side to serve as a start to the starting point of all directions possible. There in that side is only one spot is only one side applicable and one dimension present. With all the factors given one can only come to one conclusion and that is that there can be only singularity. In such a case more dividing by two will land further positions on the other side of the divide. That point serving as a position for all point and cannot allow further dividing is the smallest line or spot there may ever be. This spot is the result of a most basic process of reduction as the Hubble constant is a most basic process of doubling up during a matter of time. By reducing the line constantly the only value that will eventually remain without dispute from any party arguing about the facts is Π. By only having Π and a radius as one square (the radius effectively becomes one holding any and all sides on one point) of any significant measure as the radius it will be an evenly spaced dot. From the smallest ever possible dot will grow a line in every imaginable direction relating to a prospect of Π not favouring one direction that puts all directions at equilibrium meaning that any form of what ever might develop from such a spot will have the end and the start being in the same position, which will also have to be a sphere as the flow outward will be equal in all directions. Please think clearly, is that not precisely the commitment we find in gravity, where gravity is flowing from singularity outwards but never favouring any side? This reasoning prompted me to look for singularity in such a spot because if the prime spot from which all came was a spot holding all, then the spot must hold the shortest line but more prominent it will hold the smallest form including the smallest circle or for that matter the smallest sphere. With gravity always being in the centre of a sphere where the space is least available in the entire structure (there is not even space left to fill) one finds a flow of gravity from that centre spot outwards in all possible direction even-handedly. The fact that the original gravity will begin as a circle or will be a circle is the direction it will take when being the first spot created. All progress will be evenly in all direction because no direction will stand out or be in favour above any other direction at first.

The spot forms a full circle, but the line running through the circle is forever present because that is the future radius of the circle that will one day develop the circle, which is equal to the present diameter. The fact of the presence of such a possible line in such a possible circle dividing the possible circle into two parts makes the centre line equal to the half circle. The line forms the half circle but not only that the line presents the half circle as much as the line is the half circle. The line then is 180^0 and the half circle is 180^0 because in singularity the two factors are the same. The same value is of course $\Pi^0 = 1$.

In this half circle of the future, which is no half circle as yet because of a lack of space there are three future points indicating the space less ness that will go on to become space filled with something. On top of such a circle to form must be a marker indicating an awaiting boundary or future border and at the bottom of the future circle there also must be a similar marker that is no marker as yet. Between the two possible points that are not there yet is a future line running that is not there yet. Then indicating the possibility of a position to come that will bring about the half circle being a future distance apart from the future line indicating a diameter that will one day be there a third such a marker must be established for the future. That forms a triangle with two more sides being connected by either a line being one or half pi being one. From singularity comes about that the line is the same as the half circle is the same as the triangle and all has one value being 180^0.

From this come the most basic principles in as much as forming the ground rules of the law of Pythagoras.

When drawing a line such a line then starts of with a dot serving the spot that holds all sides equal. That means the line serving as the future radius will be equal to the half circle which is then Π. The only aspect of the point that stands in for the end of the single line forming the radius of the circle is that we then mathematically reach the single dimension. We decreased the line to where a circle being Π formed on the single dimension. This dimension also hold the circle dividing line because from there the radius must once again generate a value and by such a gesture that the extending would form the circle that forms the sphere that eventually lead to the formation of particles. This leaves a problem to investigate.

With no line possible there had to be another dot that formed since the Universe has many dots that formed lines. But let us not to get confused and lost in the range of possible diversions but let us stick to two dots. One dot was next to the dot next to the dot, but as I said we stick to one dot next too the second dot. M X ● M / r^2 ● is the first step gravity began with. That leaves us with a huge problem in as much as when r = 0 then r^0 = 0 and 0 dividing any value will leave 0 as the answer. If the particles were inseparable at the start it must bring about that gravity would not be forming since the distance will not permit any dividing. By allowing the distance separating the particles to be zero, the particles melt into a unit. Again this is Mathematics and not my incoherency as some Academics dismissed my work. Let me run through the argument one more time because I have been insulted by Academics in the past telling me I am bending mathematic rules with my applying double values to try and produce some argument. The two particles formed by an inseparable unit separated by a sharing of a spot. We know that at least two spots formed because there are many more than just two that remained to become part of the visual Universe. Let us name the spots because that is what humans do best if they do not know what to do with what they have to do. Let us call the on em and the other one spot next to em we then call emtoo. Between em and emtoo there were nothing because em and emtoo were inseparable. By they're being inseparable we would naturally be inclined to think that the separation value should be nothing or at least zero. But putting zero in that place is a mathematical excluding procedure leaving future mathematics excluded. With m multiplying m_2 and then dividing ÷ r with zero (r=0) such a procedure will leave the lot at zero and with that nothing is going nowhere. That means although we think the space between the two parts are nothing the non-existing space has to be at least one to be a future factor.

Every part of the argument is sound but was never yet used. I repeat once more if my argument reflects on inconsistencies those inconsistencies are not about my work. In order to disprove my argument replace Mass one and Mass two with any number possible, then divide such a number with to the square being zero. If there was no space then the value of the particles had to be one. If there was no space between the particles the particles then had to form a unit. But if there is a mathematical possibility of reducing a line to the single dimension then there had to be a factor representing r as a factor of one. Take $(M_1 X m_2) / r^2$ and substitute any of the factors with zero and the result coming about has to be zero. The factors in the equation have to have any and all the elements at a value of at least one. Only if r was a factor of one can gravity bring about any mathematical equation developing from this argument. That means the mass on both sides must have a factor of one being a limit, which does not allow such further reduction of r and any further reducing of r beyond the limit will not be tolerated. Only if r = 1 then r^2 can be 1 and mass can be apart. Like it or not but believing in the Big Bang must also bring about the accepting that the cosmos moved apart somewhat. The fact that r brought increase in the space separating the mass produces a problem that was solved already. About a century and a half ago Roche found just such a limit. Once again I were confronted by zero becoming growth. There is a huge hole that needs filling when bringing into a relation any forming of an alliance between a cosmos coming from nothing and filling with nothing and a cosmos growing spontaneously through balance shifting prominence. Mathematically the fact of applying nothing as a vale applying in the cosmos is not a strong and convincing argument. The minute one brings in zero as a multiplying factor forming a definite value working into the calculations of the cosmos, growth disappear. If growth was not a factor, the zero factors could be involved with some form of maintaining stability and where then further growth will accept the responsibility of zero

The Roche limit is:

The region surrounding each star in a binary system, within which any material is gravitationally bound to that particular star. The boundary of the Roche lobes is an equipotential surface, and the lobes touch at the inner Lagrangian point, L_1, through which mass transfer may occur if one of the components expands to fill its lobe. It names after the French mathematician Edouard Albert Roche (1820-83).

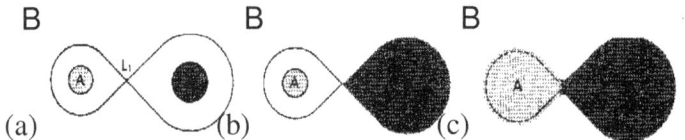

THE ROCHE LOBE: In a binary system, the Roche lobes of components A and B meet at the L_1 Lagrangian point. (a) In a detached system, neither star fills its Roche lobe. (b) In a semidetached system, one massive component, B, fills its Roche lobe. (c) In a contact binary, both components overfill their Roche lobes and share a common envelope.

The closest encounter worth noting we ever had with this law in the modern age of news and Television was the Shoemaker-Levy 9 incident during the previous decade. At the time and even in the present no one drew any similarities but after completing this book the reader should find why I could draw such similarities, which there is between this incident and the Roche limit. Even the phenomenon called the Sound Barrier became clear when applying the Roche factor with the laws governing the influence of singularity.

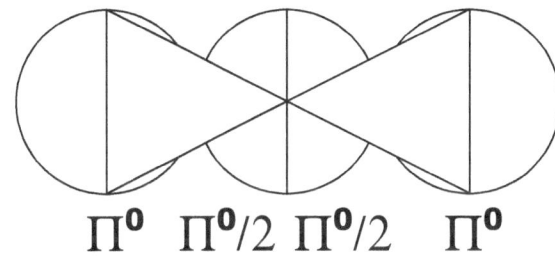

At the very first sign of any of the sides departing from the centre shared by all, all other points must also show signs of a willingness to depart. There will be one point where r still is one coming in as a factor but pi moves out from only being a factor of Π^0 = **1** and at that point pi will become a full factor of Π.

$\Pi^0 \quad \Pi^0/2 \quad \Pi^0/2 \quad \Pi^0$

But keeping Π as one (Π^0 = **1**) we keep the Universe in the first dimension.

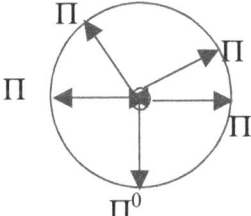

This point, which I now am referring to, is the point where Π a fully appreciated value while the diameter D still remains a dimensional factor of one. His is the dawn of the second dimension where space was there but space was sparsely shared in some cases. It is when Π^0 shifted to become Π for the very fist time.

The point without movement, the point holding singularity must have a value of Π being the eternal dot but since the dot has no dimension in having form the Π that indicates the dot must be Π^0. From such a point there has to be to the side of the centre point be a point where space do start. That point will then receive a diameter but that point will have form only in being a circle. In that point there is a shift from in relevance from Π to the centre Π^0 and for the first time it brought about two separate values for Π.

A   B$_1$ B$_2$ Extending into the distance

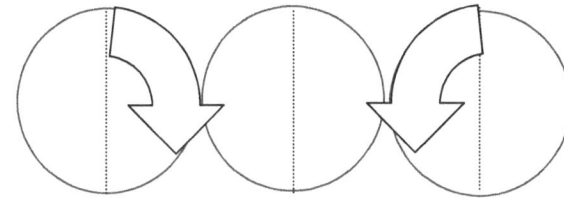

We have established the fact that em and emtoo was divided by r and then r had to be one since r could not be zero. Such a centre would then carry the same value as em and emtoo. That means whatever value em and emtoo receives has to go in equal measure to r with em sharing half of the divide and emtoo sharing the other half of the divide.

$$\Pi \quad (\Pi/2 \times \Pi/2) \quad \Pi$$

Because the three points existed on equal terms in singularity sharing a same spot the coming out of singularity will enforce that equal value comes to all. That means the circle gets to become Π, the diameter becomes Π and the distance setting the structures apart will also become Π. This is what the coming from one point brings along. Only when being part of the second dimension can there start being separate values.

While the form was still being in the single dimension from the one side of the form the dots had to establish identities apart but not separated yet. The one circle had a factor of $\Pi^0 = 1$ and the centre had to have a value of ($\Pi^0 / 2$) extending past the very next object but also cutting such an object into a square double half value that was going to come about as soon as the other dimensions came into form. In the relation at present em is extending toward emtoo by means of establishing a valid r and emtoo is establishing a valid extension to em be using r and this leads to two valid values for r being ((7+7) /10) and (10/7). The values I give here I shall explain later.

The only definite place one will locate zero is in between the starting point of the lines going in opposing direction in the position the lines hold before there was the least of directions applied, but that is only because there is no such a position, not because any line is coming from there. As I have indicated and positioned em and emtoo the two points may share a position but separation is forever a possibility and for that reason if there is no other reason we then have to put a dividing possibility at a value of one ($r^0 = 1$). The two lines are still one holding the opportunity of parting as an option but have not yet parted and therefore are on the very precise same spot. Being on the same spot does not mean being inseparable or being the same. It only means sharing a spot. The line coming from there is already there because it already has the choice of going in any and all opposing directions and when it starts running it will place filled space in that location not yet present but also holding a factor of one since it will become filled in the future. This is because the space at present is filled with a line. Where this space is now already filled with a line The line had to have a start. The starting became the line running and by running the line is filling space. That means with the line there it filled the possibility that a line could form and not with a line not being in place at all. It is again taking the r separating em and emtoo on its factor value of one and not our human visual accepting value of zero. One may not discard any future possibilities of growth by giving those possibilities a value of zero. A line might form or space may form where the line later may form. We humans tend to dish out a value of zero where ever we do not visually are able to find a value at that precise second and in doing that we also place such a value as a running obstacle into the future. But our habit of doing that is proving to be a human shortfall because with our shortsightedness we think of the here and the now in excluding possibilities while we should think of the future by including possibilities. By disregarding a positional value as zero we exclude such a position from ever being possible. By giving it a factor of one we include such a position as a future possibility. When reversing a line we might find a better idea of what is in place and where it is in place. Gravity is officially a force without limits going past and through borders and has an unlimited reach. It seems to remain even and this is conflicting with the flow of perceptions about mathematics. In as much as showing that r serving in a factor value, as one has to form a limit where em and emtoo than that of r and aligning the discovered singularity produces the Roche limit as such a dominant factor in the cosmos. With my retracing a simple line helped me to find an explanation about the Roche limit, a feat not yet done in science. The Roche factor is next to singularity the second most basic foundation in cosmology and is the starting point where singularity spawned into dimensions. As it is fundamental in all cosmic development and with that it denounces the gravity principle introduced by Newton as $F = G (M. m) / r^2$

The formula $F = G (M_1.m_2)/ r^2$ is unable to explain the principle discovered by Titius and later by Bode and in contrary to all statements to that effect made by Accepted Science policy makers the Titius Bode principle is not coincidental. In fact it is one of the four most adhered and important cosmic pillars holding the cosmos structural in place. From the two examples mentioned above comes gravity. In past few pages I proved how one could arrive at the facts that prove how the Titius Bode Principle leads us in the direction the origins of the solar system. But before we can accept the influence of the Titius Bode Principle we have to deal with "Nothing" and as such dismiss nothing from science. "Nothing" in the universe is coincidental; "nothing" in the universe does not apply. Where mathematics meat lines nothing disappears. Should any principle not match an accepted theory or change the accepted theory, the theory does not apply.

The content of my work holds a new view about Cosmology, which I have been working on for the past twenty-seven years and exclusively for the past six years. I always had a problem with the idea that space constituted of nothing, while I came to realise that lines mathematically couldn't start at zero because there is no evidence of zero as a factor in mathematics. Should you disagree with my statement the question in need of answering is this: What will the length of the shortest hypothetical line imaginable be and moreover, what would the total overall length be in that case? The shortest possible line (hypothetically) must be so short it must have an initial and ultimate point sharing the same spot. The two points must be one and only then can further reducing of any line not occur. If it used zero as a start, the zero part would not count, because the line will only start at a point past zero where the line then will start forming an infinitely small dot. I press this point in urging the understanding because there is such a point, but in my attempt to underline the fact, I have to per sway the reader to abolish four or five thousand years of accepted and practised mathematical culture and that is no easy feat. In applying the most basic method of taking the line back as far as possible brings a dot because of the equilibrium that will stem from such a position. The dot is in infinity, however small, it is not zero. Zero ultimately means not existing and then that point, as a start does not exist. The smallest line has a beginning and an end at the very same spot located in infinity, and infinity may be beyond human scope, though infinity is still not zero. Infinity may constitute of something we do not yet understand, but we may not define our human misunderstanding because infinity is not present in our minds and therefore by not sensing a value we disregard such a value as nothing whereas it is visibility nothing but in being potentially there it is one. It is the same as a person hearing a dog bark and investigate. When not sensing what the dog was barking at, the person turns around and disregards the barking as about nothing. The dog's reaction was not the indicator of the nothing the dog sensed something. The man's wits let him down and his wits produce the nothing. The dog will not bark about nothing because then the dog will not bark at all. The fact that the dog barked produces a possibility of something out there, which the dogs is getting annoyed about. The man's inability to detect what it is that the dog is sensing becomes the nothing but not the possibility of something worthwhile to investigate. We use nothing to avoid and not to valuate. In this aspect lies the difference there is between arithmetic and mathematical science where arithmetic can have position such as zero since arithmetic excludes the cosmos calculating numbers only. The nothing we see we made the nothing we find but the fact that distance is there, it is separating structures and by that is bringing in the factor of one which we cannot see. It is the way we try to disguise our inability to detect which produce the nothing we then use as a value, but still we substitute he nothing in applying arithmetic with the name as nothing and then the names used becomes the factor of one. Cosmology is not about numbers because no one can calculate the number of stars. Cosmology is all about lines and angles positioning objects, and in those there features no zero. No line can be zero long and forming a position of zero degrees in relation to another object.

A man may have that many oxen or so many sheep and even this amount of wives, (in Africa) or not have any therefore having then a total of nothing, but there cannot be nothing between the sun and its orbiting structures. The having and have-nots are part of arithmetic. Light will indicate a line flowing between the sun and whatever planet, following dot after dot thereby proving the existing of the possibility of something going about by a straight line, and any straight line in relation to other straight lines will be under the law of Pythagoras in as much as obeying the rules of trigonometry. There is no possibility of a straight line not forming in space. If there is space, there can be a straight line. The mere fact of two spots having different positions in space gives the two dots different values. If the line has the length of zero it is not present. If the triangle has one angle of zero it is no triangle because all other angles dismiss at the same time. Mathematics converts the values of integrating lines according to Pythagoras and arithmetic is about numbers to be added or subtracted. By mathematically excluding zero from cosmology a new universe opens to the human mind. With the distance between the sun and Pluto being roughly one hundred times more than the distance between Mercury and the sun, the distance must hold something more than pure vacuum filled with nothing except one atom hear and there occupying the vacuum between them and the sun. If space supposedly comprises of nothing how can nothing then become plural forming more or be multiplied by a number as to indicate a growth in something not even existing. As the one becomes one hundred the one cannot substitute a value of nothing but then must be part of something. If the one substituted the nothing, all laws of mathematics will go in disarray because when one multiply any number by zero it becomes zero placing both planets in the sun. If Pluto was one hundred times closer than it is at present was it one hundred times nothing closer? 100 X 0 + 0 = 0. That is mathematics!

By allowing the three hundred a value the nothing must form one making that which is between Pluto and the sun not to be nothing but there has to be something. This argument follows mathematics to the letter and in precise detail. With Pluto and the sun being apart that being apart has to have one of something in place forming the being apart from each other's cosmic position one time multiplied by the many ones we find in that space standing relative to other space regarding whatever the space becomes what is between the sun and Pluto. That factor cannot stand in for not one, which is the same as nothing as that is because one cannot take the place in the position that zero secures. By excluding nothing from the equation space becomes something bringing in a value lying inside the realms of the infinite that must form singularity. As the zero becomes a dot, something else becomes clear about the dot. Looking at the night sky we find darkness overwhelming the space in relation to the stars bringing across light.

My approach to cosmology shall prove to be somewhat unconventional but through the abandoning of the accepted, it enabled me in locating the precise location of singularity that forms the connecting basis of the universe (and this I say with some degree of confidence). There **are two locations** but I shall **first concentrate** my explaining effort on **the prime singularity**. Singularity did not vanish into the unknown after the completion of the Big Bang development but is in a place science incorrectly valued and classified incorrectly and in that, there is something hiding the truth. If singularity was or is where the beginning is we have to go back and see just where such a beginning was. I cannot accept that the Universe started at zero and neither does anything else in the Universe start at zero. My excluding the possibility of zero includes that the Universe is not filled to the top with nothing and neither is nothing part of outer space. The universe is about lines allowing light to flow from one point to another point and in following that line it has to continue in the line as the line has to represent something. The Universe is all in relation about lines indicating distances between cosmic structures. The cosmos is in short about lines connecting points in space being apart. It is about a line starting and continuing from such a start. But science advocates their opinion that such a start of a line flowing between any and all objects can hold zero because according to them the Universe are full of nothing. If the Universe in as much as outer space is a container filled with nothing at the present moment, and there is no place anything that was part of outer space previously could release to and there was no emptying of what ever filled it before, then it could not get rid of what was in the outer space when it started with what it started off with. We must then accept from what is not in the Universe was not in the Universe at the time during the start that at is present at present according to science because it then still must contain the same nothing and must have that same filling from the start to the present. If it was nothing it still must be nothing and that same substance being nothing is what it also used to grow using it as it grew because it filled outer space with nothing growing from and growing to nothing. Is that true? The filling of the Universe could not go anywhere so one has to presume it started off from nothing and from there it kept filling with nothing since what ever was in the Universe at the start had no place to escape to or no place through which to escape. That is only applying if it is nothing filling the Universe at large. Can nothing grow as much as a line is growing from a start of nothing? The answer is that such lines not only indicate a distance but since the Universe came from such a small space as science propagate with the theory of the Big Bang then all particles in the Big Bang Universe were rather cramped for space when the Universe started from that small line between particles and is now the same line but is now so big. In the past it seemed being so small and showing the space between particles to be awfully short at one time. It was short but how short was it? Did it start off as nothing? Is the line starting at nothing as science wishes us to believe? If it does then all lines must start from nothing so we better investigate this trend with the start of a line.

In this following I show my argument with which I hope to prove the counter part of what science believes. I am about to prove that which science sees as nothing in space and in material is the very location of singularity.

Lines mathematically cannot start at zero because there is no evidence of zero as a factor in mathematics. Should you disagree with my statement the question in need of answering is this: **What will the length of the shortest hypothetical line imaginable be and moreover, what would the total overall length be in that case?**

Locating zero

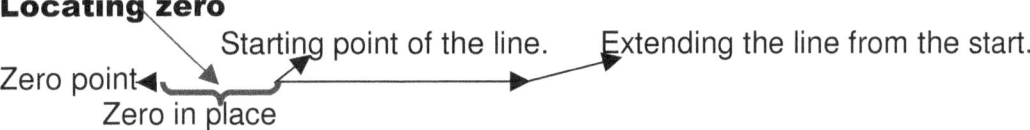

Let us duly test my statement by taking the line back as far as possible. The shortest possible line (hypothetically) must be so short it must have **an initial and ultimate point sharing the same spot.** The line that **cannot reduce** any **further** must be **so short** that **directions flowing away** from each other **are located** in the **same position**. Any theoretical line being the shortest possible line cannot have the line holding the initial starting point at point zero and advance from there. Mathematics simply will not allow it. If the point had zero as all it had to offer, such a point is not present. Mathematics is not about numbers. Mathematics is about sizes, positions, and dimensions. Expressing three dimensions will be x^{1+1+1}. We tend to look at the dimension and not the number. We ignore the indicator as being the natural and only take the dimension the indicator points to in account. We look at 1^0 and find the zero in space. In space there is space. In space there is $1^0 \times 149 \times 10^6$ km of space between the Earth and the sun but there cannot be $149 \times 10^6 \times 0$ km between the Earth and the sun. The zero means there is no such a position. If it used zero as a start, the zero part would not count, because the line will only start at a point past zero where the line then will start. Zero ultimately means not existing and then that point, as a start does not exist and where the line then stars is a point in existing. When the line **has a beginning and an end at the very same spot** and it wishes to extend the position as to further the possibility it has, which direction should it favour. Extending the line in any one direction will favour one direction without any clear reason not extending in other directions. The fact of direction being present only proves and is proved by another point established, which is placed in relevance to such a second pointing a position already established by the relevancy of two point located in a direction to one another. But if one point starts one line there is no favour of direction since there is no established direction yet. The only mathematically sensible option about extending any line starting at a pre-designated point without any other point to establish a pre determined direction will be non-bias progress in all directions equally in order to give a meaningful flow of mathematical equilibrium. Not one direction stands superior to other directions and all directions are equal with no bias anywhere. Of this statement the Pythagoras mathematical principle is proof of and that I explain later.

Let us dissect nothing, as we find nothing in the presence of the cosmos. The distance between the sun and Pluto is roughly one hundred times more and if the distance between Mercury and the sun, but both has nothing between them and the sun. The space filling the distance from the sun to Mercury has nothing more than the space between Pluto and the sun. That means the distance between the sun and Pluto is as equal in relevancy than the distance from the sun and Pluto since both is the measure of nothing. If the one substituted the nothing, all laws of mathematics will go in disarray because when one multiply any number by zero it becomes zero placing both planets in the sun. The distance between the sun and Pluto **is Pluto is 5900×10^6** kilometres of space, but in that statement we take it that the one of a kilometre is present in such a multiplication. The one constitutes the presence of fact being a statement of a value. By saying the distance constitutes of nothing we have to substitute the one factor with a factor of zero. Then the calculation must read **Pluto is $5900 \times 10^6 \times 0 = 0$.** Including nothing as to state the presence of that part contained by the calculation delivers the total of zero. By excluding nothing from the equation space becomes something bringing in a value lying inside the realms of the infinite that must form singularity. Applying this logic to the Lagrangian system and interpreting that information to the law of Pythagoras a clear pattern come about.

The reaction responding from my argument is that it is silly, but should that be your personal opinion too then test where the silly part applies. Bring the zero into the calculation, the zero that science so eagerly place in outer space and see the mathematical result. By applying the distance one accepts automatically that the figure become calculated with one as it represent one in being a calculating part of the cosmos. The calculation as all calculations normally are is in order to calculate something and the something will at least stand in as one in relation to the rest being part of the calculation. But saying that the factor of one in fact represents nothing since nothing is so much the part in the calculation being calculated, then the zero has to replace the one as the fact of being calculated.

The claim becomes obvious when observing the connection between the half circle, the straight line and the triangle, which could also promote all the qualities lurking behind the pyramid. Consider the connection between 180^0 sharing three different forms all part of mathematics where each is different in form, but equal in value and then one may realise in considering the very basic in mathematics being the Law of Pythagoras on which all mathematics are focused. The triangle stands in for one factor represented by one at a value of 180^0. So does the straight line become a factor of one and the half circle also becomes one where the factor of one equals all 180^0. All three are most seriously part of shapes in the cosmos. Revalue any one form to zero and the rest too must follow and share the same value. The Law of Pythagoras is about angles in relation to lines and not one angle can represent zero because that will reduce all the lines also to zero. The measure of angles between stars at a distance uses parsec as the indicator, but the parsec between the stars indicating an angle has to represent an angle whereby one may measure distance and such a distance cannot be zero because then the parsec will be equal to zero. Again it is multiplying the factor with the measure but if the measure is about a factor of zero, then the factor too becomes zero. That is as basic mathematics as I can present.

If the argument seems ridiculous it is not my mentioning such a fact that is ridiculous but the mere fact of the reasoning also becoming a recognising of an argument accepted by science making it as such ridiculous. If space is nothing then it has a number to use indicating just that value being zero or the capitol O indicating zero. Try and indicate what is measured and calculated in space, but not by simply not thinking about the fact and therefore simply ignoring that what is measured forming the sole value of space, but put the value of nothing as part of the distance in calculation because that is what is measured. When stating the distance between the Earth and the sun place on paper what will allow the kilometres measured to represent the factor that is being measured. If represented by one being the total of one by hundred and forty nine million kilometres of nothing put that language in the International language of mathematics that spans all dialects spoken on Earth. Put it in mathematical terminology by saying there are 149 000 000 X 1 (multiplied by the kilometres) multiplied by what it is being measured which is 0 and what will the total come too... a full zero.

149 000 000 X 1 (km) x 0 (indicating what the km are made of) = 0

Mathematics says it. If there is something to be measured then the least value the measurement can have in relation to what is used in the measuring has to be one. It cannot be zero and be measured...and we do measure outer space! It sounds as if something here is at fault. It is not with my mentioning the inconsistency one should find fault but the fault is with the fact that it is there and no one noticed! I am not to blame just because I am mentioning it, but the blame must go where it belongs.

I think it is by now little understood although I imagine not nearly accepted that by adding a million of nothing to one nothing there will remain one nothing and that is still nothing. Nothing cannot accumulate therefore I cannot accept anything holding the vastness of space being able to constitute nothing as the major component.

Mercury has 58×10^6 km and Pluto is 5900×10^6 km space between the sun and the planet. That indicates a distance and a distance comprises of something, for if was nothing then both would have equal nothing and be next to the sun. I repeat, the distance indicates something because nothing would place them both in the sun and moreover in the centre of the sun. Having nothing between

Mercury and the sun and between Pluto and the sun places Mercury and Pluto at a same position within the sun. By saying Pluto has one hundred times more zero in between the sun and Pluto makes such a statement laughable. Except if a learned Professor conducting a class does it. If I would say Mercury holds one hundred times less nothing, such a statement will make me an idiot, but used as science makes such a statement plausible. That means the more zero or the more nothing one find between cosmic structures puts such structures further apart. There can only be distance with something concrete applying the distance between it. The problem is identifying something from nothing that defines the difference there is in science. I cannot see how nothing can become plural or more sometimes. Realising this I went in search of that which nothing is substituting. The issue I went in search of is what to substitute the nothing with and fill the nothing with that something. Let us go on the interesting search of finding what prevents the Universe from tumbling in on itself. If the Universe was truly nothing, the nothing would not support the structure and the structure would disappear into the nothing not supporting it. The Universe is about lines forming angles and holding distance, that much we established so far.

When reducing the circle in size one have to reduce the radius or the diameter because the pi is the indicator of the form as being a circle. Divide the r until there can be no dividing any further and that cannot in the end indicate zero because no matter how small, in that will forever be a value in place.

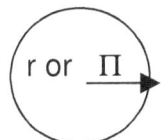
● r / 2 ● r / 2 ● r / 2 dividing r reduces r to infinity but not Π as Π remains stable, protected by the rotation of matter forming a circle around singularity.

Taking that into account it is important to recognise that notwithstanding the size of a line, there is another line (or dot) eternally bigger as well as eternally smaller than the line in question. We can never grasp the size of a line that forms the utmost or the least of possibilities and therefore size belongs to the human mind forming conceptions of big and small, but it has no place in the cosmos at large. This concept not only applies to size, but to all limits and divides we wish to create forming borders we can appreciate. When looking at the circle in the conventional manner, we persist with errors brought about in culture and not by applying some significant modern logic.

By reducing r indefinitely to the tune of half each time, r would become infinitely small, beyond human calculating means, however as mentioned in the case of the smallest dot holding one spot, r would become insignificant beyond human comprehension even, but never reaching zero and still Π would remain intact and dictating form. I believe one can begin too see where my suspicions are heading because the flaw comes about in the manner mathematics are practised for thousands of years. But before coming to the mathematics I would first like to bring your attention to the practical side. I am promoting a theory in which I am able to prove there is as much contraction going on in the cosmic universe as there is expansion and the contraction is as much part of the expansion. The universe rides on a balance and we have to locate such a balance. To prove my theory I firstly had to locate the centre of the universe. Even admitting to such a notion sounds like madness, but please give me a chance to explain in more detail. If I wish to achieve success that would depend on my ability to convince all that outer space comprises of material and as such we can locate such material even if we are unable to see such material.

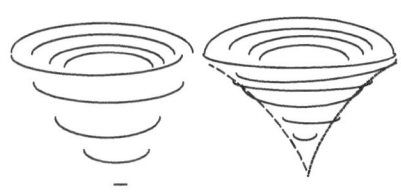
To find the invisible I had to locate singularity. I realised that my effort to locate the point holding singularity enabled me to backtrack the exploding universe to its origins.
By applying some basic effort I have located the position from where all movement came and the direction it took moving forward in time

The reversing of the circle radius is not alien to nature at all. An observation coming instinctively to mind one may recognise is that the form reminds rather explicitly of natural phenomenon as hurricanes, water whirls and even the shape most commonly favoured to express the cosmic object referred too as a Black Hole. The similarity may be more than coincidental. Let us consider the statement in the reverse. In our calculating of a circle we apply two formula methods. The one use an r to indicate the radius and the other use a D to indicate the diameter, which is double the radius and therefore needs to be divided by a four to eliminate the Newtonian inverse square law amounting to the difference there will be between the two. The one using the radius is Πr^2 and the other formula is using the diameter is $\Pi D^2 / 4$.

In any circle or sphere the size only depend on the fluctuation of r in the square as a component to the circle or sphere but that does not affect the form by indication of Π in any way there may be. The conclusion from this is that no line can start at zero because that will be a mathematical impossibility. This statement by itself excludes zero and with zero excluded one then begin to appreciate all the rest of the concepts governing corrected cosmology. A line or spot starting at zero would therefore be shorter than the shortest line possible. For obvious reasons can no line, or any line grow or extend from zero because such a line must then quit zero and become something, thus abandon its original value. That would mean the start of the line has a different value to the end and a line holds conformity through out. When any line is starting from point zero it can never leave zero because of the influence of being zero disqualifies any possibility of growth. If the line then had to grow in all directions at the same pace the line must therefore be a circle or being three-dimensional, a sphere. Flowing from this fact is that in the universe there can be no zero point or unfilled space. In the case of the growing sphere the value of the circle is Π, and that is where creation started. That gave me the clue where to start looking for singularity. One would find singularity in the value Π and the value Π will be in all things rotating in a circle. You might wonder how does that apply to the cosmos and moreover to gravity?

By accepting that there is some conductor (not the ether of old) between two cosmic structures can one accept that there are certain invisible undetectable influences on the edges outside the surrounding of material. One then can see how space conforms as it converts to liquid heat. In the same effort one can see how material confirms heat from space to material. By reducing and confining the heat drawn to the centre the space becomes more concentrated as the heat levels begins to rise. Take any bicycle pump and compress the plunger and the result will be that the heat created by such action will burn the finger you use to cover the valve hole. The heat comes about from concentrating the space, which holds the air. But the air does not concentrate because one does not bring in more air than there was before. The relevancy changes as the space reduce to change the space back to heat. This action is the very opposite of an explosion. But reducing pace the action brings about that the space turns to heat. This is most crucial in accepting because this is the precondition about the understanding as much as accepting a new concept, which I try to introduce, and at the same time I try to produce a concerned effort in dismissing myths from cosmology. Gravity is about turning space into hotter denser space as it reduces space and that is the reason why there is no visible or measurable stronger gravity in the centre of objects. The centre does not indicate more gravity because the gravity that is the measure of the accumulation of heat that the gravity produce in that space $a^3 = kT^2$. The heat increase as the gravity becomes more intense because the more intense heat is the gravity increase. By the reducing of the space coming down towards the smaller area such coming down leads to space reducing, which brings about heat increases. The stronger gravity personifies in the denser heat produced by the reduced space. As space increases heat dissipates and that we find is what happens in an explosion. The bigger the heat release the more space becomes available as winds (shock waves to use the name hiding the truth) blowing across fields. The winds come about as space multiplies through the release of heat creating new space that was not there before the time. In the explosion will the heat decrease be the decrease of gravity that was before the explosion bounding the heat into condensed space and the explosion is the release or antigravity reducing the heat as it increases the space. In my search I stumbled on two accepted but not intergraded laws and when I found and located singularity the two laws became very much plausible and factual.

Take a circle and reduce such a circle constantly to where it no longer can reduce. Reduce it to a point where only form remains part of the circle because the radius has gone beyond human measure and becomes so small it is not noticeable with what ever tools man may use, then what remains is pi since pi does not indicate size but indicate form, and form is all that then will remain.

I believe one can begin too see where my suspicions are heading because the flaw comes about in the manner mathematics are practised for thousands of years. Space is nothing because that means space is a standard fit all issued out before the time. Space cannot increase and winds are ghost blowing their breath. Winds are as much antigravity returning reduced space back to increased space. Before coming to the mathematics I would first like to bring your attention to the practical side. I am promoting a theory in which I am able to prove there is as much contraction going on in the cosmic universe as there is expansion and the contraction is as much part of the expansion. The universe rides on a balance and we have to locate such a balance. To prove my theory I firstly had to locate the centre of the universe. Even admitting to such a notion sounds like madness, but please give me a chance to explain in more detail. I realised that my effort to locate the point holding singularity enabled me to backtrack the exploding universe to its origins. By applying some basic effort I have located the position from where all movement came and the direction it took moving forward in time…and yes, even time as such.

Anything occupying space in the cube will apply r and by r I mean just a distance not using Π because Π serves as a form indication while the collective product of r will determine form as well as accumulative dimension total. Notwithstanding the name used confirming the shape or r named as length width or height, it is all just a straight line bringing about the cube with all its other names that may find attachment to specific form but nevertheless still remains only a six-sided cube with connecting lines applying different angles changing in some cases.

The normal perception is that any circle growing spontaneous would grow by the radius, which is r. In mathematics that may be true but it is not true in nature. In nature that cannot be the case because r is an indication of a straight line. By growing with the aid of a straight line from the centre to circle the influence that that would have on the circle would result in many circles following one another and not a continuous growth.

Gravity is the dimensional changing of space holding r as reference in the cube as to the sphere holding Π as the reference. In order to generate spin producing time in matter occupying space, therefore creating dimensional change, Π has to be a factor indicating the possibility of spin because implementing Π the circle sides will follow one another without establishing separation. The answer must be in finding Π, and thereby locating singularity. If singularity is in affect the original point of the cosmos birth, the reducing path we should follow will indicate the whereabouts such a point must be.

In the normal applied mathematics there are two standard formulas used to calculate a circle. The one use an r to indicate the radius and the other use a D to indicate the diameter, which is double the radius and therefore needs to be divided by a four to eliminate the Newtonian inverse square law amounting to the difference there will be between the two. The one using the radius is Πr^2 and the other formula using the diameter is $\Pi D^2 / 4$. However one looks at the mathematical expressions and Kepler's formulating of space-time there is an exceptional difference between the two scientific uses. When investigating Kepler's formula one do find it appreciably differs from the normal Mathematical equation like $a^2 = r^2\Pi$ and $a^3 = 4/3\ \Pi r^3$. In the normally used mathematical expressions such equations tend to concentrate on the volumetric aspect. In the case pf Kepler's expression it is something else that wants to surface. It is another idea that is coming to mind. In Kepler's formula a^3 stands to symbolise the third dimension and such a third dimension becomes equal to two other dimensions grouping and sharing value to equal a^3 efforts. It is not the circle of the rotation because with such a normal circle the radius is in the square and Π evaluates form. Here there is no mention must of a factor Π, which one would suspect to be somewhere applying since the circle is Π and Π is the circle and the two are inseparable. But not in Kepler's a^3, where there is no mention of Π at all. The fact that there is a radius of some sorts used to indicate a position cannot hold the square as it normally does in the case of the normal equations. In the mathematical equation the factor indicating the position of the circle edge has the square value being called the radius or in some cases the radius doubles and which then is the diameter, and the circle indicator is Π. But in this event the formula value will bring about a square value to the answer one receives. It will bring a value to the surface of the circle. In Kepler's formula it specifically does not. I am not the one that brought Newton into disrepute. Before me the cosmos did. The comets with they're not colliding did, and so did Roche and Lagrangian principles. Hubble was another one and it becomes apparent that every one that made a study about matters in the cosmos was in some disagreement about Newton. By Newton's effort to improvise on behalf of Kepler Newton made a statement that Kepler never made. In all honesty nature reacted strongly against the claims Newton made on behalf of Kepler and not about Kepler's work but about Newton's modifying of Kepler's work.

In short: how can a comet sail past the sun time after time without colliding and still apply a contraction in the manner which Newton suggested by the one claiming a freezing grip on the other? This strongly contradicts $F = G (M.m) / r^2$ How can five structures as the LAGRANGIAN POINT form around a centre structure while the centre structure keeps the five in position at equilibrium? By rejecting Newton's improvising this strongly contradicts $F = G (M.m) / r^2$

How can the Roche limit send structures spinning around an axis they establish at the time and either the lesser candidate destruct at a distance when there are in comparisons about size and gravity or they push each other into development beyond their individual means.
By rejecting Newton's improvising this strongly contradicts $F = G (M.m) / r^2$ How can Hubble's visual Universe expand while Newton's Universe must contract? I am not the one that started the dispute with Newton but it is the World of Physics that will not admit to such a dispute! By rejecting Newton's improvising this strongly contradicts $F = G (M.m) / r^2$

LAGRANGIAN POINT is another result flowing from the singularity position and most directly coming from Pythagoras principal and connecting that most basic mathematical in conformation of Kepler's formula.

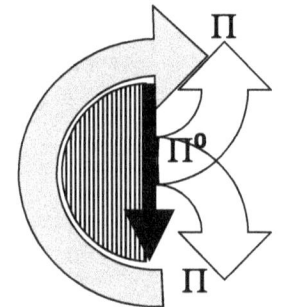

Kepler brought in a formula that came through two lifetime studies and the formula read that space is equal to time in motion. Mathematically it reads that $a^3 = T^2 k$. This formula brought Newton's claims into dispute because what this formula said was that geodesic space is not confined space and while laws apply in the confined space of Earth it does not apply in geodesic space. From Kepler on can see the precise moment the Cosmos started. It is Mainstream science that is hindering the accepting of the formula that is dampening the human understanding of the Cosmos.

When did the cosmos start is the question every one is in search of since Biblical days. I would say it had to have started with space but science has the cosmic time linked with time. There could not have been space without time and there was no time without space.
The first moment came when **k** moved a way from singularity to establish space. It was when **k** introduced individual entities apart. It was when confinement was broken and particles appeared for the first time. Understanding this comes hand in hand with accepting that there is two forms of structures other that elements confined in an atom.

Hydrogen 1	melts at -259^0 C,	boils at -252^0 C,
Helium 2	melts at -269^0 C	boils at $-268,9^0$ C
LITHIUM 3	melts 180^0 C	boils at 1300^0
BERYLLIUM 4	melts at 1287^0C	boils at 2770^0C
BORON 5	melts at 2030^0 C	boils 2550^0 C
Carbon 6	melts at 804^0C	boils at 3470^0 C
Nitrogen 7	melts at -210^0C	boils at -195.8^0 C
Oxygen 8	melts at -218.8^0C	boils at -183^0 C
Fluorine 9	melts at -219.6^0 C	boils at -188.2^0 C
Neon10	melts at -248.59^0 C	boils at -246^0 C

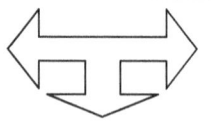

Melts (meaning it becomes a liquid) at -259^0 C **Hydrogen 1** boils (meaning it becomes a gas) at -252^0 C,

There are elements. There are liquids. There is gas. Forget about the example we find in water because it seems if any one talks about the three materials all concerned immediately thinks about water as one structure that can ice and can boil when it is not flowing. That is not even an example because water is not a true substance but is a compounded combination of volatile elements forming the most outrageous concept the cosmos could think up. By combining some of the most volatile elements the cosmos created the least volatile substance known to man, which is water. But the three forms water personifies is the structure compiling the universe and prove stages of development.

There was singularity. Singularity came first. From singularity came material as a solid substance. From elements comes liquid, a soft substance. Then comes gas when liquid turns to space. Again we can see a pattern coming about. But it involves Kepler more than any Mathematics.

All lines would form a duplication of another line sharing value since there will always be a possibility of yet another line in the realms of singularity lying between the two lines in question reducing the size infinitely to either side of the divide we humans create. Boundaries therefore are human and as man made substances it does not belong to the cosmos outside the influence of man and must be discarded. The understanding of insists on one vital precondition. We have to abolish our instincts about things being big and small, high and low, hot and cold, tall and short. All the human measures are truly unfitting to the cosmos. Those are measures made by man. In the cosmos the tiniest object is the key to that particular Universe. By going smaller one is going bigger. By going within the atom we go out and meet the Universe. In singularity we find the cosmos because every singularity not only represent the cosmos but also in fact is that Universe. No mathematics will ever measure the thickness, because as the line that is standing still it cannot have a width at all. The moment a width appears which one can measure or calculate, the line will become part of the factor forming the divided and not the divide. The instant when space connects, the spin direction will produce the partisanship of space and spin. Any form of space including even in the most minute will produce a favouring of direction. Through such motion gravity comes about because the motion is gravity and that produces the time aspect T^2 thus thereby changing the direction by rotary motion. The moment there is an area there is a measurable rotating brought about and no longer a non-interfering divide. Such a line holds space in a position that runs far beyond the boundaries and limits of the three-dimensional. Another factor of such a line would be that the radius (let us substitute the radius r with the using of Kepler's **k**), **k** would be immeasurably small. The factor **k** cannot be zero because infinitely close to that first **k** is the start of the third dimension where time plays the part as the fourth quarter. The presence of **k** is undeniable and recognisable yet it is not visible. The fact that **k** is there albeit stripped of any influence disqualifies it from being zero and therefore not being there. With **k** already beyond any measurable space, leaving a^3 as a factor of one and not being able to pin any volume measure to that one **k** will have to be to the power of 0 being k^0. In Kepler's formula $a^3 = T^2 k$ the area a^3 would be one because of the dimensional non-existing of measured sides in any direction. If $k^0 = 1$ and $a^3 = 1$ the only alternative T^2 could possibly have is also one. The factor of T^2 identifies the time in the formula and when the formula indicates time as one, the time component must therefore be eternal. Only time in eternity does not change

The more commonly used mathematical formula applying when the calculation of the sphere comes about is $a^3 = 4/3 \Pi r^3$ where it places one third dimensional but lesser factor in direct relation to another third dimensional relation and all that in relation to the form that is applying. By using $a^3 = 4/3 \Pi r^3$ it is definitely not what Kepler said when he produced his formula. The square he allocated to time as well as the cube he allocated to space does not find representation in the mathematical expression of $a^3 = 4/3 \Pi r^3$. In mathematics we find a deliberate lack of the time factor when using the equation $a^3 = 4/3 \Pi r^3$. This must prove the differences in what Kepler found to express and what Newton thought Kepler tried to express. However using Kepler there is no criss-cross matching of dimensional accumulating. If I am reading the situation correctly Newton saw Kepler's mathematical skill somewhat below the dimension of Newton's genius and that spurred Newton on to bring changes to Kepler's formula but in doing so many things went missing through incorrect translations and miss interpretations of genius. Kepler was not referring to a mathematical space on a flat sheet of paper, which was what Newton saw. Kepler produces a value linking space to gravity being time in as much as calculating space in the geodesic and measuring time in the geodesic and bringing about factors in formulas through figures that geodesic space informed Kepler about. If Kepler was mistaken then the Cosmos was wrong about the cosmos and if Newton was incorrect it was about mathematics that Newton then was wrong about. Kepler places time in the square directly in relation to space in the cube in association where time shows two distinct qualities. The one factor is time in the circle rotating while the other is in the linear or the straight line implicating the position that the other would have. This proves that time or gravity (with time and gravity being the same thing)

In all instances of measuring the distance the orbit travels around the sun as the space displaces or space covered by travelling in the time it is covered and dividing such a ratio one find the distance of the orbiter from the sun the in relation to the other factors form one or very close to one.

Planet	Period T years	T^2	Distance k	Space a^3	Ratio
Mercury	0.241	0.058	0.39	0.059	0.983
Venus	0.615	0.378	0.728	0.381	0.992
Earth	1.000	1.000	1.000	1.000	1.000
Mars	1.881	3.54	1.524	3.54	1.000
Jupiter	11.86	140.66	5.20	140.6	1.000
Saturn	29.46	867.9	9.54	868.25	0.999
Uranus	84.008	7069	19.19	7067	1.000
Neptune	164.8	27159	30.07	27189	0.999
Pluto	248.4	61703	39.46	61443	1.004

At the first glance Kepler's formula seems to be numbers and positions applying between the sun and specific but different planets in the solar system.

In that it is just about numbers but about that it is much, much more than just numbers. The numbers paint a picture and tell a story. Kepler produced a formula from the numbers and not numbers from a formula. The numbers brought about answers but we fail to ask the correct questions. By seeing a mathematical circle one miss the total picture and the story the numbers tell. The figures explain dimensions working in conjunction and together they combine dimension where the picture behind the story becomes a colour spectacle in comparison the mathematical grey that a mathematical circle produces. It is relevancies carried from the sun and the sun is the governing singularity representative for the entire solar system. This is about relevancies applying throughout the Universe. This balance is much, much more than what the figures say. It underlines and it explains gravity as a life form in the cosmos other than what we consider our life to be.

The German mathematician and astronomer KEPLER, JOHANNES (1571-1630)
German mathematical and astronomer became Tycho Brahe's assistant in Prague in 1600 A. D. where he undertook to complete the tables of planetary motion Tycho had begun. Kepler first calculated the orbit of Mars. He spent much time trying to reconcile Tycho' s accurate observations of the planet with a circular orbit, but concluded (in Astronomia nova, published in 1609) that Mars moved instead in an elliptical orbit. Thus, he established the first of his laws of planetary motion. A theory that the Sun controlled the planets by a magnetic force led him to the second and third of his laws, which were published as part of his treatise on theoretical astronomy, Epitome astronomiae Coernicanae (1618-21). The Rudolphine Tables (named after Tycho's patron, the Holy Roman Emperor Rudolph II) of planetary motion appeared in 1627 and were still in use in the 18th century. Kepler also wrote De Stella nova, on the supernova of 1604 and Diptirce on optics and the theory of the telescope. The overall view followed in this book **Matter's Time in Space** places the true significance of his work in true contents. In KEPLER'S EQUATION is the equation that relates the eccentric anomaly of a body in an elliptical orbit to its mean anomaly. The equation is $E - e \sin E = M$., where E is the eccentric anomaly, M the mean anomaly, and e the eccentricity of the orbit. It is important as one of the mathematical relations enabling the position of a planet about the Sun, or a satellite about is planet, to be calculated from the orbital elements for any time. However this only relates to the solar system, and KEPLER'S LAWS only apply in the contents of the solar system. The three laws governing the orbital motions of the planets, discovered by J. Kepler is as follows:

The first law states that the orbit of a planet is an ellipse with the Sun at one focus of the ellipse. The second law states that the radius vector joining planet to Sun sweeps out equal areas in equal times which as it says refers to time and not the circle. The third law states that the square of the orbital period of each planet in years is proportional to the cube of the semi major axis of the planet's orbit. The first law gives the shape of the planet's orbit; the second describes how the planet must continuously vary its speed as it follows its orbit, moving fastest at perihelion and slowest at aphelion. The third law gives the relationship between the planets' average distances from the Sun and their periods of revolution. Instead of placing, the true value to Kepler's laws I. Newton placed his own interpretation to Kepler's laws, and in doing this, he wilfully destroyed the principle working of the Creation. Through Newton's tunnel vision, he applied his own miss interpretations to the correct presumptions of Kepler. Newton reduced the implication that Kepler findings hold by introducing to the law of gravitation. He then went about and changed it to three laws of motion. I. Newton generalized Kepler's first law, verified the second law, and showed that the third law should be amended to the form; $4\pi^2 a^3 / T^2 = G(m + m_p)$. In this, the value of T and a are the period of revolution and semi major axis of the orbit of a planet of mass m_p about the Sun of mass m, and G is the gravitational constant. The major aim of this book is to correct these misgivings of Newton. I shall return to the statement about $4\pi^2 a^3 / T^2 = G(m + m_p)$

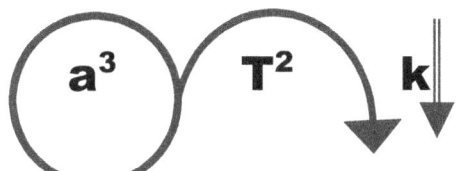

What Kepler saw was more of a dimensional nature than the practical mathematic symbols and values. On the one hand was a value to the third dimension, which equalled two-dimensional values one the second dimension, and one to the first dimension.

In the argument Kepler made he had hide so much more facts into one formula than what I think even he realised. Well, it is much more than that the Accepted Policy Protectors Of Science ever came to realise. He officially formulated space-time, he officially coined not the name but the origins of the Universe being the Big Bang and he was the first to put the speed of light in relation to cosmic development…and all of that with his rather simple formula. He said the space a^3 not the circle (a) or the circumference a^2 but in the circle a^3… where such a circle represent a factor in the third dimension.

The formula he compiled was not rather but very specific about the area being a third dimension area and to prove it beyond doubt he placed it in the relevancy of the formula in a ratio of presenting the third dimension in space. He said a^3 is equal to $T^2 k$. Newton and Newtonians came afterwards and played with mathematical toys as to challenge their mental capabilities. Newton introduced a $4\Pi^2$ to indicate the presume circle on the one hand and on the other hand he brought this lot equal to $\{G(m + m_p)\}$ which he then presumed to be the general Universal gravity constant (G) and the sum total of the two structure mass. Newton saw a ring circling around a centre having $4\Pi^2$ to indicate such a ring outside a centre and he positioned $\{G(m + m_p)\}$ where the two mass factors combine the gravity effort in the general grand gravity constant in space. I have had so much resistance in the past from all Academics but that is not what I see what Kepler saw. With what I saw what Kepler's saw I shall trace that back even as far as to the centre of creation.

In their eagerness to calculate they calculated a formula to measure the circumference a^2 of a circle being Πr^2. I have seen an Astro physics examination question paper where they use $4\Pi r^3 / 3$ as the formula to calculate the sun and other stars volumetric space! They formulated the measuring procedure of the circle being in the third dimension that will show how big the volumetric space is of a sphere at a^3 being measured with the procedure being $4\Pi r^3 / 3$. This too was a fanciful devise allowing mathematicians to be much superior to the rest of the commoners and to dictate to the lowlife how and what they should think when they think and if they indeed can think of anything to think of. Then some Mathematician and an Englishman of Substance came onto the idea of gravity. Being a mathematician the Englishman placed the Universe at the feet of mathematicians. He saw circles where Kepler saw three dimensions. He saw three dimensions where Kepler saw nothing. He knew time had to be somewhere as something and then covered it by denouncing the circle cycle as nothing.

What then is it that Kepler saw as he formulated $a^3 = T^2 k$. At the normal flow of time it takes the electron a certain time to spin around the atom. The atom uses space a^3 and the atom is a certain length k that forces the distance the electron has to travel in one cycle period T^2. The atom a^3 connects the electrons travel k to gravity T^2. The relevance k produce to support a^3 is to point T^2 to two positions the electron will be in the duration of one specific time. The electron travel will be cyclic and periodic in relation to the space the atom holds. The space stands related to the gravity with which the Earth reduces space and with the space and speed with which the atom travels through space.

The perception is that light travels as fast as time flows but I disagree with that idea. Nevertheless the perception is there that the speed of light is as fast as travelling of any sorts can reach. We accept the electron's travelling speed imitates the speed of light as much as it is permitted by gravity to do so. By this imitation the electron come as close to time as it can ever come. The electron rotates around the atom nucleus indicating an atomic border of some sorts. From the centre one can draw a line pin pointing the position of the electron during the duration of a time period. In this time will indicate movement of the electron through space. The time indicated must be T^2 should space be in the third dimension a^3 and singularity will connect through k. By linking space a^3 to singularity, which produces time T^2 it will have to indicate the influence of singularity through the single dimension connecting of k. In the relation only k will be representing the single dimension factor since that places the Universe in space and time. Should one place the time factor but presenting time as t in a single dimension role the two dimensional time have to disappear with the three dimensional space that also will disappear. In that space and time must disappear. We all can accomplish that task by taking a photograph and print the image on paper and call it still photography. Then the paper will hold time in the square while the paper is in the cube all indicating individual and complimentary k pointing to individual and complimentary singularity producing space-time.

The one position in space will place the electron in time where the electron then will be below...behind and above the atom sub particles with which the electron shares frequency. The position indicates implicate the electron T^2 in the environment a^3 establishes.

By taking the line to k back to where the line or k cannot reduce further $k^0 = 1$ establishes such a value where k then finds a position in the single dimension. But in that case a^3 also is equal to one and so is T^2. In fact k still has to produce a line and we find that k represents a^3 to the full as well as T^2. The point where k forms the most slightly distance the area a^3 establishes a value outside the single dimension because T^2 adds a value. The fact that T^2 comes in as a factor in the presence of the first sign of k appearing indicates the start of motion taking a^3 from one location to another specific location. It indicates the travel of the planet during a month or a day or an hour. It does not indicate a circle except at the end when completing one cycle. T^2 is the distance in time a^3 will take k from indicating one point to indicating another point. The formula points to a referring of the very time space was indicated by position location and time. The astonishing part is not as much the way Kepler formulated his formula to cover the movement and the position of the electron in relation with the rest of the atom, but the brilliant way the mathematicians neglected to see the fact. Kepler saw a three dimensional a^3 something in a specific position in time T^2 relating to a specific density k of the atom. With space in a cube as it cannot ever be otherwise the time too has to be in a square because placing time in the single dimension of t the time then becomes part of a single dimension such as one may find in a photograph picture. One can justly use the same formula to implicate the electron taking time to complete the distance between two points indicating the area from the centre of the atom.

The time frozen on paper in a single t is effective in remembering the viewer of an event but that is not the event in the present any longer. That was how the event occurred during the time from where the camera shutter opened T_1 to where the camera shutter closed T_2 and the time frame T^2 was then during the open period of the camera shutter. But afterward it represented **t** when looking at the picture and the looking of the picture became an event during a specific T^2 that went from where one is taking the first look to where one is looking away from the paper carrying the first dimensional image of an event gone by and that is at that stage a representation of **t** in another milieu of $a^3 = T^2 k$. The **t** in the single is when mathematically presented as only **t** indicating a mathematical single flat dimensional view of time and is then correctly applied because it represents a reminder of a four dimensional event $a^3 = T^2 k$ that went single dimensional because the moment in the fourth dimension was then frozen in a single dimension on paper while the fourth dimension $a^3 = T^2 k$ soldiered on and time will always be representing T^2 as Kepler stated in the square allocated to space having a cube $a^3 = T^2 k$ at a time even before gravity got a name. But reducing the dimension of time to a single **t** one will find the ability to mathematically design the paper on which the photo image will be printed in time T^2 using space a^3 in the third dimension to apply the ink in the third dimension. Printed on the paper is an image that is not part of space-time while the ink used is space-time and the paper is space-time. The ingredient all hold different and k indicating different singularity connecting forming individual as well as group space-time.

In $4 \pi^2 a^3 / T^2 = G (m + m_p)$

$a^3 = T^2 k$

$a^3 / k = T^2$ but at the same margin is

$k / a^3 = 1 / T^2$

$k = a^3 / T^2 =$ singularity

$a^3 / T^2 = G (m + m_p) / 4 \pi^2$

and $a^3 / T^2 = k$

then $k = G (m + m_p) / 4 \pi^2$
But I showed that $k = a^3 / T^2$ and Newton's claim is that $a^3 / T^2 = G (m + m_p) / 4 \pi^2$

> a^3 forms the space the atom claims while travelling in the Earth spinning all the way and travelling with the Earth around the sun

> **k** positions the electrons travel in the space relating to the space the atom holds while travelling with the Earth in the Earth around the sun in relation with a specific position **k** will indicate in relation to the sun.

> T^2 is the time it takes the electron to be relevant to the position the Earth places on T^2 while the Earth captures the space of the atom by providing the space for the atom to be within while the Earth travels from one point T_1 to another point holding T_2 in frequency to the atoms T^2 relating to the Earths T^2 in perfect harmony with the sun having another T^2 relevant to all the other factors we call cosmic particles. Big or small it is only about cosmic particles holding space in time in relevancy.

The only definite place one will locate zero is in between the starting point of the lines going in opposing direction in the position the lines hold before there was the least of directions applied, but that is only because there is no such a position, not because any line is coming from there. The two lines are still one holding the opportunity of parting as an option but have not yet parted and therefore are on the very precise same spot. The line coming from there is already there because it already has the choice of going in any and all opposing directions and when it starts running it will place filled space in that location because the space was already filled with a line starting and not with a line not there at all. When reversing a line we might find a better idea of what is in place and where it is in place.

In the action of the inseparable drawing closer and moving closer gravity finds the dual value of linear and circular gravity. There is no separation of the two factors acting as one but both have different application and values in the unit. This is the result of singularity having three parts acting as one but giving three distinctions in application. Gravity is as much part of dismissing space as it is about making contact with space in time. Since the connection comes about as a circle, the connecting points will relate to Π as the value. Due to the spinning nature of such a point with all surrounding the point will be alternating direction favouring change every second and in that the value to such a point can only be Π because of its constant changing. Using r would specifically oppose another r from every angle because the use of r will bring about a static relation to the previous and following instant and therefore it will cancel the constant spin flow.

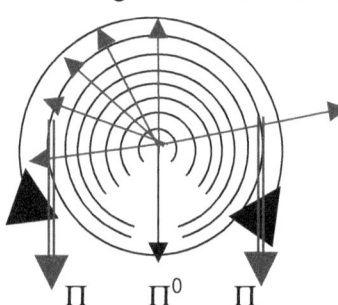

$\Pi = r$ in constant directional change as time flows through rotation

Pinpoint positioning of singularity Π^0 with Π positioning space to either side forming the border set by singularity

The new direction pointing to a new location in relation to the previous point will oppose the previous point it had in relation to direction considering the centre point.

Em B₁ B₂ Emtoo

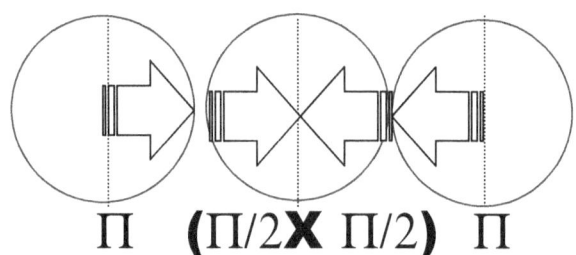

Π ($\Pi/2$ **X** $\Pi/2$) Π

We return to the fact we established before that em and emtoo was divided by r and then r had to be one since r could not be zero. Such a centre would then carry the same value as em and emtoo. That means whatever value em and emtoo receives has to go in equal measure to r with em sharing half of the divide and emtoo sharing the other half of the divide.

The single dimension is a dimension covering everything into the dynamic of one. This brings about that **k = 1 a = 1** and **T = 1**. That is the first dimension and the first dimension is a dynamic of one being the result of a dimensional 0. $k^0 = 1$ $a^0 = 1$ and $T^0 = 1$. The factor k was at no stage zero. Only the dimensional factor is 0. The extending that k was capable of was zero. The factor k was never zero. The factor **k** can never indicate zero point from zero or point to zero because just one zero will dump the entire Universe into zero.

Then a moment arrived where **k** developed from k^0 to **k**. This brought about a revolution of cosmic proportions. Matter divided from singularity as matter claimed space. The growth of **k** had to produce a^3 which is an interpretation of the space k will bring in place. The most tiny and slightest of growth established **k** coming from k^0 to **k**. By establishing **k** that very second a^3 also came about. But through Kepler we can see how singularity achieved space to come about. It came through spin. Kepler said that $a^3 = T^2 k$. Space broke away from singularity by applying spin. Gravity is spin. When **k** extended it secured a^3 being space but through the spin of T^2 the space separated between particular singularity, to individual singularity. It secured individual space from space by rotating space. Through the rotation came boundaries, which we now consider to be particles in time. It is the time (from T_1 to T_2) that it takes **k** to swing into a different relevance to a^3 space. It established movement in the area holding the least space. The relevancy started when k claimed space by motion from singularity. That is what Kepler claims. It does say how Kepler incidentally forgot to improvise for the claim of a circle but accidentally forgot about including $4\Pi^2$ on the one side and G (m + mp) on the other side. Kepler placed the growth **k** directly relating to the area a^3 separating as the spin T^2 provides the space. Kepler showed with his formula what gravity is. Gravity is the least space (singularity) **k** claiming space a^3 through spin T^2. Kepler announced space-time, the Hubble constant, the Big Bang theory and all other later cosmic developments.

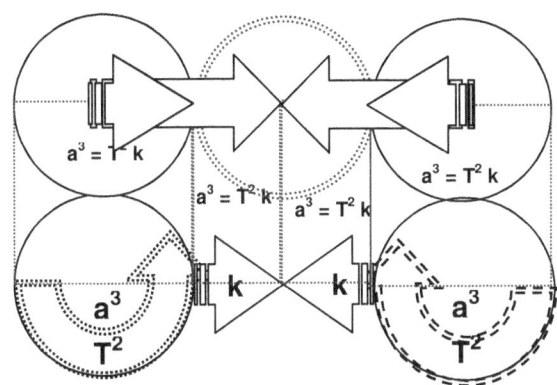

Space developed sectors through time applying differences as singularity changes the universe from T_1 to T_2 through a^3 and T^2. The factor **k** positions the centre of the universe and then sets rules applying in that Universe as far as setting space in time by applying space-time. Space = a^3 and time T^2 coming about from singularity pointed by **k**. **Space-time $a^3 = T^2 k$**

If the sun held a relevancy of 10 relating to seven with one or two or three of the planets…well yes that might be coincidental, but when it shows such a relation with all of them where all of them includes planetary fragments being between the Earth and Mars making such a coincidental claim on ten structures perfectly distributed is more than ducking the truth. To honestly be honest about the finding of scientific truth and discarding such evidence as coincidental is being unfaithful to you. While considering yourself as all scientist that should be to be level minded and not thinking you to be a participant of bias but in the same breath blow away such clear evidence should be honestly considered as gross clear dishonesty. To go and dismiss this certainty as coincidental because it does not fit into a Newtonian Universe is stretching the truth to beyond the accepting norm. One should not try to focus on an image of such a spot or dot because there is no image. The line dividing the cosmos and that run through every particle, no matter how large or small is beyond our vision. Such a small line, so small it is not even noticeable to the cosmos in the 3D is large enough to part the cosmos into sectors. It splits the biggest there is into particles and we are not even able to notice the precise location of such a split. In truth there is no top or bottom that we living in 3D can see. We shall have to use a general conception brought about by intelligence. Your intellect tells you about such a spot, but that is all because that spot is on the other side of the universe (quite literally). From the centre of the dot there is a top and a bottom spot. From those points there is connection with four quarters. That produces six connecting points that are all aligning to the centre.

In this singularity there can be no sides and without sides there can be no drawing showing the explanation by means of illustrations. Where I do just that, I ask for your forgiveness because being human means I have no capable means of performing an explanation. Yet I am forced to do just that and I have to allow the means of sketches, well knowing the implication that such act is not allowed.

There was the dot. The dot had no borders therefore there was no separation and still we know there were more than one in a group of one. The evidence of this is very present in the cosmos at present and one can find such evidence all around us.

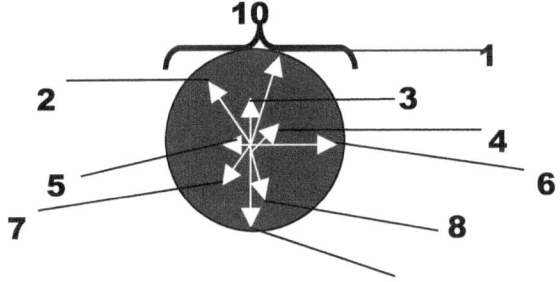

With the cosmos still in a single dimension there were no limits as we know limits to form in the Universe we use and no borders indicating limits because after all it is the single dimension where there is only one dimension holding so much diversity. The borders were part of development because we can witness the legacy of such borders in the present day holding the 3D in place.

The overall picture resulted in a ring and all rings hold Π to secure the form. The only form that existed then was Π and therefore even today the borders use Π to indicate positions. But in the single dimension such definitions were far from clear and the only distinctions came from securing singularity in preserving the position of singularity to apply gravity and thereby absorb all anti - gravity. But anti gravity could not control expansion by counter acting contraction through gravity so the overheating continued forming non-existing borders in some thing infinitely solid just as Einstein predicted because this took place before light came about and therefore before the speed of light became part of the cosmos. The cosmos formed a partnership with one side overheating forming antigravity by expanding into space through the applying of the overheating and the other side formed gravity or contracting of space

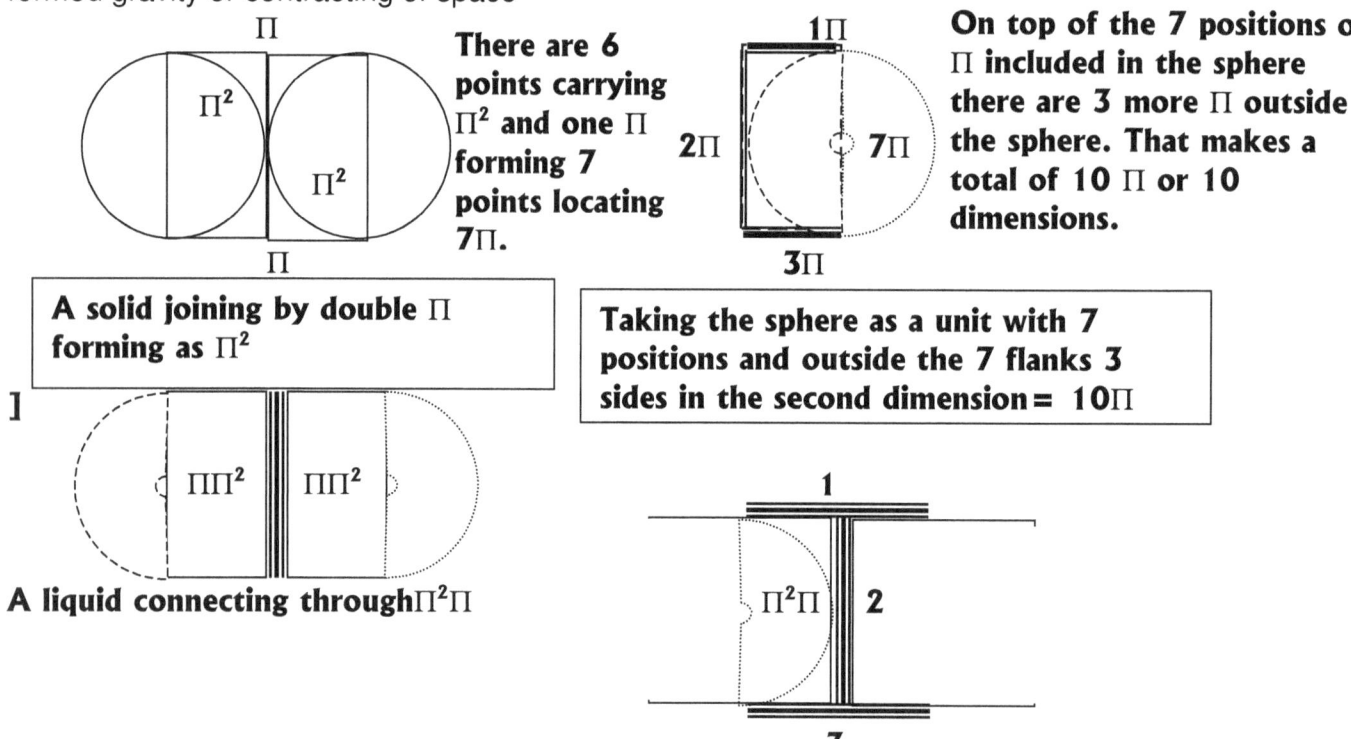

Total connecting relevancy of the sphere forming matter connecting to space
= Π²Π 3
There are four locations indicating values inclusive of the space of the atom. That is the base of gravity in the atom and in the atom is the base of all gravity.

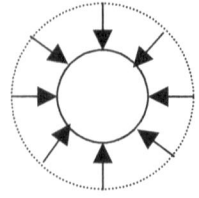 Gravity is about reducing space

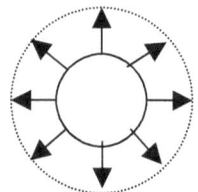 Expanding is all about heating. Heating takes up more space and gravity reduces space.

This says it all and yet every person with a position of influence in science is missing all there is to see in Cosmology! Greatness in Cosmological terms is not in size, but the measure goes by intensity of density and lack of space. A smaller (a^3) result in a larger T^2 where (a^3) is the space the object holds and T^2 is the sizable gravity the cosmic object has. It confirms Kepler and disagrees with Newton. According to Newton Betelguese should be formidable when applying gravity but the Black hole is the true undisputed giant! By taking the diameter, as the measure is clearly no solution in a method to calculate the gravity because it solve not one thing. The circumvent this failing the Academics changed the approached by applying the usual r measuring gravity but instead of having the radius holding the square they brought in the speed of light and further disrupted the truth by applying the square that should fall on the r in the normal calculations to the C that is supposedly there to bolster the gravity. The speed of light is the worst or best form of antigravity depending on which way one look at matters. Light is the strongest antigravity there can be and to throw that into the Black hole by the square to hide the insufficiency of the methods applied to calculate gravity is once again another cover up to hide Newtonian not functioning in cosmology. By producing C^2 in an attempt to bolster the gravity figure they supposedly are able to calculate in the gravity of a Black hole and placing C^2 in conjunction with R symbolising the radius is just the way not to improve the incorrectness which their theory quit deliberately bring about as a measure to determine gravity. Then what about the measuring of the gravity in the Neutron star and how will they explain the Neutron star because the Neutron star will either be stronger in gravity than the Black hole which is clearly not the case or it will be pathetically weak. If the Black hole has a diameter of 10 kilometres then the Neutron star has a diameter of twelve kilometres. By using C^2 in the Neutron star the Neutron star suddenly have a larger gravity than the Black hole has. It is either that or the neutron star is so weak it has less gravity than a comet. This ridiculous scenario developing just shows how little mathematicians have any grasp about cosmology. They should keep to building dams or skyscrapers where the mathematics of them is useful and is appreciated and stay out of cosmology. For this remark they will get back at me as they usually do.

Mass is the result of gravity and gravity is not the result of mass. Gravity brings about mass but mass does not produce gravity as Science wish to advocate. Gravity creates mass but mass does not establish gravity.

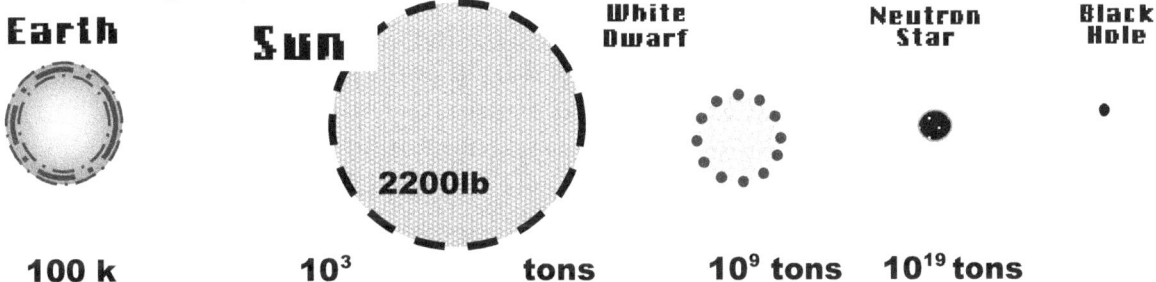

The idea is that gravity pulls material closer and more so in bigger stars. But that gravity pulling can only come from the accumulative effort of every individual atom according to proton mass (number) as a unifying effort of all the atoms in the star in accordance with mass applied. The idea is that mass is the same everywhere and is never changing. Why would there be such huge mass increases in the bigger or should I say smaller stars. What would entice the material inside such stars to grow more massive if mass comes about from the pulling of one particle closer to the next particle. If it was about pulling on each other the mass of the particles could not increase through that. Even by combining the mass of two individual atoms the increase is already in the equation. By locking the two into a unit should not change the mathematics because 1+1 = 2 whether the two share one unit or two units cannot be any mass increase in a star because all material is within the unit. By fusing the star cannot become more massive using that specific method since it gains no further mass and two hydrogen atoms plus one oxygen atom can at best be equal to if not less massive than one neon atom. The mass cannot grow because the star does not produce mass, if we stick to the accepted views. By fusion protons will only join without further rising the mass they have apart or combined. If the particle has a mass as two units, and the units join in volumetric occupation, they still have the same mass. Some facts about science are too astonishing to be real!

The above facts are part of Accepted Science and accepted facts, but my theory about gravity dismissing space to the advancement of compacting matter is not accepted through all my trying to introduce my ideas to Accepted Science. One hundred pounds in mass will be equal to a mass of one ton in the sun. One cubic meter being one ton on earth will hold ten thousand tons of material in a star one class more developed than that what the sun is. In more developed stars the figures rise above Human comprehension. But it is so clear that the space diminish as the mass becomes denser. By compacting matter the space reduces in the same process. I have been trying for years to get any professor to admit to that and the rest of my theory. Understanding star development must lead to understanding the Big Bang.

The heat available during the Big Bang explains the lack of space at the time. Space is heat expanded and heat is space contracted. The one is the reverse of the other. To expand is to bring about excess heat and to contract is to produce gravity by eliminating space. The Coanda effect backs my argument. Where gravity is applying the strongest heat is the most and space is the least.

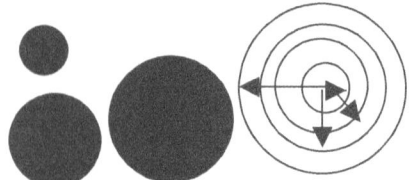

Looking at the affect of gravity it shows the precise quality of no distinctive point, as gravity never seems to end at a point but flows all over affecting all that holds a position in its sphere of influence. The gravity coming from China meets the gravity coming from America at no particular spot but intermingles without distinction.

In the sketch above the circle to the right would come about from a straight line r growing influencing the appreciation of Π, but to influence Π would lead to a breakdown in r as Π and r are different entities. The circles to the left shows a continuous growth by extending Π every time and since Π is the same part as the previous Π, only extending that billionth of a millimetre or many times smaller each time, the circle will be truly continuous without any signs of a break. In the context of dimensions one find coming from the centre Π^0 an established eternal flanking of Π to six positions since Π^0 forms the centre to the six sides and all six sides not having a diameter yet must apply Π to indicate specific value. What I try to say in this elaborated effort is that where **k** for there very first time extended from Π^0 to the edge of 3D where Π begins it had a certain distance. We humans are incapable of ever measuring that extending growth but be as it may it is there. Such expanding is beyond is beyond human thought but it still had enough value to raise a Universe. By allowing this expanding of **k** personified as Π to continue uninterrupted as one flow of a continuing growth of Π a line of ever so small but still productive will follow one line upon another line producing a cover of the full area of a^3 the time factor T^2 or for that matter gravity Π^2 will be flowing constantly through out a^3. The factor a^3 will be improvised by the singularity measure of Π^3 and the time factors T^2 and **k** will then be in singularity terms $\Pi^2\Pi$.

In the very centre, which I am referring too, rotation must end or start depending from what vantage point the relevance is placed.

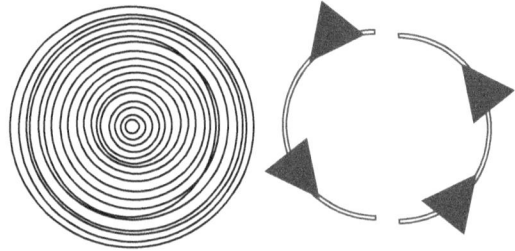

The very centre form an eternal divide that will not allow what is on the one side to present an influence on the other side. It divides spin. It divides direction of spin. It divides all rotation from the outside that one may detect and such divide is there because at one point spin will run to the left coming from the right and just immediately next to that point must run a direction from left to right.

It cuts without contributing or participating in movement. It divides without any favour.

That is singularity not having a dimension of space and not having a dimension of time, or a radius connecting the rotating distance to Π. It is the point where T2 breaks down and a3 stops to exist. It is a point in all-rotating objects. Every rotating object holds a centre from where the rest of the rotating direction will differ at any and all given points. Not one point is exactly the same, but in the very middle, the centre no one can draw, measure or see is a point not in motion.

In the centre runs an axis line that forms the division of rotation. No one human will ever be able to indicate the precise line, but such a line must exist because of our logic telling us about such a line. In the centre one will always find one more line smaller than the outside but forever also always bigger as it is towards the inside.

Locating and finding Singularity

In the **precise middle** of all **objects in rotation** is a precise centre dividing the object in sectors that will **start the spinning initiation** from that centre point. Thus, the spinning object **will have a middle point**, a very specific **centre point that does not spin** and only holds Π as a specific value because no radius can apply. But also the one value such a line **cannot have is zero** because the line **is there and holds contact** to the rest of the material bringing about that **zero does not start any** line and therefore the **value of the line must be infinite**, just as described in **accordance** and by **the definition of singularity**

As I am introducing a very new idea, I whish to explain in better detail what I try to convey.

While the toy top is spinning move the rotating line progressively to the middle by reducing the length the line have from the edge to the middle. At one point all further reducing ends.

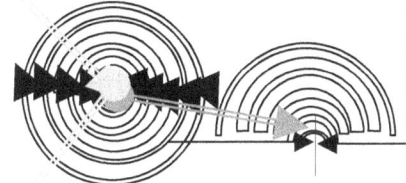

As the rotating direction moves inwards, the rings will become smaller and smaller.

That point albeit hypothetical, is also as much a reality none the less and is placed where that point **must be standing still** because every line **running from that point** in **opposing directions** are also **in opposing directional spin the other or opposing side.**

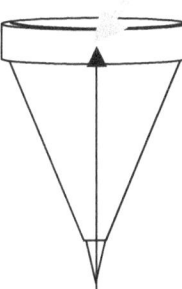

In considering the spinning motion in the fraction of time in the detailed instant every aspect of rotation will turn in every instant of change in time. Although the points had the same characteristics only one instant before, they oppose the characteristics it had just before and just after the very instant in which they are and to which they relate by similar points also in rotation. The fact of the graph proves my point in quarterly opposing dimensions and values,

Such a pint will also establish a line we call an axis. There must come a point where the ring is infinitely small, where it can reduce no more, where it reached its ultra limit, but at that point it cannot be zero, because the point is there for all to realise but nobody to see.

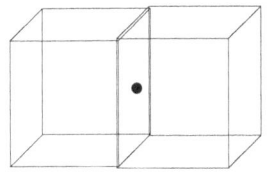

Taking the outlook from the point the sphere is holding from that centre out into space there are ten points connecting to the centre. In that are the dimensions of singularity connecting to space where five connects to space in the second dimension of singularity, and five connect in the third dimension of singularity.

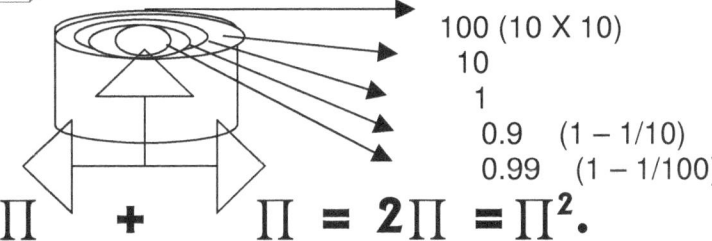

100 (10 X 10)
10
1
0.9   (1 − 1/10)
0.99 (1 − 1/100)

$$\Pi + \Pi = 2\Pi = \Pi^2.$$

Understanding all the following is connected intimately and all conditionally to the fact of accepting that all individual particles in the universe use motion and therefore spin. On the border where k becomes Π and step away from singularity into the very first point proving to form 3D a factor of Π^2 comes into place.

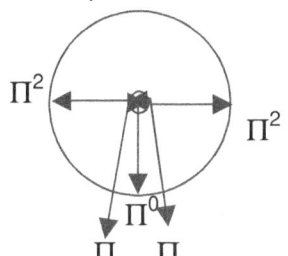

In dimensional terms, which explain later on the value of 2Π relates to Π^2. Then that relation extends to the next value where Π^2 relates to Π , which relates to Π^0. The first space in the circle will then be $\Pi^2\Pi$. From the centre being in infinity one can realise by applying mental power the single dimension factor not seen but present all the same. Extending that into the 3D comes six Π and any one of the six will further extend to form a seventh point as Π^2

In the circle as a factor distinction precise circle and using r as a there is no in outer space floats detectable interfering using $r^2\Pi$ the r has to have distinctive qualities placing it apart from Π. Where the growth shows no separate but a continuous flow from the precise centre to the edge the flow would become in relation with Π depicting the Π replacing r as reference to any point on the circle. By distinction in the circle division is possible but by using Π distinction possible making it a solid flow. Any object being and such floating is seemingly random with no specific favouring a movement in a particular direction. Such a devise is depending on influences not in our scope of detection. But then the object comes closer to the Earth and reaches on specific point where the six dimensions that influences the object in the cube suddenly changes. At one point one of the six dimensions fall away as it disappear and the object quite literally falls to the Earth. it changes a stance from floating to falling. The support of one side disappeared and the centre point of the sphere took over the control. At that point the object is under the influence of one centre point in the sphere where the sphere in this case is the earth and we also know that in such a centre point one will always find the strongest or the controlling gravity.

The sphere holds six sides in relation to form as unit and as does all other shapes and forms. All forms have to have at least six sides indicating different exposures to the Universe. But with gravity having a free choice, gravity always chooses the sphere. As I shall prove later on gravity is the strongest where the form produces the least evenly distributed space. The first condition for gravity is even-handedness through out the sphere holding the applying gravity and the second is to have most or the strongest gravity located where the space is least. That gravity then has a position in the very centre of the sphere and from that centre the gravity produces all the edges or borders that the sphere consist of. In the case of the sphere this factor makes the sphere much more dominating than any other form does. From the centre point controlling all sides is gravity and with gravity applying control the sphere has seven sides to the square in any other possible form having at least six sides.

The cube can come in whatever form there may be but the sphere adheres to precise measure and behind this principle is all that forms the Universe. The cube has six sides connected loosely and can change form just by changing the relevancy between one side (or more) in relation to the distance brought about by the other sides. The sphere being a complex circle stands related where the sides has to apply precise measure in equality. This becomes a law because in the precise middle one will find the strongest gravity as that gravity holds the object in form and true to form. If there is even gravity spread in all directions the form must be a sphere and the sphere insist on seven points relating to sides or borders.

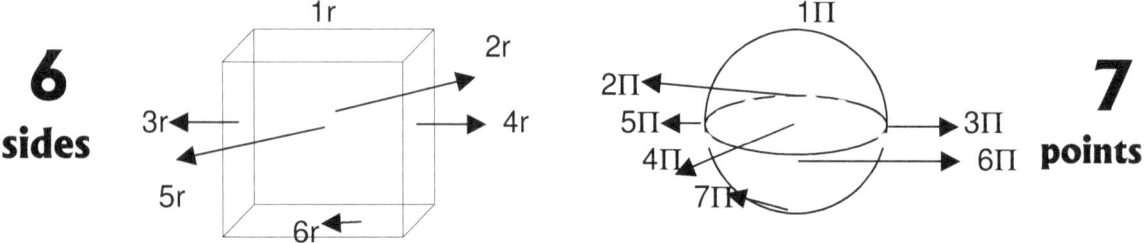

6 sides — **7 points**

In the sphere, which I am referring to there can be no radius but only the extending of Π from the centre Π in six opposing directions relating to one another by the square but remaining Π because of the unity the matter holds in relating to space. In the event of using r the individuality of r as a concept will bring about a specific ring at any specific point where every such a ring will define every time various on going circles. One cannot use such a definitive line because such a line will have to cut through atoms at some points. That is not the effect one get from gravity. Gravity includes and joins all aspects within the field but such definitive line will have to exclude some and include some because of the definite points and rings developing. The gravity influence we are in search of is something like a woven cloth covering the area and covering all in the area. It is like a silk blanket covering all the aspects. That means the value Π is running into and past the entire surface as if the surface is one consistency without different particles. That is what motivated me to look for gravity as being heat because everything consists of heat and if gravity is condensing heat gravity then will be all including. It is not possible to draw a precise line that would form a precise ring and not cut some atoms in parts. Because where r is used there will always be an atom disallowing the precise positioning of the circle the circle continues on a solid basis holding Π as a positional reference and not r. In every sphere there then are the seven Π relating in precise dimensional and positional equality forming equilibrium to the centre Π as well as to one another by 90^0 and 180^0 implicating the dimensional positioning. Therefore the sphere holds 7 $_{by}$ Π and the cube holds 6 X r^2. I use the symbol r to define the idea of a line weather the line might be used as a radius or to be indicating length breadth or width I do that to avoid the pitfall of using names to hide truths. Where space comes into contact with the sphere the cube loses one of the six dimensions it has to the more dominating seven dimension of the sphere whereby the seven dimension in equilibrium will dominate the six dimension loosely connected by r bringing about that the cube then has 5 sides to the seven of the cube. Because the space surrounding the sphere takes on the shape of the sphere and not the other way round where the sphere resolves in accepting the form of the cube, one may presume the form of the sphere is the most dominant of the two choices.

In the centre of a sphere there is a definitive position where the strongest influence of gravity is located. It is the centre of the sphere where the space is the least. From that centre point gravity extends in keeping the edges of the sphere perfectly true to the form singularity has being Π in every aspect. But also such extending continues beyond the specific edges of the sphere as it influences the space surrounding the sphere. We know this by another name given to divert every one from realising no one has a clue why that phenomenon is applying. We call it the Coanda effect where the process shows that even liquids submit to change of form. The condition for such a submission is that singularity carrying Π is in place and committing the influenced space. Only by applying singularity through using Π does the Coanda effect apply. With the sphere being defined by singularity to confirm singularity with the use of Π as form the sphere is influential enough to remove one side of the square cube, which is, loosely connecting sides to form the cube. The cube is very loosely formed and the sphere is very defiantly controlled by singularity therefore the sphere is able to dominate the space in the cube by removing the nearest side. With the removing of the side the cube in form loose one supporting dimension and therefore will not be able to secure what is in the sphere to the form of the sphere. That is gravity. It is reforming space to the requirements of singularity.

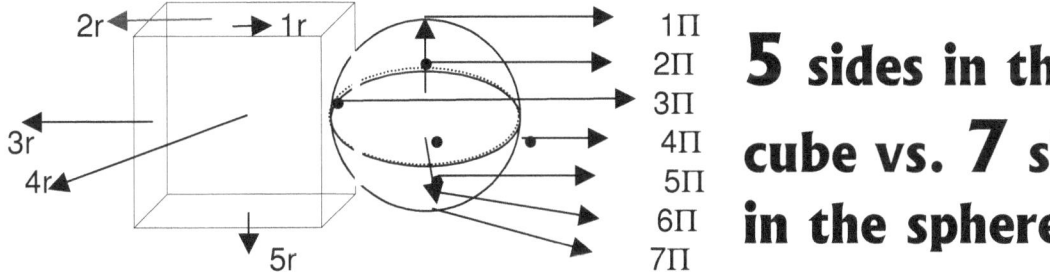

This means that in the cube at the point of contact between the cube and the sphere the cube experience such a contact point as if the "bottom falls out" of the cube and without a "bottom" to support objects they fall to the sphere as objects does fall to the earth. Remember that a body "floats" in space, but at one specific point it starts to "fall" to the earth. That is gravity and it is a dimension change much more than any force. I shall explain this last remark later on. That too is the Lagrangian system with five cosmic structures holding relevancy to the centre structure where the centre structure stands in for seven positions diverting from singularity and the orbiting structures standing in for five positions in space.

Gravity is all to do with dimensional changing and reforming of forms to re-affirm alliances supporting singularity. It is the reforming of space converting space to more concentrated heat.

The Universe is in the three dimensions using twelve dimensions that is visible to us and indefinite number of stages in size differences ranging from the immeasurable small to the immeasurable large where mathematics become a short fall to the next and the previous dimension.

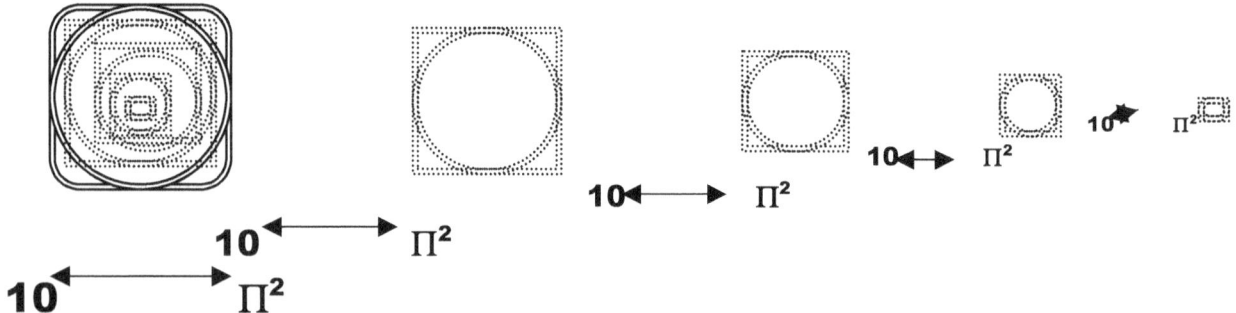

There are always 10 in positions running smaller and running larger. Even to us thinking we are at the edge, because we use light as an information source will find that the Universe are infinitely bigger than what we see and infinitely smaller than what we see.

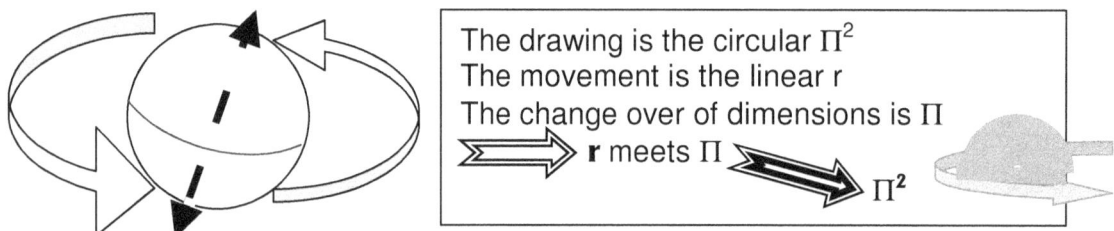

Gravity is the dimensional change of space taking space from 10 to Π^2. This happens by means of applying the Titius Bode configuration of space adapting form through the seven dimensions interlinking ten dimensions to reform the concentration of the space to heat.

The drawing is the circular Π^2
The movement is the linear r
The change over of dimensions is Π
⟹ r meets Π ⟹ Π^2

The result of the five to one relation comes the Titius Bode principle.

From this line of reasoning I dismissed the theory of the presence of a force being gravity but rather consider it as a dimensional changing contributed by the spin of the earth and the spin comes from singularity located in the centre of the earth. It is all about dimensional changing that influences space as a factor of ten to reduce to Π^2 on a continual basis from point forming new dimensions through billions of such points.

Space-time is a four dimensional position of the universe where the position of an object is specified by three coordinates in space and one position in time. This evidence we find as matter grew into the dimension we now share with billions of stars in the cosmos.

With the dimensional change from space in the cube to space in the sphere a relation of 5 to 7 comes about depicting gravity on one side of the divided Universe. The principle of 5 sides in space relating to 7 in the sphere holding matter forms the basis of the Titius Bode and the Lagrangian principles.

The German mathematician and astronomer BODE, JOHANN ELERT (1747 – 1826), published a formula in 1772, now known as Bode's law, which yielded the approximate distances of the six known planets, from which he predicted the existence between Mars and Jupiter of an undiscovered planet. His major publication was Uranographia (1801), a comprehensive atlas of the entire sky showing over 17 000 stars and nebulae. For fifty years, he oversaw the publication of astronomical data in the Berlin Academy's yearbook.

BODE'S LAW
A numerical sequence announced by J.E. Bode in 1772, which matches the distances from the Sun of the six planets then known. It is also known as the Titus-Bode law, as it was first pointed out by

the German mathematician Johann Daniel Titus (1729-96) in 1766. It is formed from the sequence 0, 3, 6, 12, 24, 48, 96, and 192 by adding 4 to each number. The planets were seen to fit this sequence quite well – as did Uranus, discovered in 1781. However, Neptune and Pluto do not conform to the 'law'. Bode's law stimulated the search for a planet orbiting between Mars and Jupiter that led to the discovery of the first asteroids. It is often said that the law has no theoretical basis, but it does show how orbital resonance can lead to commensurability.

BODE'S LAW

Planet	Mercury	Venus	Earth	Mars	Ceres	Jupiter	Saturn	Uranus
Bode's law distance	4	7	10	16	28	52	100	196
Actual distance (10^{-1} AU)	3.9	7.2	10	15.2	28	52	95	192

Bode's Law:

A numerical sequence announced by J.E. Bode in 1772, which matches the distances from the Sun of the six planets then known. It is also known as the Titus-Bode law, as it was first pointed out by the German mathematician Johann Daniel Titius (1729-96) in 1766. It is formed from the sequence 0,3,6,12,24,48,96, and 192 by adding 4 to each number. The planets were seen to fit this sequence quite well – as did Uranus, discovered in 1781. However, Neptune and Pluto do not conform to the 'law'. Bode's Law stimulated the search for a planet orbiting between Mars and Jupiter that led to the discovery of the first asteroids. It is often said that the law has no theoretical basis, but it does show how orbital resonance can lead to commensurability. The importance that becomes known is the sequence the Titius Bode law saw in the number arrangement of 3; 6; 12; 24; 48; 96 etc. The incorrect application of the Titus Bode law lies in subtracting the figure of 3 from 10 leaving 7. The other way of reasoning is to add four each time to the firs value of three starting with 3 and so on. The true significance of the Titus-Bode law is that it points directly to a circular growth of 7 stages. The 7 relating to 10 is a precise derogative of the Roche limit or the Roche limit is a precise derogative of the Titius Bode principle because he two systems interlink.

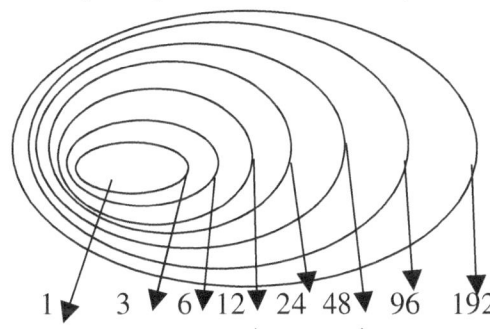

Planet	Mercury	Venus	Earth	Mars	Ceres	Jupiter	Saturn	Uranus
Bode's Law dist.	4	7	10	16	28	52	100	196
Actual dist.	3.9	7.2	10	15.2	28	52	95	192

The TITIUS BODE Principle Outside the sphere

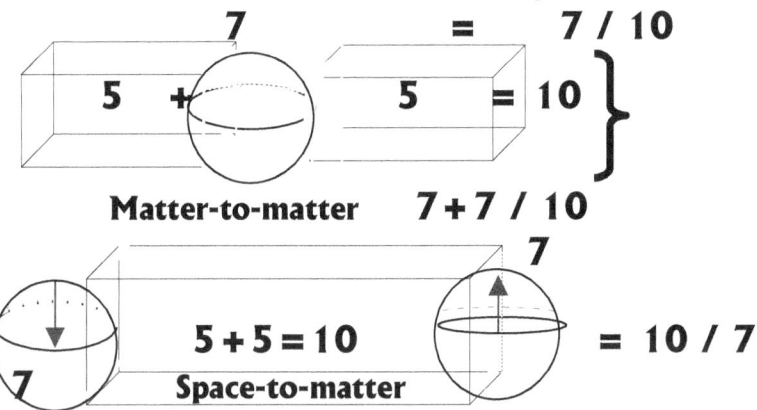

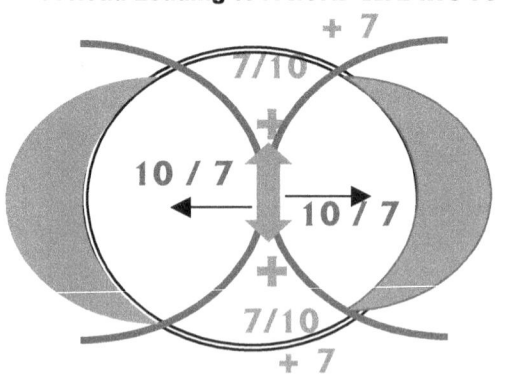

Gravity is motion in space and motion through space. Gravity is a relevancy between space travelling and the time it takes the space to travel. Gravity produces or reduces space during a certain period of synchronized spin of material in motion. Gravity is $a^3 = k T^2$ where it then becomes $k = a^3 / T^2$ In the light of this all other explaining fails the test of accuracy. It is no force because mass depend on gravity and gravity does not depend on mass.

The Titius Bode principle is a relation where space is the ten factors and material is the seven factors. By space being diminished by material one relation comes about and where material dismisses space another relation of seven to ten comes about. The Titius Bode Principle is equal to gravity @ = Π^2 = 9.8696 Proving that the Titius Bode Principle is a product flowing Directly from the growth of singularity forming space-time The Titius Bode principle directly valuating TIME to SPACE = Π^2 = 9.8696 = MATTER HOLDING THE SECOND PROTON COUPLING

Matter in relation (part of) to the total dimension of space.
(10 / 7) \ (7/ 10) = 2.04

1.4285 / 0.7 = 2.04 Taking from both orbiting influences
SPACE DIVIDED INTO TIME

(7/10) / (10/7) = 0.49
.7 / 1.4285 = 0.49 Taking from both orbiting influences
SPACE MULTIPLIED WITH TIME

7/10 / 7/10 = 1 and   10 / 7 X 7/10 =1 Therefore not influencing change
THE PROCESS PARTED USING THE ROCHE PRINCIPLE

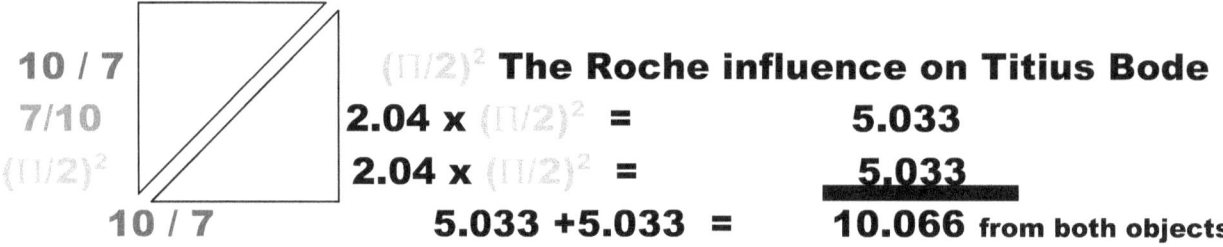

$(\Pi/2)^2$ **The Roche influence on Titius Bode**
2.04 x $(\Pi/2)^2$ =   5.033
2.04 x $(\Pi/2)^2$ =   5.033
5.033 +5.033 =   10.066 from both objects

SPACE DIVIDE INTO TIME

7/10 / 10 / 7= 0.49
0.49

10/7 / 7/10 =.49   10/7 / 7/10= .49

.49   +   .49   = .98
.98 X 10.066 = 9.8 =Π^2
TIME SPACE = Π^2 = 9.8696

TIME SPACE = Π^2 = 9.8696 = Space and time in a dimensional implication.

In every sector the directional flow will provide a distinct meeting of Π linking r to Π² and this allow the time component in the rotation.

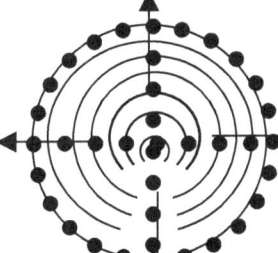

As the meeting of r points to a very distinct different r in direction such a point of meeting opposes the other points in meeting and will lead to destruction of the form Π in any the event of any value changes by Π changing Π² and r.

Keeping these factors in mind it is clear that Π² are the choice of gravity and not r².

Every quarter provide a distinct value that indicates the progress of the flow of time from the one point Π to the next point Π.

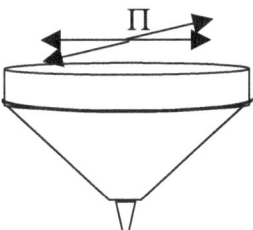

Any changers occurring in Π will lead to a an unequal triangle providing two different values to r and will alternate the link between r and Π² bringing about different form (Π) and time (Π²). When singularity forming the lines of the triangle is not in equilibrium the triangle will destroy the matching of half circle.

It is motion that brings about 3d as much as it is motion bringing about gravity Π²

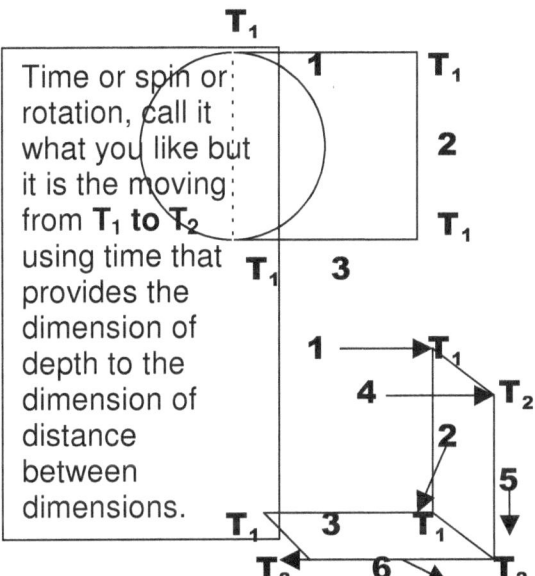

Time or spin or rotation, call it what you like but it is the moving from T_1 to T_2 using time that provides the dimension of depth to the dimension of distance between dimensions.

In the motionless Universe there will be on point in time and that point will represent k more than anything else. Every point being T_1 will only show the extending of **k** from singularity to that specific point. The fact that T1 indicates no motion brings the universe to a stand still and to a flat Universe.

It is that which give **k** the coming from the first dimension and by only extending from singularity it the forms two more positions becoming a^3. $k = a^3 / T^2$ but remove T_1 to T_2 from he equation and only **k** remains $a^3 / T^2 = k$. By the effort of spin or motion the universe becomes the three-dimensional object it all seems to be.

That too forms the answer about the question concerning the Titius Bode gravity implicating of cosmology. The seven sides are linked by rotation nothing changes because there is a steady linking to the inside centre of the sphere. But it is to the outside that this rotation brings about dimensional complications. There are five T_1 points moving to five T_2 making contact with five moving points. The moving non fixed points is the point before reducing by five to the point after reducing by five that bring along the ten points in stead of the five to one point as it is the case with the Lagrangian system. Two points relating to seven points coming from by continuing in the same direction it is going to remaining seven points as going. That means in matter there are five times two points relating to seven in a moving constant and seven fixed rotating points

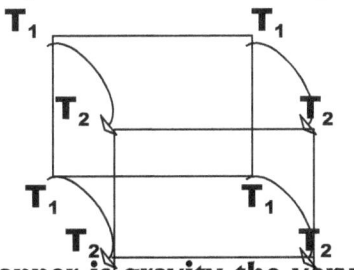

In this very manner is gravity the very same as speed where gravity is the moving from one position to another position and the duration it takes to complete the distance is time.

The reason why man can never fully create the complete 3D is because mans' inability to recreate motion that we find in time. The duration it takes any one point T_1 to move to any other point T_1 will have at the time of the arriving of T_2 produces the 3D Universe we are in. By such means does the Titius Bode changing five relating to seven to ten relating to seven bring about gravity or time or moving singularity. Use the name you like but it is all

Gravity is the very same but it is the recalling of the space by creating motion in the space.

By recalling the space it is also reducing the space because it is counter acting the time expansion provide. That then is clarifying the reason why gravity will always on the limit be stronger than light. At a point it slows the time component down to such extend the space reduces faster in that time than what light can produce motion.

The Titius Bode law is an extending dynamic deriving from the law of the gravity dimensional factor where the space factor in a square of ten relates to a matter factor in the square by half (half since nothing can be in two places in the universe simultaneously) of the matter factor of Π^{7+7} or the square of space (10) relate to the matter factor of 7. From such a point every other point will be opposing any other point not pointing in the direction to which the first point is pointing, whereby it extends the direction it holds. No matter what the point is or where the point leads, such a point holding a specific direction will be unique in the direction it is rotating because at that or any other specific point wherever, it will be directing not in the direction it spins but in the direction flowing from the centre point outwards.

Gravity starts where gravity started in the very first instant. The very place where **k** left singularity and stepped into the 3D **k** extended ever so slightly but never more influential. The point is where k formed a line outside singularity and went from Π^0 to form Π.

The length of the extending is so small it is beyond any manner of human conception or mathematical comprehension but so vital the Universe was the result from that action. With that action of extending k by the very utmost most slightest of margins the Universe we know came into being what it is. The factor **k** went from Π^0 to Π. By the extending of **k** space a^3 came into place. But that space only came about by T^2 producing motion. By **k** extending space came into place, but only through the motion of gravity and antigravity. The motion T^2 produced brought **k** the independence which a^3 secured. The factor $k = a^3 T^2$. Seen from the point singularity holds every aspect there is in the cosmos including other singularity is space-time. From where **k** starts expanding **k** produces space a^3 through time or motion T^2. It is $k = a^3 T^2$. To one point holding singularity that point of singularity holds everything there is and there is only the space-time extending by **k** producing $k = a^3 T^2$. There cannot be space without motion. There cannot be motion without gravity and there cannot be space without gravity. Gravity is motion producing time forming space capturing gravity. It is a confined unit belonging to every singularity extending.

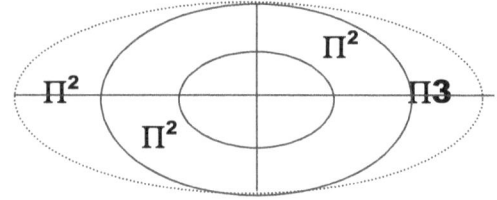

That is the relation there is At the point of cosmic birth the cosmos is Π^3 as Π extends from Π^0 becoming 3D through spin. Where spin ends space ends and gravity ends. It is a unit undividable. Linking Π^3 to Π^0 is the motion Π^2 brought about by spin or rotary motion.

There are four time sectors in the Universe coming about from singularity and singularity fills every one. In the one sector there is a proton Π^2 connecting to singularity Π^3, which is connecting singularity in the form of Π to singularity in the form of Π^3. In that there is the Kepler formula Π^3 (a^3) = $\Pi^2 \Pi^0$ (k T^2). Then there is the other proton filling the second opposing quarter of time implementing the same procedure with the same result coming about. In the third quarter is the first neutron connection following an identical path but linking to a proton forming the space-time. In the forth quarter the motion divided even further by splitting the motion as it conforms space-time from **3 to Π**, which in the end is the motion of Π^2. In all instances space a^3 is confirming singularity k^0 in time T^2 except in the one that produced the Big Bang. The k^0 confirms T_1 to T_2. Singularity is and remains Π therefore Kepler takes on Π but the dimensional impact remains the same. This connecting happens both sides on the Universe Time has four parts to fill and singularity criss-cross fill it be connecting directly through the proton to proton ($\Pi^2 + \Pi^2$) or the proton to neutron (Π^2) or the link of the neutron to electron link ($\Pi3$) Adding this total confirms the mass difference between the proton and the electron at 1836 times. The proton has a mass of $1{,}673 \times 10^{-27}$ which is 1836,12 times greater than the electron's mass of $9{,}109 \times 10^{-31}$. $(\Pi^2 + \Pi^2) \times (\Pi^2) \times (\Pi 3) = \mathbf{1836}$.

It is the mass that space generates where the space has to reduce size by becoming more intense and concentrates 1836 time more when entering the point of singularity.

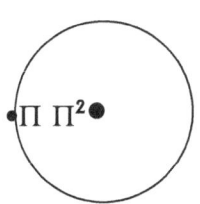

In single dimension seen from one aspect, with single dimension contacting the edges forming the sphere it will still keep the seven positions because the sphere remains a unified structure though apart because of singularity. In the core of the sphere connects the proton alliances $\Pi^2 + \Pi^2$ with the solidity of the neutron holding Π^2 as a second forming value. From the centre to the outside is a connecting of Π, in relation to the Π^2 that brings about the liquid or neutron form. In the centre of all the atoms space is relinquishing a position through the dissolving of heat by means of maintaining singularity. But through it all, another singularity forms in the very centre of the structure, claiming the position where space is the least available that bind the singularity of all the atoms sharing space as a unit. With that evidence I realised there are a connecting of singularity and that connection is electricity. In the cosmos all objects form a sphere. Some solids do not seem to be a sphere and space is no sphere, but the truth is hiding in the way of connecting. At the centre connects Π^2 forming the base of the solid. At any one specific given point forming the surface of the sphere is another marker holding the connecting relevancy of Π. When there is no sufficient heat to form space that will part Π from the other holding of Π, the two will combine in a solid joining connecting as 2Π that translates to Π^2.

In the way space and the sphere connects the sphere will have 7Π points holding a relation to 3Π points not within the sphere forming the 10Π that creation started with. This will mean there is a division forever, and such a division may run smaller everlasting. With fluids connecting it is simple to recognise the sphere as Π for the form will indicate Π as the form of the sphere. By gas forming the connection there are the three points of space being apart and not forming Π, but still holds a relevancy to Π^2 through the value of Π. The gravity applies as much to material as it applies to form installing form. In this way stars are spheres are just more cosmic atom and all rules apply as much to stars being just cosmic atoms as the rules apply to atoms being just individual stars. The Universe ends with singularity and starts with singularity where from this size is just space-time, The rules in the cosmos are the same applying to all in the same manner

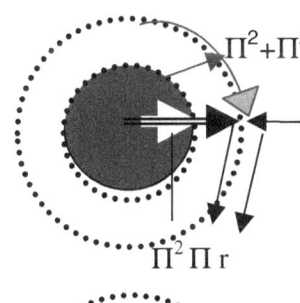

In the circle $\Pi^2\Pi$ which consists of the atmosphere the space surrounding the rotating object will also extend by Π as the concentration of the spinning motion draw or drag on past Π^2 extending the influence of Π^2 by the value of Π. Very clear evidence about this one can see in the Coanda effect. This extending of Π^2 to accommodate Π we refer to as the atmosphere, but physics apply to this extending in the normal fashion. The soil of the structure represents the solid proton being $\Pi^2+\Pi^2$. From the spinning motion Π does not stop at the end of the solid structure but the influence of Π extends and this then becomes the atmosphere. The influence of Π^2 stops at the end of the solid structure but the influence of Π extending plays a most dominant role in the cosmos, although not yet recognised and that factor is most crucial to a better understanding of the implications of laws governing the cosmos.

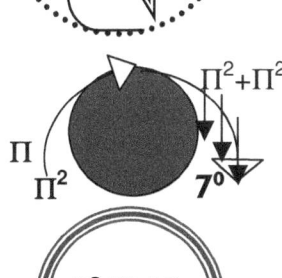

With the circle being $\Pi^2\Pi$ the Π^2 will reflect the circle in the square with Π forming the extending of Π^2. This is an extending of the six Π forming in alliance with the centre Π. This produces that any extension of 6 forming material one further extending goes into space and relates to a seventh dimension. The extending of Π will not end immediately but will carry to the surrounding space the circle influence through rotation. The influence immediately above the circle will have the biggest influence and reduce gradually as the value of Π reduces in the leverage that the space has on Π and a gradual but definite change from Π to r will affect the extending of Π progressively more. The decline of Π will follow the same contour of the circle at 7^0. Every one of the dimensions indicates an individual significance as I shall show later and the increase into space runs by 7^0.

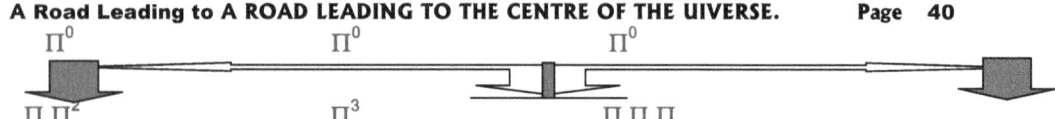

Singularity split the universe into two parts that under no circumstances can ever meet. The one side of the universe perform a balancing act to the other side of the universe that duplicate but never double. The dot started overheating while the dot remained cool by activating gravity

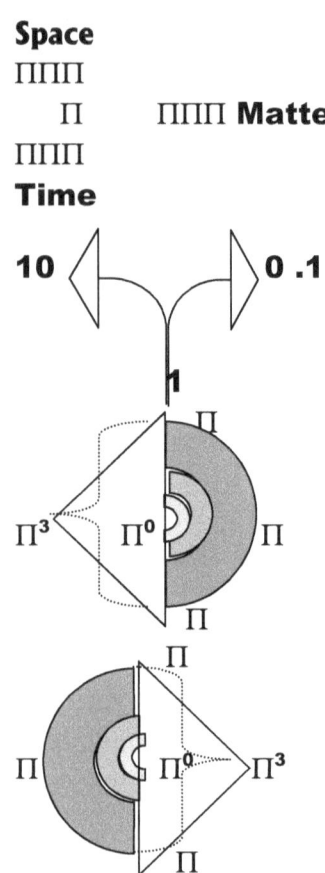

With the first dimension came matter, but also came space and came time splitting the universe in segments of matter relating to space filled with matter and time influencing the spinning matter.

Taking the queue from the numbers line that runs in opposing directions Π^0 going larger as well as smaller. But the centre takes a value of one. It is a private choice preferring $k^0 = 1$ or $\Pi^0 = 1$ but that splits the universe into two part, being smaller and being larger.

It is apparent that one cannot substitute the correct formula used to measure the area of a circle by using $a^3 = \Pi r^2$ because if k is the diameter then the formula must be $k^2 \Pi$. But k cannot be Π because in Kepler's formula k takes the value of the radius. In that case what will the value be of T^2? That places the formula outside the normal use of mathematics practised in the normal sense of $a^3 = \Pi r^2$.

By using the Kepler Formula $a^3 = T^2 k$ it is good to change the values to Π and see what pans out. If k = 1, k at the same time would be k^0. By replacing $a^3 = \Pi^3$ then on the other side of the universe $k = \Pi$ and $T^2 = \Pi^2$. But to secure this k in the centre must be 1 leaving $a^3 = 1(1 \times 1 \times 1) = 1$ and $T^2 = 1(1 \times 1 = 1)$ That complies with Einstein's definition of space-time being: Space-time is a four dimensional position of the universe where the position of an object is specified by three coordinates in space and one position in time.

If k is the middle being $k^0 = 1$ then $a^3 = k^0 = 1 = T^2$ When time is in a shift freezing then $a^3 / T^2 = k^0 = 1$ In order not to overstep my limits by changing valid formulas I changed Kepler's formula to $R^3 = T^2 = 1$.

But the book being written in Afrikaans the R stands for Ruimte meaning space and T is time. From that I deducted that the space used in a specific location will equal the time meaning the density of the heat in space. That brings the proof that space equals heat and space is the same as heat. Heat deforming or exploding is the equal to the space created. Also it confirms the substitute between Kepler and Π is correct

The sphere was the first to come about and only after the sphere could not produce the gravity required to suppress the overheating forming the antigravity did the Universe try other options. Then the atom came in to form…did the atom clusters form…did the material clusters form…did the antigravity apply enough to allow overheating bring about expanding where the relevancy will produce one softer and one firmer structure…from which the antigravity expanded into space. When the heat turned to apply antigravity that eventually produced space, as we now know space to be the Big bang was happening. But before the Big Bang there were some mighty jerks, cracks and jerks announcing the Big Bang to come. It was the proton $\Pi^2+\Pi^2$, which afterwards connected to other protons which became independent Π^2 connecting yet again to other Π^2 particle that with he help of further antigravity overheated and expanded to $\Pi 3$. One can see there were many stages to produce what we now have. However most important is the fact that the laws that once implemented the stages of developing those laws still apply and still present our Universe to what it is.

How many dots was there is a question no person can answer because everything was un-dividable solid and yet it did group together to form every atom located in the 3D.

Individual singularity and governing singularity and group singularity enhancing the gravity every time singularity find an accumulation.

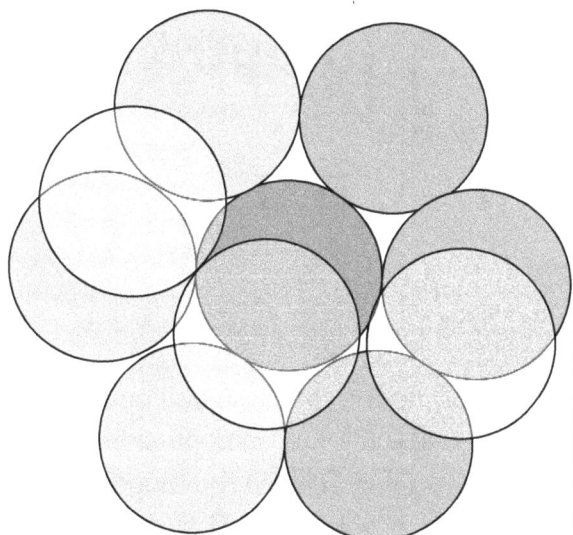

On the one side of the Universe in relevance to all the dots that came before, three dots landed forming one side while three dots formed the second side and three dots formed the third side, all relating to a centre dot which in turn related to the original centre dot from which all the dots came and developed.

The universe came into position by deploying dots supporting other dots and some dots remained dots while other dots went on to become dots of hybrids as it was supporting dots through claiming dots of lesser density and pass that on to dots with larger density.

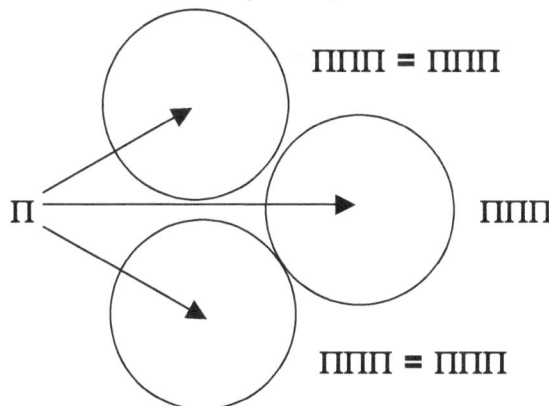

Matter formed where matter had to have $\Pi\Pi\Pi = \Pi\Pi\Pi$
space to occupy since it was to be in some space $\Pi\Pi\Pi = \Pi\Pi\Pi$
therefore $\Pi\Pi\Pi$ **met with** $\Pi\Pi\Pi$ to form the proton in $\Pi^2 + \Pi^2$ because the matter is within the space it holds and another Π^2 employs Π as a representative of singularity. This then placed the seven positions of singularity as the ending of matter and the three squares ($\Pi^2 +\Pi^2$ **and** Π^2) of singularity as the limit of material. The last $\Pi\Pi\Pi$ became $\Pi^0\ \Pi^0\ \Pi^0$ and that became the space producing heat without occupying matter in order to allow heat to be restrained inside the dome singularity provide.

When I refer to an atom it includes all unified cosmic structures holding an excluding formation such as an atom or a star or a galactica does. This is where **k** defines many but as a whole also one a^3 / T^2 determining space within time holding space within time to the value of unifying the lot in one cosmos container. There is little difference in the cosmos to say a lead atom or a giant star or a large galactica It is all space confined to time extending singularity because in the cosmos there is no big or small.

Pythagoras shaping the cosmic beginnings

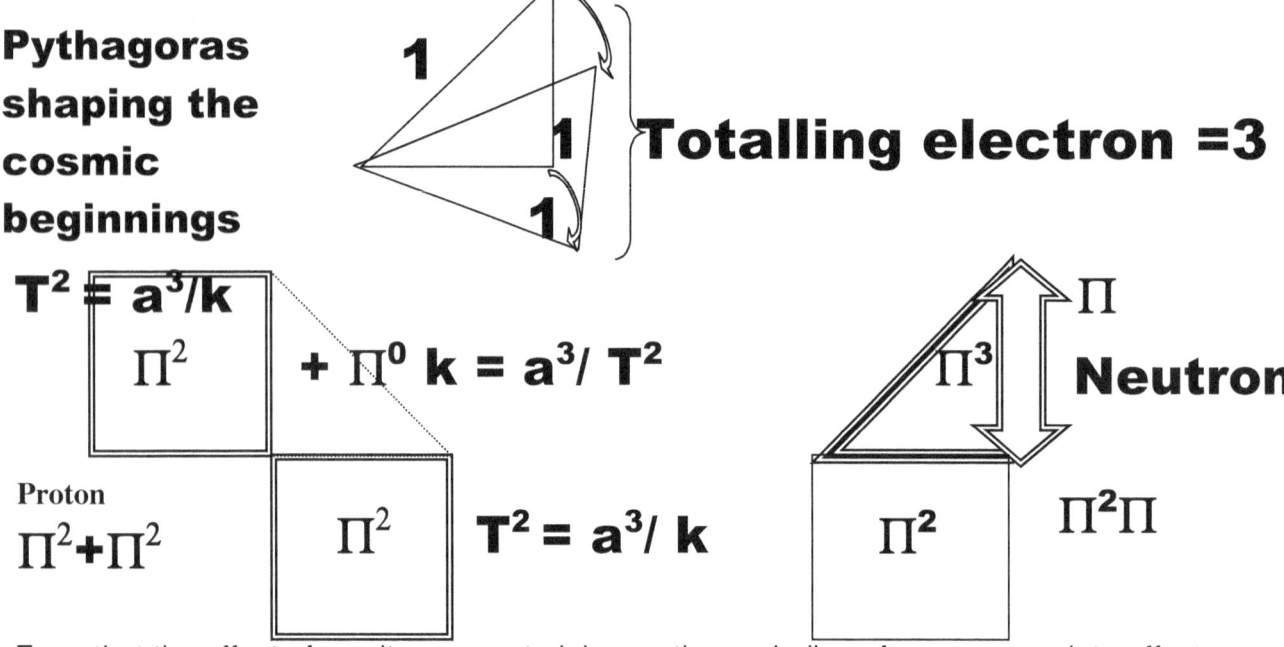

From that the effect of gravity as a restraining on the exploding of space came into effect.

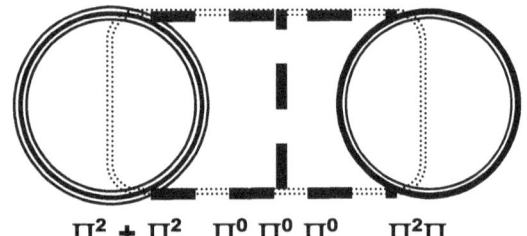

$\Pi^2 + \Pi^2 \quad \Pi^0 \Pi^0 \Pi^0 \quad \Pi^2\Pi$

It is all about relevancies applying the relations gained and lost through relations. If one place $\Pi^2 + \Pi^2$ on one side then $\Pi^2\Pi$ is related form where $\Pi^2 + \Pi^2$ is in the other side of the Universe being on the other side of the relevancy. Then $\Pi^0 \Pi^0 \Pi^0$ will again relate to the other two factors forming the "outside" of the other two being the "inside".

The universe divides into two separate issues because of singularity. Nothing can be in two places at the same time where as all the rest in the Universe has to confine to the law applied by singularity. Objects can only be in one side of the universe holding three parts or in the other side of the universe holding three parts. From the totality three will be a double with six sides too shows, but that forms 3D. From singularity it is flat with three sides forming on either side of singularity.

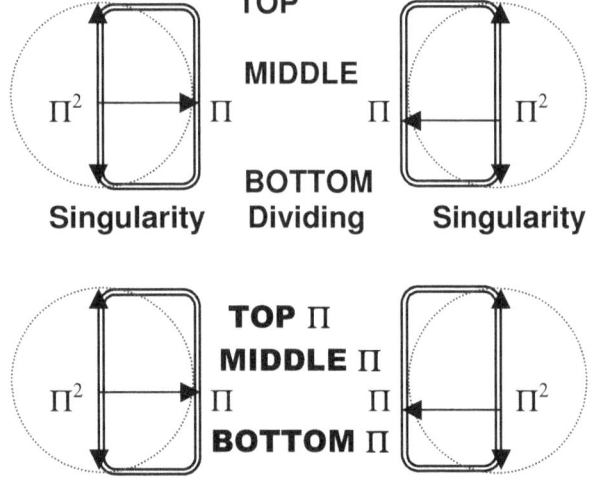

But when the Universe was in the single dimension, all values were Π, therefore every value related to $\Pi\Pi\Pi$ forming three of the same that was very different because it was where Universes met and formed relations. Every dot formed an individual Universe and every dot

The names I use in TOP, MIDDLE and BOTTOM must not be viewed as sides but merely as terminology using names to implicate divisions. Direction depends on positions and positions form a value only when the observer forms part of the cosmos and not part of the observing.

At first when material presented one side of the Universe matter had three sides to show. Matter had to have space to keep matter somewhere in some part of some universe and that made up three positions. Between the two universes **k** and T^2 placed a value but since only singularity applied any values the value therefore was $\Pi^2\Pi$ where $T^2 = \Pi^2$ indicated time coming from 7/10 in relation to 10/7 and $\Pi^2/2$ (proof of that is somewhere in the book) and $k = \Pi$ valued by singularity. When space-time developed 3D the dimensions falling outside the sphere becoming space-heat formed as $\Pi^0 = 1$. The electron holds a relevancy of 3 relating to the Neutron being $\Pi^2\Pi$ and the three keeps the electrons in different universes relating to separate or individual singularity.

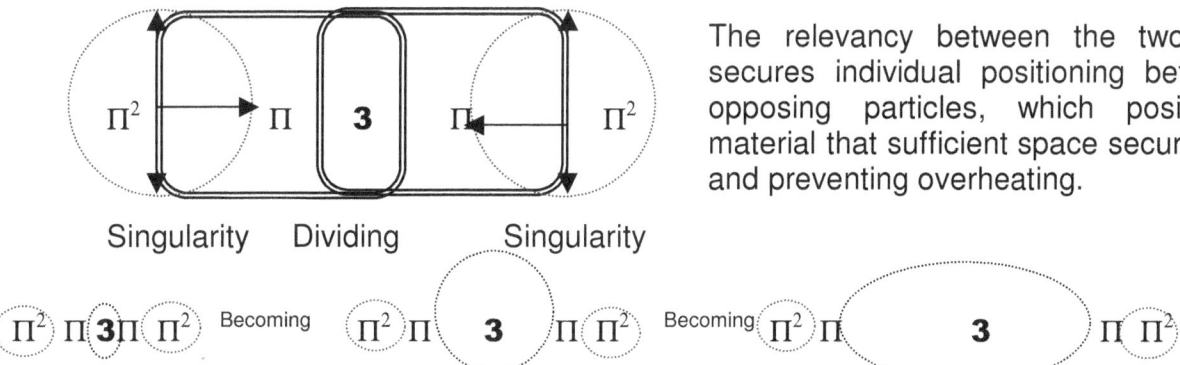

The relevancy between the two particles secures individual positioning between the opposing particles, which positions the material that sufficient space secures cooling and preventing overheating.

Singularity   Dividing   Singularity

As the relevancy between the particles promote overheating or applying antigravity (overheating) to the responding cooling or applying of gravity, the one repels material into space-time while the other is collecting material into space-time. The one loses material and ensures a model of preventing overheating while the other gains material and sustain a model of overheating is prevented The one principle we named the Hubble constant where overheating produces space and the other one we called gravity where gravity is demolishing space, but both phenomenon is at present dominating the flow of time in the Universe and will do so until equilibrium again comes about.

During the Big Bang two things happened. Particles all overheated. By overheating material enable the securing or claiming of more space. Only by overheating or increasing heat can material claim more space. In order to supply **k** with any reason to grow into where **k** then become the fibre of material there had to be a way fitting natural processes to do such extending. The precondition of material to grow and claim space is to accumulate heat, and that must have happened because we can trace the excessive heat there was even today. If k grew, the temperature had to rise. But all temperature was the same in singularity. Singularity is homogeny in all areas including heat. Singularity can generate heat on one condition and that is that is by producing more spin. The Coanda effect is vivid proof of my statement. But to heat it must spin at a higher rate and by spinning singularity then produce more heat. That is in motion and all proof still exist that singularity had and has no movement. Spinning the top will bring the top to become more existed and then attempt to elevate the motion of the top to separate from the Earth gravity. In singularity there was no motion yet and there was no heat yet. Still to get k to extend the temperature had to rise. Let us have a good look at gravity.

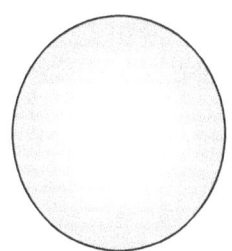

Why would a water drop floating in a space capsule in space, in micro gravity always form a sphere when left capture free form? We all accept that the true cosmic form would be and most probably will be the sphere...but why would the sphere form as the original form when matter is not pre-cast to have any specific form and therefore take on by cosmic pre-cast the sphere as form? We know gravity is there, but qualifying gravity as a force lets the process of investigating science a bit off the hook. Thing become rather simplistic in the modern age when all else is so highly investigated but gravity is merely defined as a force influencing matter. Why would we find in space, where there is supposedly nothing, something we named micro gravity and would bring about micro gravity when gravity is not present? What would cause gravity up to one point and from such a point in the area there supposedly is nothing there is micro gravity. By extending k froze to set the standard for heat. By freezing k came about and set the Universe into a concept other that one unified lump of not being anything of sorts. The freezing action brought on **k** and **k** brought on space through spin. By the extending off **k** did the Universe obtain a^3 / T^2. But it had to produce **k** by freezing **k** into existing from where k produced a^3 / T^2. This we have to understand about fusion. Fusion is about creating a freeze in the deepest of heat and where all is engulfing in heat surrounding all, from that freezing must come about to apply fusing. It is this first action ever that has to repeat to establish fusion. It is in gravity attempting to secure the most heat under the prevailing conditions where gravity eliminates the most space to establish a freezing centre in creating fusion. But that does not solve the indicating action. How did the Universe liberate material and heat/ space from singularity because with singularity comes eternal non-changing-everlasting in conditions remaining in absolute equilibrium. This equilibrium maintains because all development extends form equal equilibrium through out. What evoked change and that is the question the Atheist will never answer but that too is the most basic and ever-lasting fundamentals of the Universe. This split second start before the start, but there are other books I delve into this matter. From the deep freeze came the Hot Big Bang bringing about to Universal displacement relatives being $10 \div 7(4(\pi^2 + \pi^2)) = 112.795$ ands then a second one established 3D by introducing the six to seven sides Universe at a density point of $7 / 10 \, \pi^6 / 6 = 112.162$. There is of course a lot more about this than what I mention at this point. But what made the Universe freeze to form the Universe in space and through time. It had to start with a specific reason applying. Once the process started there was no stop to it, but there is no chance that the initiation of the start was spontaneous by nature. With $k^0 = a^0 / T^0$ all stood still in singularity and that factor is still with us controlling and generating the cosmos. It is there for all not to see and for every one to establish. The effect coming from $k^0 = a^0 / T^0$ and about $k^0 = a^0 / T^0$ is beyond denial. That centre spot in the rotating of objects is up to this point of cosmic development as incapable of having space-time and can only secure space-time as it was capable of at the start and as it will be capable of up to the very end. In only $k^0 = a^0 / T^0$ being present it was as stable as it still proves its stability. There were something external from the Universe that was outside controlling the $k^0 = a^0 / T^0$ Universe that set the lot into space being motion. That is the one aspect natural physics will never answer. What forced the cosmos out of the stability of singularity as $k^0 = a^0 / T^0$. There are two controlling cosmic principles it control of all. The one is gravity and the other is the light performing as the example of the epitome of antigravity.

There has never been any explanation offered about gravity being able to bend light except mathematical calculations producing the factor prevailing and the principle produced. Einstein declared that large objects producing massive gravity could bend light. Human abilities are meaningless in the totality of the cosmos therefore what we perceive as massive and what we perceive as nothing is in cosmic standard amounting to about the same. It is our norm we create bringing on our true Human incompetence in realising limits prevailing in cosmic standards. If large gravity can bend light all gravity can bend light be affecting the flow of light. Gravity does not bend light because light is not a solid that can bend. By dismissing the space through which light travels such dismissing must lead to rerouting or redirecting the path light will displace space-time. The closer the light travel or pass the centre core of any object the closer it comes to the main gravity within the structure and that alone will bring on a greater effect the more the space conforms to density. Gravity is bringing space in relation with time through applied motion. It is about space becoming reduced or redundant because of motion either by space moving or by matter moving. The duration of the time in relation to the space affected bring about the gravity applied. Gravity is space in relation to the centre while at the same time it is the centre coming in new relevancies to space surrounding the material structure. There are always two components to gravity in space being $a^3 = k \, T^2$ It is k in the way the centre stands in the space changing in relation to a centre a^3 but at the same time it is a centre T^2 changing relation of that centre to another centre k applying control. It is a constant re-matching prevailing centre influences on space occupied by material and space not occupied directly by material. It is the time the space takes to bring about new positions in space occupied by motion and through motion that takes a certain duration time to move from point to point. Gravity is motion of space towards a centre and the time such motion takes while that centre is attempting motion away from a controlling centre and the time it takes to complete such attempt. Gravity is speed and speed in space in motion through time duration. Light is the attempt to establish motion not controlled by any centre and the time it takes to establish space between such a centre of space control and the light finding an ability to dispense the space by reactive motion.

Gravity produced space by allowing as much as producing overheating. But what is heat if there is no cold to set the standard for heat become the other end. How did singularity have parted principle of unifying? Singularity froze in applying gravity bringing about particle separation within singularity.

In the investigation of light and gravity and objects and gravity, the mathematical rule of the invert square law must apply without question. But according to the observation of Roche that is not the case. From what one gather through the Roche limit implicating two orbiting structure the opposite is applying. One must accept that although k proves as an indicator it is also much more when complying the thin influences brought about by singularity in the values carried on by singularity.

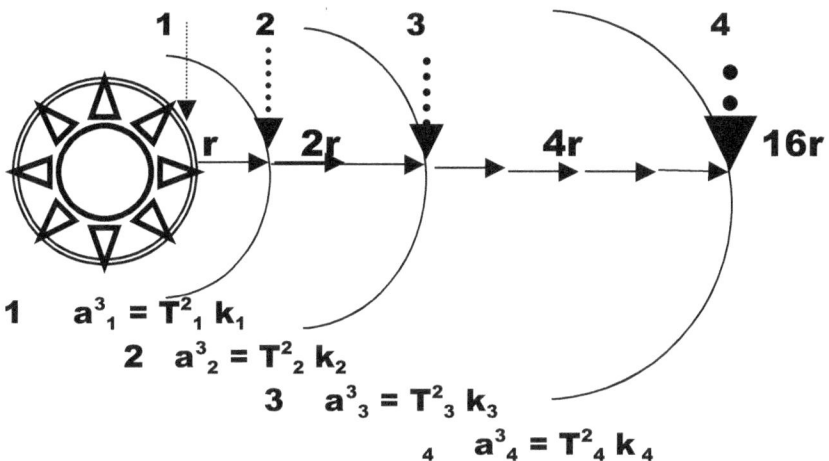

1. $a^3_1 = T^2_1 k_1$
2. $a^3_2 = T^2_2 k_2$
3. $a^3_3 = T^2_3 k_3$
4. $a^3_4 = T^2_4 k_4$

As a^3 increases, so does T^2 as well as **k** increase and with that the influence of gravity per space unit increases with the concentration demise of a^3. But why would that be and what are we missing? Light shows there is an influence out there in outer space, that redirects light's route through space when passing large gravity fields. It is about the relevancy of **k** influencing the a^3 to allow the T^2 of light to divert in route because of influences established by **k** on a^3 and slowing down or increasing the line diverting. In this measure one may also find the Roche limit applying, but to truly understand how the Roche limit comes in place and how the Roche limit work one have to replace Kepler's factors with singularity and singularity extending being Π^3 $\Pi^2\Pi$ and **3**.

There cannot be light without gravity and there cannot be gravity without light but that is on condition that the light we perceive to be light is only the symptom of what the Universe use as light confirmed. Light is not what we think light is but that explaining will too extended to explain in this part. Light is heat concentrated to a form almost material and space is heat dismissed. Therefore light, heat and space is the very same thing and all form the extending of antigravity applying. But gravity is more the redeeming and the re cooping of light where light then is antigravity. Light is space concentrated in motion at speed and gravity is reforming space by motion forming speed. Gravity is speed and light is space at speed. In drawing a most basic picture of light passing the gravity lines extending from any structure, I felt it was most insightful that the brains in cosmology was not able to see why light does *not bend* in the presence of increasing gravity. More surprising was that I found the mathematicians had to call on Einstein for advise on a most ordinary problem. Light does *not bend* when passing large objects. It is Kepler's formula applying, and the evidence is clearly in front of the searcher for truth. But one has to go back to Kepler to re-apply what Kepler formulised and change the significant from Newton's significance.

In the Roche limit the space factor provides space to a solid structure and therefore the value of r is replaced by the value of Π bringing about a square in half of Π. The cube holding 5 to either side removes allowing the extending of Π to indicate position to space.

5/2
Five sides divided by two spheres.

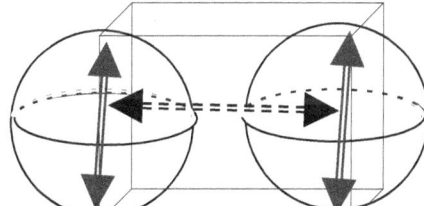

Where Π extends to lock onto the next sphere's extending indicator, Π has to connect to Π forming the square of space and translating that to the half of Π being $(\Pi/2)^2$.

The Roche limit 5/2 becoming = $(\Pi/2 \times \Pi/2) = 2.4674$ as singularity interferes

The space between the spheres divide in half, but because of the extending of Π and not applying r as ordinary mathematics will suggest where Π replaces r the singularity extending from Π^0 will be half of Π in the square of Π = (Π/ 2)² = **2.4674.** In this lies the dynamics why planets have a positional (be it rather a dimensional) relation of 7/10. Half of the five of the Lagrangian points is the Roche in conjunction with singularity. With singularity coming involved singularity will enforce the value change to fit Π.

The Lagrangian ratio is normal dimensional particle layout brought with development through time as a result of the Roche principle. One can see the Lagrangian system by looking at the elements developing clusters of five forming such characteristics of similarity in groups. Looking at how singularity applied connection in 3D there will always be five relating to a centre where that centre carries half of the value. The will be singularity taking a centre with one point to the top of the centre and one point to the side of the centre. The centre then becomes the one side in corresponding to the side the centre takes. From the centre another connecting will be to the bottom the front and the back. Every point will form a position where it will support the centre in displacing space by providing heat in an attempt to secure the prevention of overheating. The function of the liking of five to a centre on as a group effort is with the group work individually to secure the survival of the individuals forming the group and being the group but as a group as well as the securing the group as a unit by a mutual concentrating of space. By halving this effort in the Roche proves the motive behind the forming of the five in the Lagrangian points because of the destruction or development that space less ness then in Roche bring about. The five to one forming the centre of the governing gravity by establishing a principle singularity such a centre improves the gravity effort by all taking part to dismiss heat and create a concentrating heat flow to the centre whereby the group as individuals and as a structure will survive through mutual gravity. This is the effort of the whole cluster in underwriting the centre spot control. It is plugging five spots same as electrons do and in that accelerate the space flow by reducing the space flow as a motorcar carburettor does. In producing mutual support it underlines the individual support required by all participants to prevent future overheating or antigravity. The extending of space collecting and the extending of space reducing prevents overheating by establishing matter in six by connecting matter in six to a centre one. That centre one has the factor of less than one forming the Alfa singularity to the cluster forming the gravity.

Where the Roche factor is the singularity influenced half of the Lagrangian system, the Titius Bode is the dimensional duplication of the doubling of the Lagrangian system in space occupied and space not occupies by material.

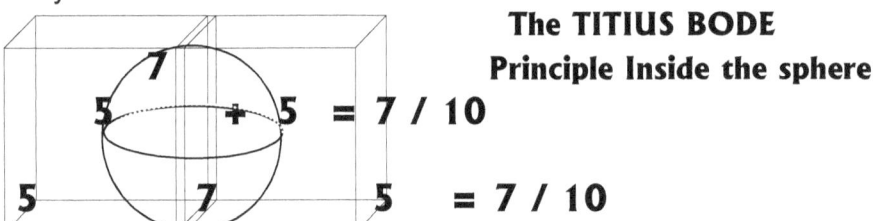

The TITIUS BODE
Principle Inside the sphere

Space-time is a four dimensional position of the universe where the position of an object is specified by three coordinates in space and one position in time

By rotating around a centre that is standing still such a centre forms a divide that separate the unified unit. **Any point will be opposing itself** within the **rotating of 180°** where it **then change every aspect** of its **previous flowing** characteristics it had or **will once again have** in 360° from there. While in rotation from the view point of a bystander it all may seem static and never changing but to the object in spin every next instant in time will be diverting from every aspect it had every second passing, and the direction it held in relation to the direction it held the previous mille, mille second as it will totally be incompatible with the direction it holds the very next mille, mille second of rotation. This is why we can use degrees measuring the circle by (6²) (forming the square relating to matter through singularity) X 10 (square if space) = 360° however it is always in motion. That proves no point can be static or constant, though it may seem that way to outsiders. Although matter is matter, matter can also be anti-matter and moreover form its own anti-matter at the same time. This degeneration of structure is very likely to occur with overheating. Revaluing Π to Π² will bring about a new contact point where Π meets **r** forming another relation in Π². Every time material swap sides it also qualifies as anti matter to matter because if it goes out of orbiting rotation frequency, it has the ability to collide with the same matter it forms union with but is located on the other part of the spin. It then becomes in a situation where Π **revalue to r. Time is** the **changes in relation where Π contacts a different r** not withstanding the many r points there may form because **every r constitutes a different value** to the universe through other ratios and relevancies brought about **by heat and light. Time is the duration it takes Π to rotate between any two given points of r** and therefore must always amount to **a square (T^2)** moving from point to point through the **cube of space (a^3)** in that **duration of time (k)**. With that it proves **Kepler's a^3 (space) $=T^2 k$ (time in the instant of motion)** but motion must continue through a specific value in space where the space-time is maintaining relevant equilibriums throughout singularity connecting.

When concerning a sphere and we establish the Laws of Pythagoras to provide a centre principle all lines running through the centre will be effectively related by groups forming 180^0 and 90^0.

We take a line running between two points as being 180^0 and the rest of the explaining is saved in the accepting part of mathematics. Any one of the two points the line start or ends at is a point in infinity, The start and the end depends on the viewer putting the relevance to favour the side of choice. That puts the point of end or beginning in the spectrum of choice and not fact. Any direction is as equal as all other directions.

A straight line, triangle and half a circle will always have equality in dimensional capacity providing equilibrium being 180^0 because each one shares a common denominator in singularity to the value of Π. As the straight line averts a zero it holds another straight line in place to set about such an averting where the two lines will always carry a relevancy in elation to progress (the triangle) and a common denominator in the start from singularity. This concept we apply as the graph or the vector. By going back to a line, any lines and all lines, the line is a connection of dots in infinity, running from one specific to another specific and avoiding zero or dots. At every point in infinity it dips into infinity coming out on the other side by choice of direction and the direction is unforced and change presents any angle including the straight line, which incidentally is just another angle.

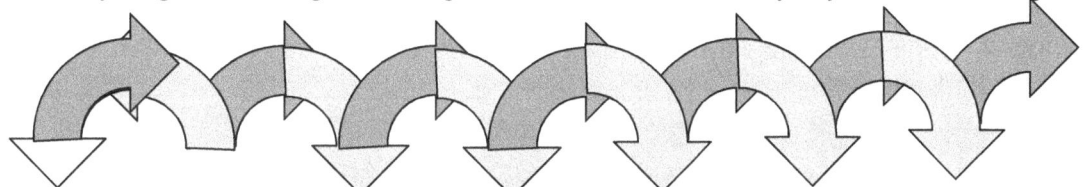

Following the flow of any line such a line is an extension of the previous dot in infinity to the next dot in infinity without any ability to skip or bypass any of the other dots in the connecting line. Any direction change including the remaining of travelling in the same direction is in relation to a line travelling all being the very same. Change does not affect the line.

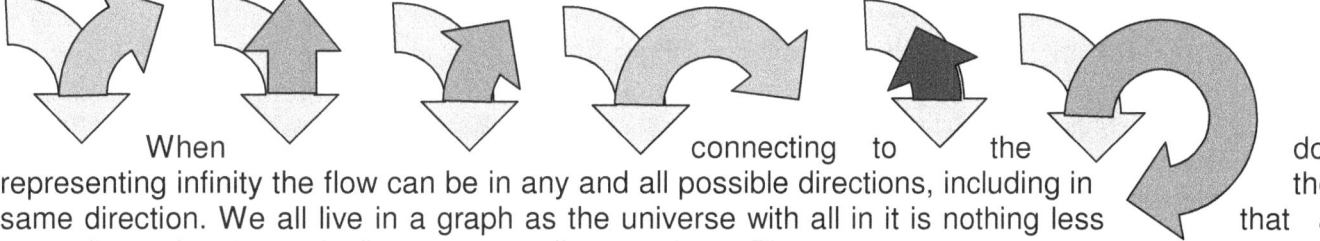

When connecting to the dot representing infinity the flow can be in any and all possible directions, including in the same direction. We all live in a graph as the universe with all in it is nothing less that a three-dimensional graph flowing according to time. That means in the case of Pythagoras the mere fact that the line shows changes in direction does not implicate or affect the line as a tool of mathematics. Whether the line changes into a half circle meeting at the other end again or meeting in a triangle in forming a half square by joining the point where it began, the result still indicate a line flowing between points.

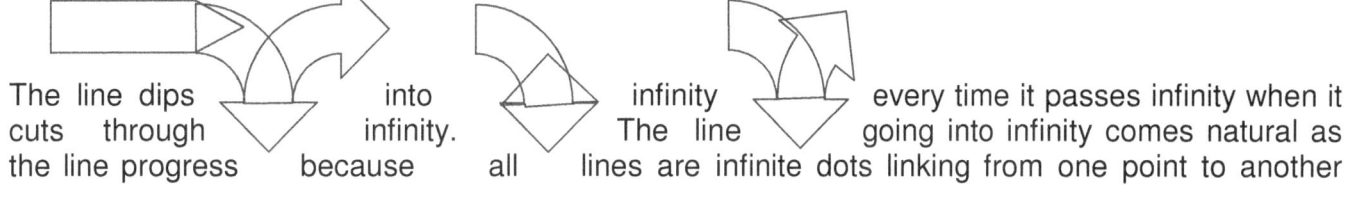

The line dips into infinity every time it passes infinity when it cuts through infinity. The line going into infinity comes natural as the line progress because all lines are infinite dots linking from one point to another point. That brings about that coming from infinity might change in angle bit that directs the route and not the form. The form is all the same

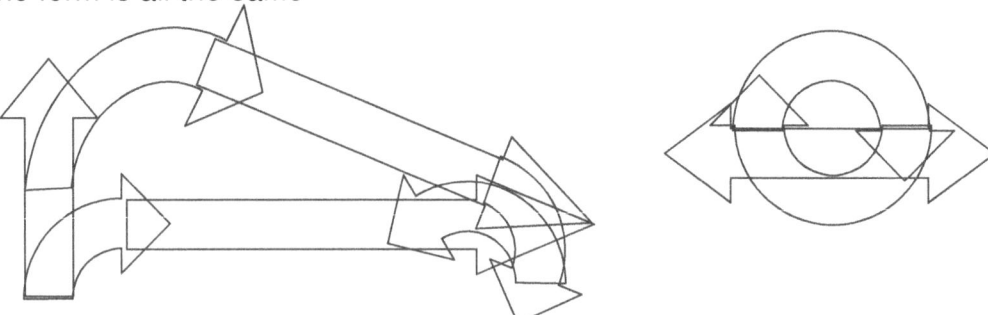

In that way a circle is a straight line following a loop as it comes out of singularity at a different angle and a triangle is a straight line that dipped into singularity but at three stages changed the angle with which the line then left to follow different directions at specific points. From the point singularity observes it still remain a straight line because there is no direction alternation in the first dimension and in that dimension it still remains a straight line in which we on the outside may experience as three forms but are in fact one single line. Only when the direction changes completely in reverse the line doubles in value but comes from multiplication for instance 2Π become Π^2.

Not long after the law of Pythagoras was understood where Pythagoras introduced mathematics Eratosthenes of Syene made as big a discovery as Pythagoras did. But in the one instance the world took notice because the world could see and understand and the other instance the world disregarded the findings because the world did not see what the implications was. The same apply to aircraft flying and when the aircraft wishes to escape the earth's singularity hold it has to comply with the laws laid down by the earth. The seven becomes as big a part of the concept as does Π as it all interacts

When we dissect the sphere to the bone of singularity we find Pythagoras at that bone. Because the sphere centre is the infinite point such a point will cover the infinite spot of the outside of such a spot where singularity produce matter in the centre extending. From the family of lines there will be three lines being close relatives crossing three lines at the centre and pointing to six edges at the birth point of material. That is then six points forming a^3. The factor of **k** is extending results in 3D lines with six points being material. From the centre of the sphere six but that six is standing in relation to the next six point which are they in a different location in time. From Kepler we know that space in 3D or a^3 is produced by time or motion in T^2. **k** $=a^3 / T^2$. Matter then holds six singularity positions where motion puts the six 6 effectively in 3D by the square of motion T^2. $a^3 = 6$ and $T^2 = 6^2$ $6^2 = 36$. Then space comes into play. Where six holds the material position the very next position is where space start and that then must be seven

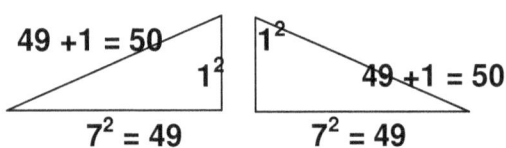

50 + 50 = 100 but that is in both sides of the Universe.
$100^{\frac{1}{2}}$ = 10. That places space unoccupied in the factor of 10 being also 7 + 3 = 10

That will bring about that a circle is an infinite number of six points in the square and in the square of double space. From that very simple mathematics comes the proof of the Lagrangian system, the Roche principle and the Titius Bode principle.

In the Roche singularity apply all three components

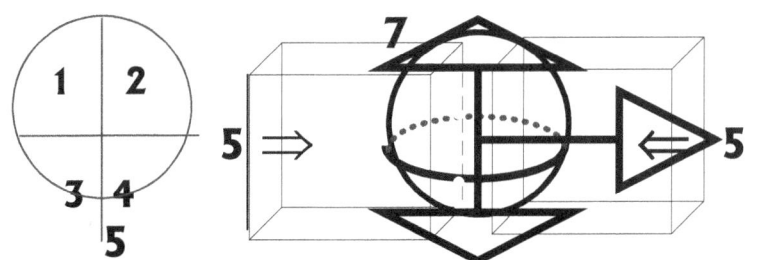

In the Roche limit the **straight line** forms part (1) and the **half circle** is part (2) and the **triangle** forms part (3) to singularity (4) Holding 5 points outside singularity

The influence of singularity as the extending of Π into space links Π^2 to r and forms 2(5)+2(5) =10+10=20

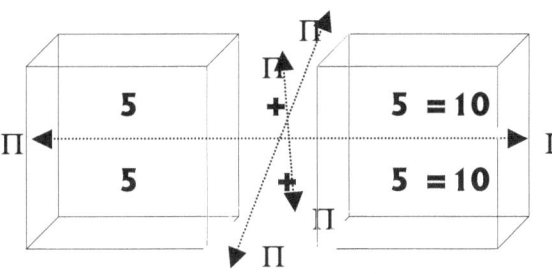

From the position of singularity there are different values in Π where each indicate a position. The value it represents being $\Pi\Pi\Pi$, Π^3, Π^2, Π and Π^0

From there it influences singularity in the triangle flowing through to the half circle. It is an interaction between circular and linear motion as the value of Π continuous past Π^2 (at the end of the solid) and every cosmic structure holds an individual and specific singularity. The field where Π extends we call the atmosphere having a value of 21.991 / 7, which is Π.

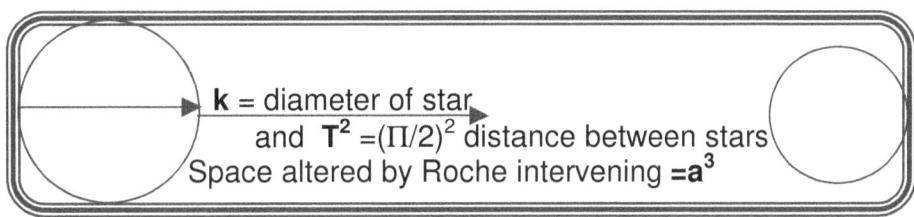

k = diameter of star
and $T^2 = (\Pi/2)^2$ distance between stars
Space altered by Roche intervening $=a^3$

Being at... Π **Going too...**

Singularity: a mathematical point at which certain physical quantities reach infinite values for example, according to the general relativity the curvature of space-time becomes infinite in a black hole.

Coming from $=\Pi$

Singularity $=\Pi$

Where singularity holds position in the centre of any and all rotating objects as a value of Π merely applying movement (in the form of atoms) qualifies all matter to be space-time. It does not only fit the description of space within Black Holes but it fits all stars where singularity becomes part of all the stars from the minute to the largest cluster of matter.

With no line starting from zero because there is no zero as a mathematical fact, then all particles hold the point of infinity and not merely the Black Hole of nothing,

From that argument one may conclude that all stars will become Black holes depending on the gravity increase they may generate.

Through rotation encircling the point of singularity and matter is (1) coming from, (2) being at, (3) as it is going too in one movement in relation to the specifics of the centre point being singularity, all matter then qualifies to form space-time.

That confirms our vision that expresses Kepler's formula $a^3 = k\, T^2$. a^3 holds three Π^3 point in time T^2 from point to point Π^2 relating to $k^1\, \Pi$.

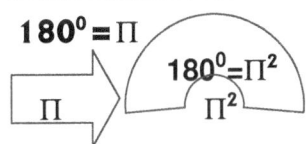

$180^0 = \Pi$
$180^0 = \Pi^2$
$180^0 = \Pi^3$

The triangle, the half circle and the straight –line has two things in common, they share 180^0 as a mutual value and they are part of singularity.

Using the concept that gravity applies Π as the circle factor Π as well as Π² replacing r² the replacing by Π brings two values as Π and Π². That I found is the case with gravity and will be apparent when explaining the sound barrier as well as the Four Cosmic Pillars. In order to create a distinction I remained using r as the indicator of the cube or non-circle that has vacant space and by vacant space I refer to non-solid structures. In the solid structure I use Π as a value for reasons that will become apparent in due time.

Gravity does not apply mathematical equations to the letter as we would like, but rather use Kepler's thinking by enlisting an average gravity applying through out because it never favours and is equal every where. In gravity one find the extending of Π implementing Π² on average as a unit and not the radius r as a specific.

Looking at the affect of gravity it shows the precise quality of no distinctive point, as gravity never seems to end at a point but flows all over affecting all that holds a position in its sphere of influence. The gravity coming from China meets the gravity coming from America at no particular spot but intermingles without distinction. This takes mathematics back to another fact beyond normal explaining. But the Lagrangian system proves much more than dimensional interlinking, it proves Pythagoras in principle.

No object can be in two spherical quarters in the same time, but has to alternate in aliens to the space in accordance to time rotation.

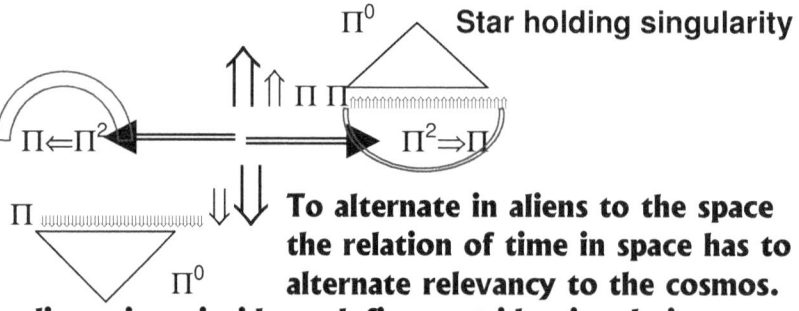

To alternate in aliens to the space the relation of time in space has to alternate relevancy to the cosmos.

Singularity holds five dimensions inside and five outside singularity as matter and space forming space-time. The ten dimensions I named the atomic relevancy is also showing the double value of singularity as singularity extends into as well as beyond space. The atomic relevancy is $(Π^2+Π^2)(Π^2 \times Π \times 3) = 1836$ that is the mass relation between the electron (3) and the proton. Proton = $(Π^2+Π^2)$ Neutron = $Π^2 Π$. The atomic relevancy holds the dynamics of singularity control.

LAGRANGE (-TOURNIER), JOSEPH LOUIS DE (1736-1813)
French mathematician, born in Italy. In celestial mechanics, he studied perturbations and stability in the Solar System. He examined the three-body problem for the Earth, Moon and Sun (1764) and the motion of Jupiter's satellites (1766). In 1772, he found the particular solutions to the problem that give rise to the equilibrium positions called Lagrangian points. Lagrange also studied the Moon's liberation. LAGRANGIAN POINT One of five points at which small bodies can remain the orbital plane of two massive bodies; also known as liberation points. Three of the points lie on the line joining the two massive bodies: L_1 lies between them, while L_2 and L_3 have the two bodies between them. These three points are unstable, slight displacements of a body from then resulting in its rapid departure. the fourth and fifth points (L_4 and L_5) each form an equilateral triangle with the two massive bodies, 60° ahead of and behind the smaller body in its orbit around the larger one. A well-known example of bodies flying at the L_4 and L_5 Lagrangian points are the Trojan asteroids in Jupiter's orbit. Among Saturn's satellites, Telesto and Calypso lie at the L_4 and L_5 Lagrangian points in the orbit of the much larger Tethys. In similar fashion, tiny Helene precedes Saturn's satellite Dione, keeping 60° ahead of Dione. The Lagrangian points are named after the French mathematician J.L. de Lagrange, who first calculated their existence.

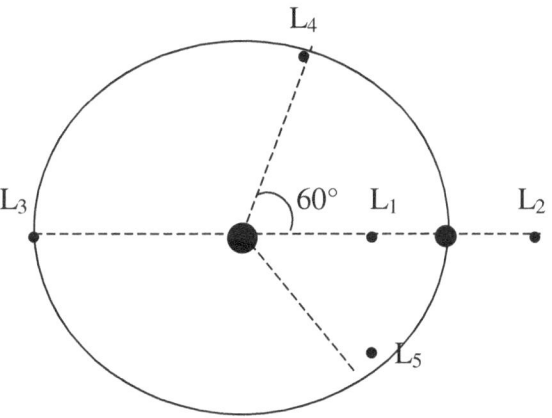

LAGRANGIAN POINT:
The Lagrangian points are five equilibrium points in the orbit of one body around another, such as a planet around the Sun

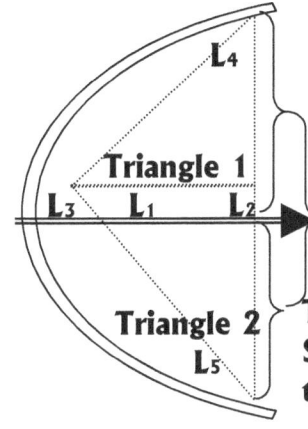

The Lagrangian System implicating the five positions extending from singularity

Singularity dividing the cosmos	1 Half circle = 180°	L_3 L_4 L_5
	2 Triangle 1 = 180°	L_3 L_4 L_5
Each triangle claiming a side of the universe	3 Triangle 2 = 180°	L_3 L_4 L_5
	4 Straight Line = 180°	

The half Circle = 180° combining as a Sphere when comprising Singularity in the matching of the value of the straight line forming the half circle and combining as the triangle and all are equal 180°

A I indicated previously there are many stages where singularity connects either through space-time implicated or more directly. Π Connects Π whereby Π will join > by 1/10 and 10/7 standing related by the linking of matter and space and space and time with singularity applying motion to space creating gravity in the process. The fact that $(\Pi^2 + \Pi^2)$ being the proton forming time is because the gravity created is the proton in motion. It must be clearly understood that only by the motion of the proton in the square is 3D created and not by space. The contact that the first motion makes with the rest of the Universe forms time in the splitting to fast to repeat. By the proton applying time that action of spin creates the duration of time being so close to eternal that all other time following will be much faster but seem much slower. By motion duplicating space can 3D establish space and by connecting to space through time can 3D in slow motion become reality. The motion of heat in space is a result of the motion of heat bringing space. But since that is so close to singularity that is hardly a human concern. By alternating alliances with singularity and on the other side of the Universe by aligning with space created can space establish the time component within the realms of time-eternal. One should remember that such time in motion at that point where the proton is meeting singularity such time is 1836 times longer in duration than the speed of light represented by the speed of the proton. The mass is a result of gravity and not the cause of gravity. The gravity is a time discrepancy coming about from space depleted in size. I essence it is a time zone that much slower. The proton is filling the four quarters of time and the gravity the proton produces fill the space next to time or if you wish the gravity filling the space. If we go back to the Roche with the second singularity being to clues we find gravity Π^2 being halved in thee square of the half by space filled to close to singularity being confirmed.. That means the gravity being Π^2 is filling the space with gravity. That filling by gravity will extend to a marker pointing singularity as the extension thereof which, then will be Π. The gravity committed in the space immediately following time $(\Pi^2 + \Pi^2)$ the proton commit once again to other time or gravity factor applying to other space-time. That means the space of material in the atom or in the star which is just another cosmic atom holds the seven positions of which six is in material $(\Pi^2 + \Pi^2)$ and (Π^2) and one indicating space within the atom being the seventh marker as (Π). This fills the atom that was about before the Big Bang. Then came the part that concluded the Big Bang. Three dimension being part of the atom falls outside the atom but is part of the atom as it produces a relevancy attached to the atom forming a value of space in between particles. That concludes the atom as $(\Pi^2 + \Pi^2)(\Pi^2)(\Pi 3)$ = 1836 time difference.

The network of individual singularity not only provide spinning through governing singularity in the sphere but also provide spinning in the geodesic through out the cosmos linking all matter to matter in a network no one will ever come to understand in full. In the sphere the four squares forming the triangles linking the lines to the half circles holds space in time maintaining singularity of different assortments. In view of the matter-to-matter Roche factor where the factor consists forming relation between particles occupying densified space-time of where ($\Pi / 2 \times \Pi / 2$) relating to the foursquare triangle the value of gravity Π^2 comes in position as $\Pi^2 / 4 \times 4 = \Pi^2$.

By implication the Lagrangian five-point system is time formed with securing space displacement. Every four points form a securing proton relation of holding position to secure ($\Pi^2 + \Pi^2$) in relation with centre singularity Π and the fifth is space filled with gravity Π^2. This is a relevancy applying all around and implicates any position to all other positions. This is gravity converting space to motion. It is space relating to time. It is the five of Pythagoras where the three of the triangle relate to the square of the line. It is $\Pi^{3+2=5}$ in relation to k or Π^0 being one as it refers to singularity. This is directly forming the basis of Mathematics or the basis of the Creation is the principle of space in motion. Our interpretation leads to out mathematical straying from facts.

The second one also fits in the singularity influence on the Universe.

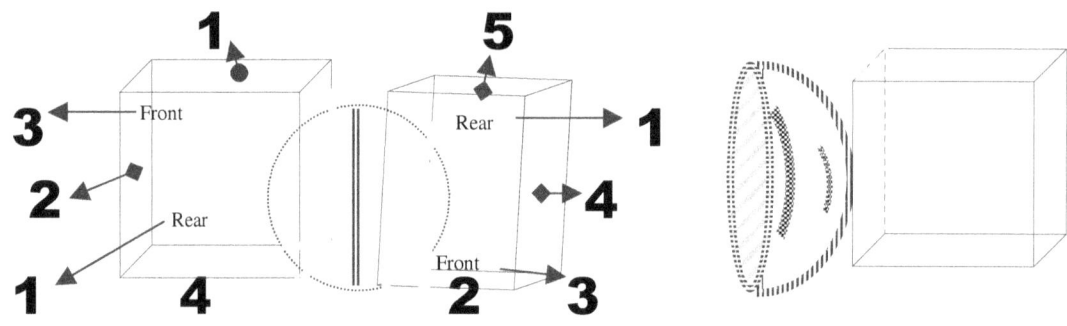

1 Relating to 5

Once again that points back to Pythagoras and Kepler where $a^3 = k T^2$. The line forming the square proves the space included. $a^3 = k T^2$ presents $\Pi^3 = \Pi^2 (T^2) \Pi(k)$

The value of singularity stems directly from the law of Pythagoras or Pythagoras is the result of the average of singularity. With the shortest line being a dot, all lines must start from a position implicating Π. A circle is a square without corners implementing Π and a half circle is therefore a triangle without corners. The corners are the factor that confused every one in the past. When replacing the value we normally attach to circle being r with Π, the law of Pythagoras becomes quite meaningful and mathematical.

By placing a connecting circle on the sides of the triangle half a circle forms. By implicating Π as a relevancy and not the straight-line r, two values of Π applies to each circle, and the straight line is no longer r, but is Π^2. This will bring about that each circle holds half the square value implicated to the allocated conditions applying to Π in that specific instance. By adding the two half squares forming the two half circles and then calculating the square root of the total that then forms the average diameter, an average of Π in the connecting line will come about. As both lines are the straight line forming singularity coming from one line being Π, the connecting line then must be the average of the two lines as Π^2.

A straight line a half circle and a triangle always have equality in dimensional capacity with of all sharing a calculated use 180^0. The reasons why this is the case is not apparent from the 3 dimensional positions we find ourselves using. Taking it back to single dimensional concepts it is very sensible indeed. Place the concept in line with Kepler within the boundaries of singularity where a line and a point and a position is inseparable but at the same time being the major divide. Leave out the symbol used and the dimensions are 1+2=3. The relevancy in Pythagoras is that the square plus the value of one line will produce the space within a square. It is two squares indicating one line but also a squire relating to a line giving an including space.

That is what **the law of Pythagoras says.**

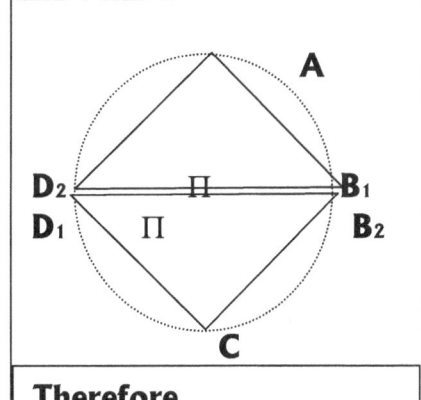

$(D_2 A)^2 + (B_1 A)^2 = (D_2 B_1)^2$ (PYTHAGORUS)
$(D_2 A)^2 = (B_1 A)^2$ (EVEN SIDED TRIANGLE)

$2(D_2 A)^2 = (B_1 A)^2$
$(D_2 B_1)^2$ (DIA. OF CIRCLE) AND ABCD EVEN SIDED SAUERE WHERE $AB = BC = CD = AD$

$(D_2 B_1)^2 / 4 = (AB)^2 + (BC)^2 + (CD)^2 + (AD)^2$
$2(D_2 A)^2 = (D_2 B_1)^2$ BUT $(D_2 B_1)^2 = \Pi^2$ (Replacing r^2)

$(D_2 A)^2 + (D_2 A)^2 = (D_2 B_1)^2$ $[(D_2 A)^2 = (B_1 A)^2]$
THEREFORE $4(D_2 A)^2 = (D_2 B_1)^2 / 4 = (\Pi/2)^2$

**Therefore
The Roche lobe is
$= (\Pi/2)^2$**

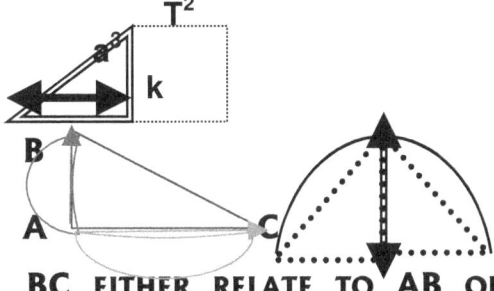

BC EITHER RELATE TO AB OR AC AT ANY GIVEN TIME OCCUPYING SPACE AS MOVEMENT DICTATES DRECTIONAL CHANGE THROUGH DIRECTIONAL FLOW

By having **k** in relation to any other side in the square T^2 will produce that area a^3 in three lines meeting committing three lines to space included. Mathematicians only see the two squares adding and when rooted it forms a line again. Yes that is the case but that is the proton part Pythagoras gives to the cosmos. There are so much more than that.

Placing singularity in the position of filling a centre it is two positions added to three positions will conclude the five points coming about. With the normal extending of singularity it will always form the triangle in a half circle whereby Π relates to the cube by 5 points to either side of the line singularity forms.

Thus there are 10 standing related to seven and visa versa.

By calculating the 4 squares in the circle with the dimensional changing of space (5) becomes the twenty. It is a triangle forming an area through two squares matching and it is a triangle receiving a three dimensional position in the flat universe by connecting three adjoining lines, It is space coming from time coming from the centre of singularity. The normal flow will allow singularity extending to 10Π but when singularity blocks another sphere in singularity the two will form a joint value and by this joining the larger will dominate the space as well as the time of the lesser taking control of the surface and the atmosphere. Through this the Roche lobe comes about with all its other dynamics I describe farther on in the theses. The principle is the same, which we know as the conducting of lightning and Jupiter uses it extensively to implement this action.

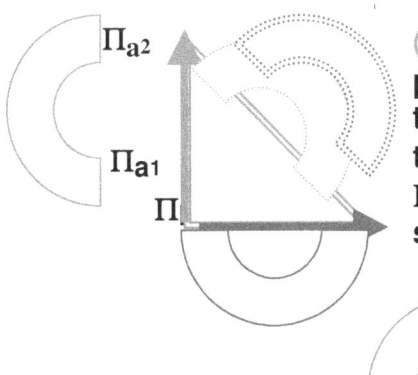

$(\Pi_{a2} \times \Pi_{a1}) + (\Pi_{b1} \times \Pi_{b2}) = (\Pi^2_a + \Pi^2_b) / 2 = \Pi^2 =$ **gravity and that is proven by Pythagoras. Gravity is the average movement of matter through space in time determent from the position where matter in the sphere meets space in the cube from a point of Π to a point of Π^2 In this the figures of $2(5) = 10$ (space) stands related to 7 from singularity as (matter)**

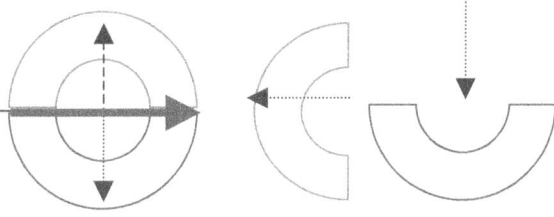
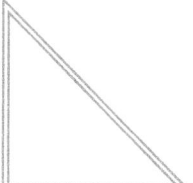

From the star holding a dominant point or most valued point in singularity it affirm all three other structures, each holding singularity individually and in a compliment of 5.

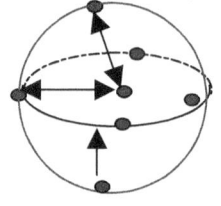

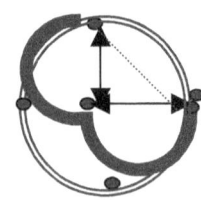

In the sphere there are never only one direction implicated in movement. Movement are always in relation to the centre position because as a line goes up it also goes in or out. When a line goes north or south, it also comes towards the centre or going away from the centre. There is always relevancy present in movement. As this moving indicates direction it also apply Π^2 for indicating value forming the time factor. Because every moving line represents one quarter of the sphere in relation to the rest of the sphere and the line also indicate the relevant position between the point indicated and the point in the centre it is a relevancy of singularity in progress. By connecting the line, as Pythagoras will suggest the singularity within the sphere become a specific value indicated representing one half circle.

The TITIUS BODE
Principle Outside the sphere

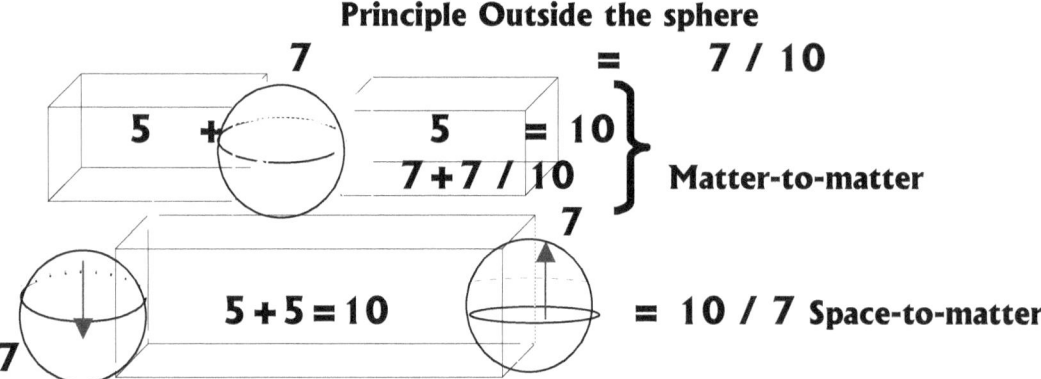

From the dimensional implication comes about, not only the Doppler's effect, but many more of phenomenon not yet understood. The dimensional relevancies formed between matter as six, matters end at seven and space at ten, comes the value of Π.

The process is all intermingled and stands in relevancy to one another. The relevancy compliment holds such attachment that none of the factors can even stand-alone. It is the way that science places every aspect in the cosmos as individual and not related to each other that launches the problems of miss understanding. The Value of singularity appreciates or demises by ten fold. For instance, the value of Π will increase by ten every time singularity applies another layer.

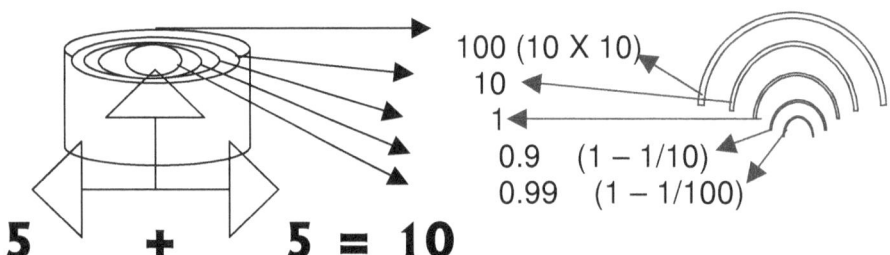

The normal flow will allow singularity extending to 10Π but when singularity blocks another sphere in singularity the two will form a joint value and by this joining the larger will dominate the space as well as the time of the lesser taking control of the surface and the atmosphere. Through this the Roche lobe comes about with all its other dynamics I describe farther on in the theses. The principle is the same, which we know as the conducting of lightning and Jupiter uses it extensively to implement this action. In the Roche limit the straight line forms part (1) and the half circle is part (2) and the triangle forms part (3) to singularity (4) Holding 5 points outside singularity. Every aspect connecting to the universe changes everything it holds totally and becomes the anti-matter to which it was matter $180°$ previously.

Gravity is a relation not bound by borders and the influence stretches seemingly indefinite. As expansion comes about the expansion will have to follow what gravity produced initially.

A Road Leading to A ROAD LEADING TO THE CENTRE OF THE UIVERSE.

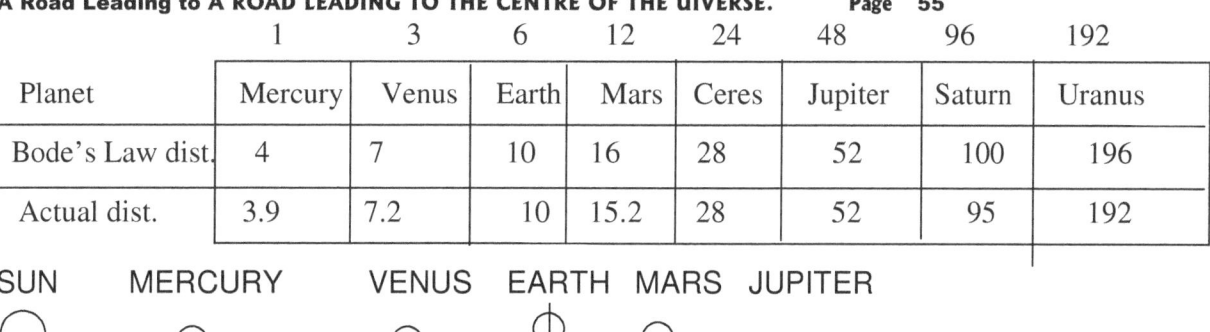

	1	3	6	12	24	48	96	192
Planet	Mercury	Venus	Earth	Mars	Ceres	Jupiter	Saturn	Uranus
Bode's Law dist.	4	7	10	16	28	52	100	196
Actual dist.	3.9	7.2	10	15.2	28	52	95	192

SUN MERCURY VENUS EARTH MARS JUPITER

B_1 B_2 → $\Pi^2/2\Pi$ Π^2 10Π $\Pi^2 + \Pi^2$ Π Then the rest.
(This I shall explain)

The spherical positioning layout forming the Titius Bode Principle

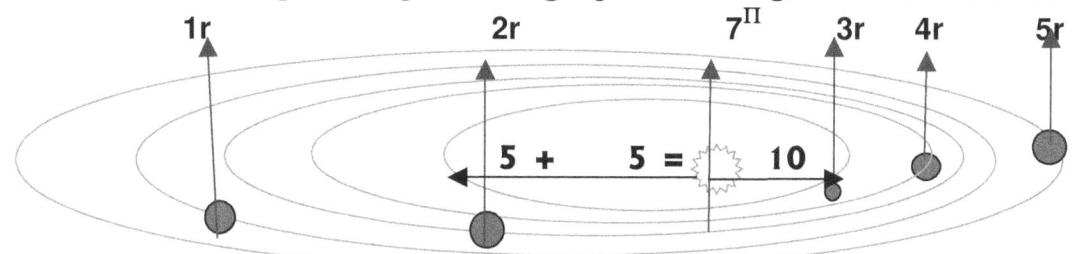

From the matter-to-matter relation in the Titius Bode configuration there are 7 / 10 + 7 / 10 = .7 + .7 = 1.4

From the space-to-matter relation in the Titius Bode configuration there is 10 / 7 = 1.42

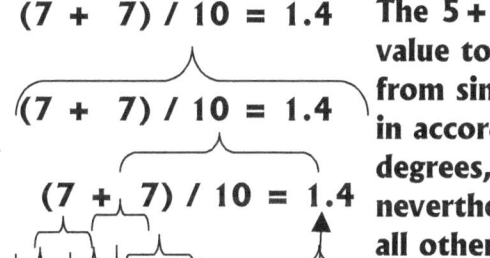

$(7 + 7) / 10 = 1.4$
$(7 + 7) / 10 = 1.4$
$(7 + 7) / 10 = 1.4$
$10 / 7 = 1.42$
$10 / 7 = 1.42$
$10 / 7 = 1.42$
$= .7 /\!/\!\backslash\!\backslash 1.42$

The 5+5=10 is a position of dimensions as space loses value to singularity. The 7 that matter diverts in points from singularity may seem, as coincidental but is valid. Still in accordance to our perception valuing the number in degrees, it seems coincidental but if it is coincidental, it is nevertheless a figure of diverting proven as accountable in all other calculations and plays a most dynamic role.

The Lagrangian 5 point system results as much from the Curvature of space-time as does the form the Black Hole holds. The Galactica is the opposing equivalent of the Black Hole and has identical but opposing similarities being the five points positioned to singularity. The galactica is generating space and the Black hole is degenerating space.

= 1.4 /\\ 1.42 Because the space-to-matter is in the square at 10 placing the matter-to-matter at a square of .7 + .7 = 1.4 the space-to-matter forces the matter-to-matter to double the distance by number as structures are place father from the mainΠ^0 maintaining singularity.

Reasons why this does not fully apply to the solar system I give in book # 7.

In this maintaining of cross referencing of singularity located in individual atoms providing spin to the governing singularity that maintain structural form in solids, many factors of singularity all form a close knit network and being inseparable as one unit, by the same margin it also is strictly individual to a point of destructing. From the inner or governing singularity outward all is concerned as spade-heat. From the dividing singularity only one reference holds a matter value forming the position next to the governing singularity and therefore 7+7 becomes a factor and not all the dividing singularity between the point of reference and the governing singularity. That way the star to the outside takes a position doubling the distance every time. In balance everything in space to the outside of the governing singularity is space be it space or matter that makes no difference therefore that is 10.

The extension of Π is well received as a dimensional implication to matter holding seven positions from singularity and space having four quarters through out the rotation of singularity forming the centre to the five dimensions (one side lost to the cube's six sides connecting to the five remaining sides) making the total sides facing space from the point holding singularity at any given instant at a value of twenty (4 X 5 = 20). Then adding the singularity cross of Π being (1+1) = 2 the relation becomes 22/7. This is crude because in more precise calculations it becomes .91 + 1 = 21.91/7 = Π

The sectors provide individual singularity as a means in sustaining governing singularity by which provision comes through maintaining governing singularity the required spin in maintaining cooling. If this process did not apply, there would be no connecting individual singularity to major singularity. The sectors provide individual singularity a means in sustaining governing singularity by which provision comes through maintaining governing singularity the required spin in maintaining cooling. If this process did not apply, there would be no connecting individual singularity to major singularity SINGULARITY BY DIVIDING SPACE INTO MATTER AND MATTER INTO SPACE, ANG ALL OF THIS ACCORDING TO THE TITIUS BODE LAW OF 10 / 7 AND 7 / 10 IN CONJUNCTION WITH THE ROCHE PRINCIPLE OF $(Π/2)^2$

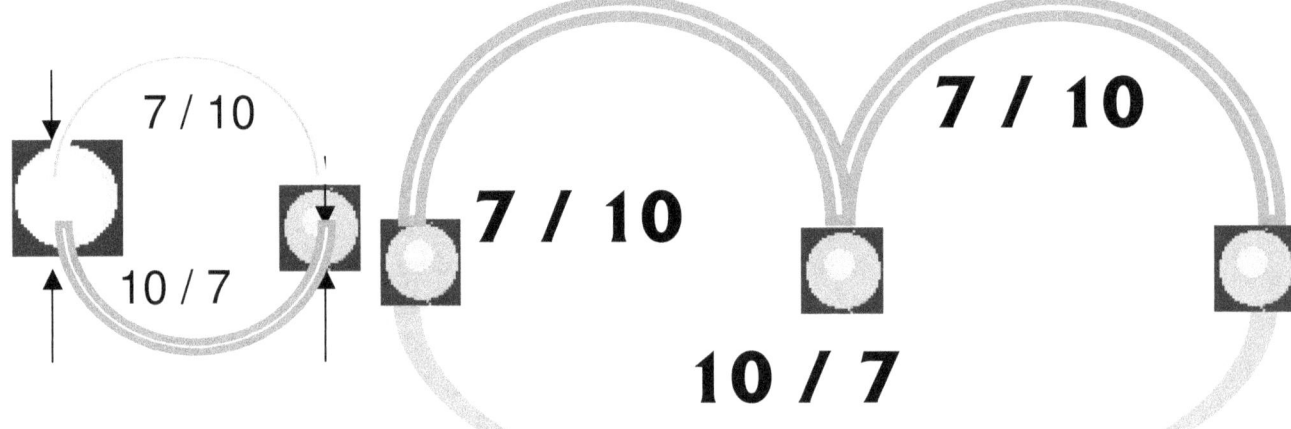

Time started at zero, eternity, whatever you wish to say, as long as you say time did not move at all. Then the command came and time overheated for the first $Π^2$ in time. That brought space into play.

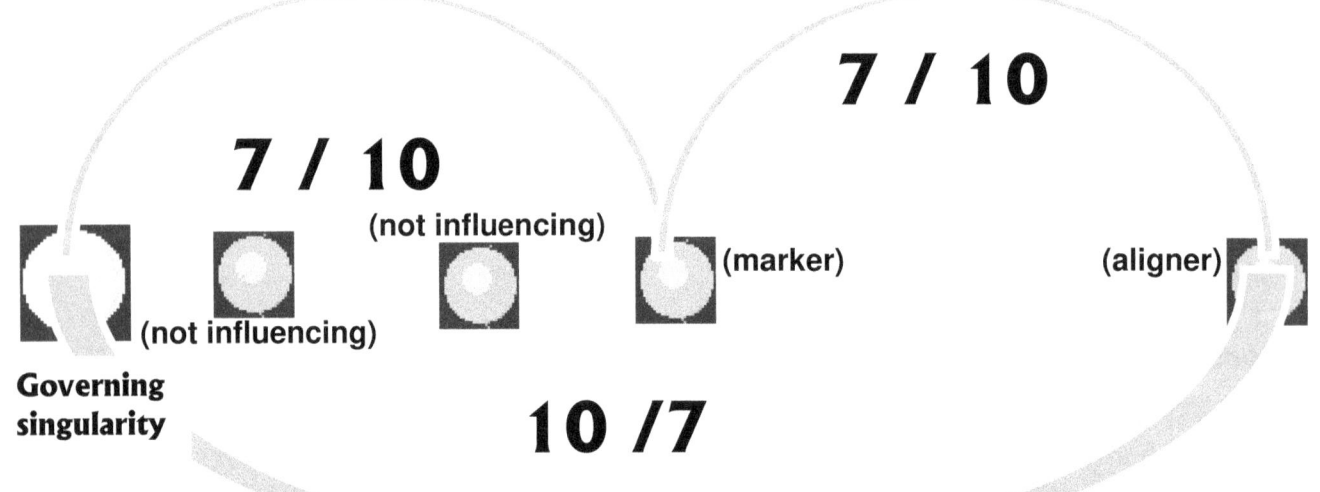

From the orbiting structure (planet) aligning singularity only one structure the very inside singularity applies as a position of reference and that is reference to the distance applied between the governing singularity. From the sun (governing singularity) the matter marker is 7/10 = 0.7 with the only one other forming a marker 7 + 7. The two form 14. From the sun (governing singularity) the outer planet forming the marker in search of position holds space in the square 10 / 7 = 1.42 in aligning with the 7 forming material of the sun. Therefore there are two sevens relating to ten forming the material positioning of the structure in orbit and from the governing singularity all outside the sun is the square of space (ten) aligning with one particle (seven) and not one of the other structure to the inside or the outside holds any value. Because 7 + 7 = 14 and 10 / 7 = 1.42 the distance doubles every time there is an aligning of three orbiters. In this there is definite proof of influences coming about between particles sharing gravity. But then again the entire Universe shares gravity and as such then all will influence everything.

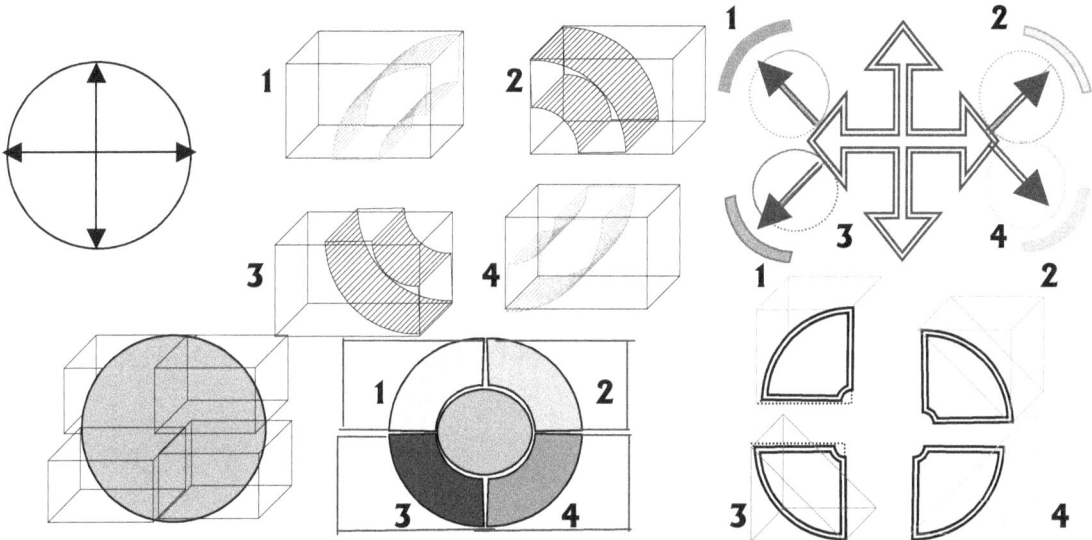

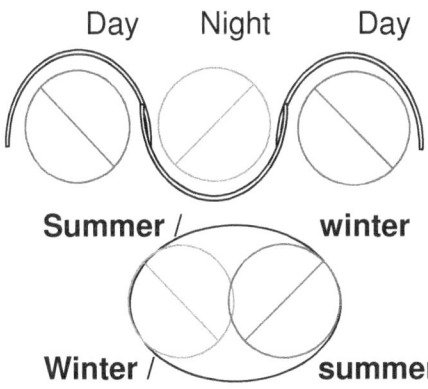

I do realise science do not recognise a relevancy between the rotation or orbit of the earth and the position of the sun as Newton claimed and I shall come to that in a brief time. On monitoring the rotation of the earth to a graphic display one find that the earth movement displaying in accordance to change in positional location does indicate a relevancy that imitates the flow of current to an almost exact. Seasonal change has all to do with the graphs influence derived from the cosmos and little to do with the position of the sun and the earth.

It is the position singularity holds in relation to the universe and the Milky Way forming currents and seasons moreover than the sun shining brighter or not. The sun in size over dominating the earths in comparison disqualifies any positional influence that can alter the earths heat standings. Through shear size the sun can shine at the top and the bottom of the earth simultaneously without effort from all normal possible angles. I show a relation between singularity in different positions maintaining seasons and north/south polarity, not only as far as concerning the earth but also outside influencing polarization. This has to do with the second position singularity holds in accordance to matter and space and is an "*electromagnetic*" (used for the lack of a better word) sustained positional opposing derived precisely from the graph in the manner when calculating electricity.

In this it is clear why the Titius Bode ([10 + 10 + 1 + .991] / 7) and the Lagrangian 5 \\ 7 systems part their ways when applying the different processes they hold. With all the differentiating, the observer must also consider the dual massage that light uses in travelling through the vastness of universal space. The thought of nothing is just what it is, a thought of nothing and although it is in the human mind common nature to present nothing as a value in the recalling of something, nothing is a presentation of the figment in the human mind. There can be no number such as nothing and that was (possibly) Newton's biggest error. Nothing represent non-existing and that is just what nothing is, it is non-existing.

The Titius Bode influence in a manner that on the one side holds the matter-to-matter relation of 7+7/10 whilst on the other side during the same time holds the space-to-matter relation of 10/7 forming equal and opposing values. From this the orbits of cosmic structures are always oval favouring the singularity dynamics of the one structure at one point and switching the favouring to the other structure on the opposing side. Because the structures can never be equal in size (singularity will not permit that where the Roche principle will intervene) the shape is always "off centre" as well. This influences coming about as the Titius Bode principal manifest in other ways proving Kepler's time relation with space through distance from singularity controlling the factors. **Once again the following proves that mass is a result of gravity and gravity does not come about through mass, because by using a new a^3 it can establish a new k, which will convert that gravity T^2 to apply to the new a^3. However it fluids, which smoke and dense heat also are. On that and other grounds I maintain that the sun on the inside is liquid. Gravity and the establishing thereof is not a God given write of birth bestowed on all the heaviest to create.**

The gravity it develops is a "*cosmic life*" not to be confused with carbon life we find and have and are on Earth, but that which makes the Universe alive. That establishing or creating or exciting of singularity by applying new heat in space by separating distance by spin is a new cosmos entity standing apart from the rest of the Cosmos. Every singularity is a Universe that can apply new values and rules by changing any of the Kepler Factors.

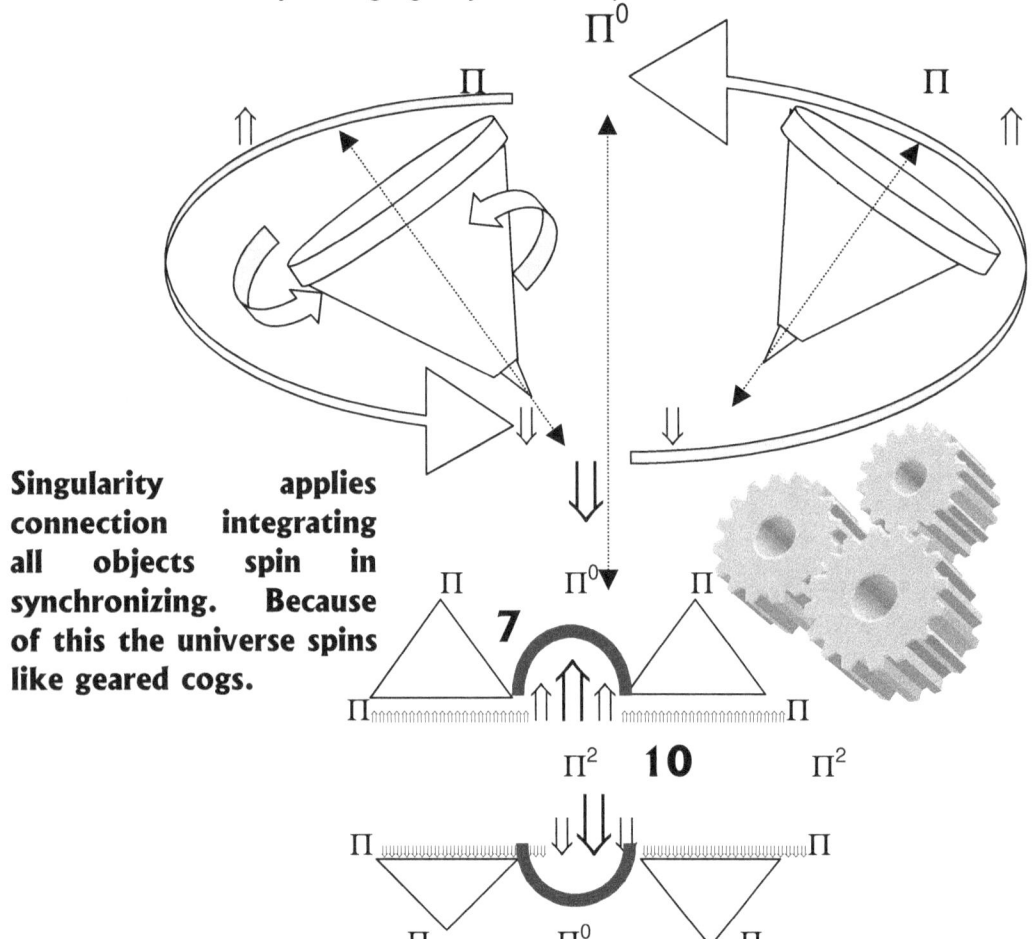

Singularity applies connection integrating all objects spin in synchronizing. Because of this the universe spins like geared cogs.

By altering any of the three Kepler factors space-time can establish a new significant gravity in the midst of gravity applying. By creating a new spin in the presence of the Earth gravity, the spin creates a gravity that will encourage in example the spinning top to try (it can never happen but it is trying all the same) through a newly charged singularity to develop a gravity that will produce such vigorous movement T^2 that will take the top to a position apart from the rotation it normally has with the Earth. When the speed of the rotation exceeds the limitation Earth the Earth allow the spinning top will start wobbling from side to side indicating a maximum effort to create lift and go in a separate spec at a separate distance from the Earth. When the top slows down the wobble will become present again, as the gravity established through the spin will fight to stay alive and apart from that of the Earth. But it shows that in Kepler's formula new space comes about from establishing a new T^2, which the spinning then forms in the alliance of space created through the manifestation of a new $a^3 = k\,T^2$ in the boundaries of the Earth. Make no mistake about the fact that the spin is new gravity that comes about in the area the top occupies and the **k** is now the rotation coming about from the centre of the new spin. The wobbling at the bottom and at the peak of the spin effort of the rotation is a gravity struggling for independence either to maintain independence or at the top to establish ultimate independence.

The ten dimensions I named the atomic relevancy is also showing the double value of singularity as singularity extends into as well as beyond space. The atomic relevancy is $(\Pi^2+\Pi^2)(\Pi^2 \times \Pi \times 3) =$ **1836** that is the mass relation between the electron (3) and the proton. Proton = $(\Pi^2+\Pi^2)$ Neutron $=\Pi^2\,\Pi$. The atomic relevancy holds the dynamics of singularity control. In the ratio and dimensions we find in the atom, all space-time derives from the atom, whatever the atom is.

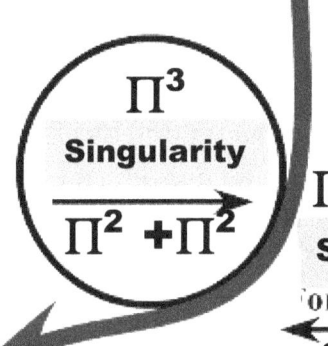

The Coanda effect is the perfect example of the curvature of space-time brought about by the extending of singularity influencing due to the shape that imitates or duplicates the value of singularity and again conform Π. By establishing a new value of singularity as Π, singularity can once again take control and establish a new Π^2 as gravity in the new Π^3 forming space

Singularity extending the influence

'orce on water

The Coanda effect is creating gravity. It is not replacing gravity it is not recreating gravity it is not enacting gravity it is forming gravity

By reducing the propeller in size and boxing in the airflow it is directing the flow to a centre where the spin will intensify as it accelerates. Gravity is created in such a way. The protons perform the spin creating the gravity and the proton number does not necessarily prove the strongest gravity.

The turbine engine is a star in the little. Massive space reducing brings on heat increase and with the accelerator of fuel added the heat increase generate a singularity enhancing equal to a star. It shows gravity come about by altering the space a^3, applying with heat added a new T^2 where that then produce the thrust to use the **k** coming about to elevate new movement, which is gravity

By reducing the propeller in size and boxing in the flow it is directing the flow the spin where it will intensify as it accelerates. Gravity is created in such a way. The protons perform the spin creating the gravity and the number does not necessarily prove the strongest gravity. The protons accelerate the moving of space whereby that spin will reduce the space volume. By accelerating the flow of space the volume per time unit decreases in relevancy

In the reducing of the intake the gravity becomes artificially constructed because gravity is the reducing of space. By injecting a fuel the situation intensifies as the possibility of raising the heat levels become much more prudent. The reduction of space will bring about heat. Injecting fuel in a place where such reducing of space already increases the heat the fuel will "spontaneously" ignite. The igniting created a heat level one can only find in the stars. Fuel will establish conditions that according to laws of cosmology only apply to stars where such heat levels will indicate the enormity of the gravity present that is generating the massive gravity accumulated in the absence of space. Gravity is the concentration of heat in the utmost reduced space thus a star is born in the gravity on Earth. In this example **k** increases as thrust pushing space a^3, which the **k** creates to a new T^2

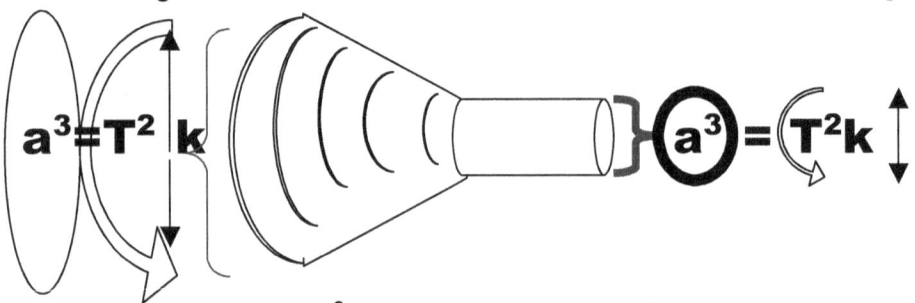

By reducing the space a^3 with the shortening of k a new standard comes about establishing a new T^2. Then moreover a new T^2 created by injecting fuel into a newly established k sets the groundwork for an a^3 coming about that can challenge the best star centres there is. With the ability of life to manipulate space-time man confuses nature in accepting there is a Tiger loose and this young star can challenge the space-time set by the gravity of the Earth. The earth will allow such rebellion just to a point and a fight will ensure that will either release the spacecraft or down the aeroplane. But all this comes about by the artificial creation of a singularity miming singularity by presenting something that can respond and produce gravity. It concentrates space as gravity does, it concentrate space in huge quantity as gravity does and therefore nature takes the action as coming from well established singularity. But never forget that life established the manipulation of the cosmic phenomenon. The action itself is not cosmic. The action is artificially established by life's ability to manipulate the cosmos. The action becomes as result of cosmic life increasing but it is not normal and it is not cosmically natural. That stands in direct contrast to the very same application of the very same conditions but where the turbine is as artificial as money, the Roche factor is as natural as water.

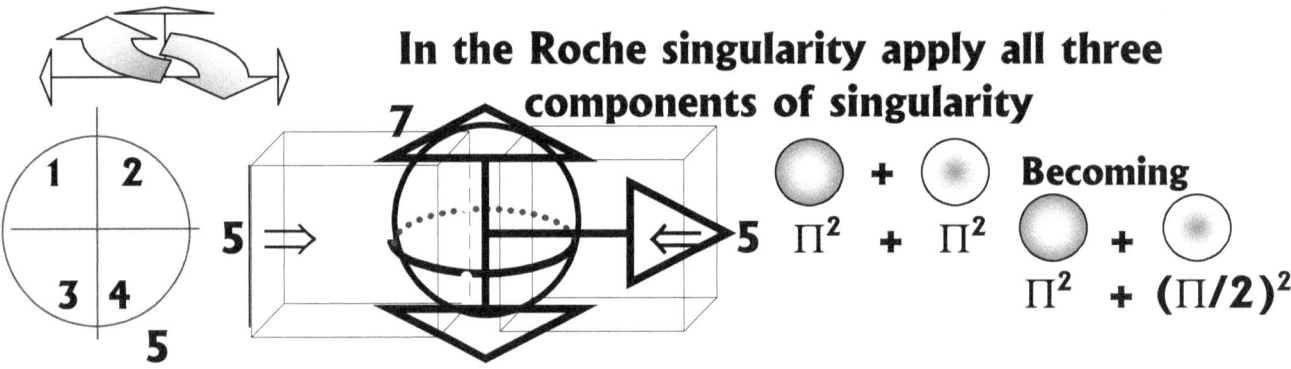

The influence of singularity as the extending of Π into space links Π^2 to r and forms $2(5)+2(5)$ $=10+10=20$

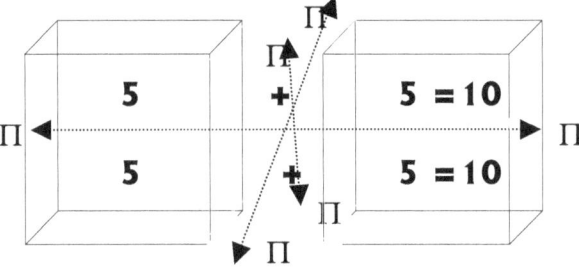

From the position of singularity there are different values in Π where each indicate a position. The value it represents being $\Pi\Pi\Pi$, Π^3, Π^2, Π and Π^0

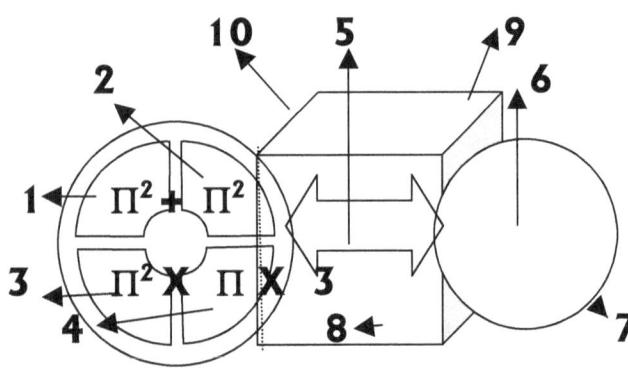

1 Singularity $X\frac{1}{4}$
2 Singularity $X\frac{1}{4}$
3 Singularity $X\frac{1}{4}$
4 Singularity $X\frac{1}{4}$
5 Singularity Π Extend

(6) Matter (7) Matter to space (8,9,10) Dimension 1,2,3) in the cube's six sides

Gravity is about a relation established when time begun between particles we know as material and particles we know as free or unoccupied space. Gravity reduces space to apply to fit the form of the sphere and later accept the form of the sphere.

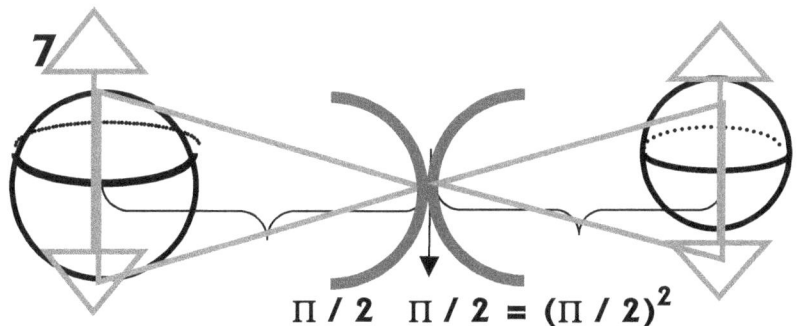

$$\Pi/2 \quad \Pi/2 = (\Pi/2)^2$$

SINGULARITY MEETS AND COMPLIMENTS EACH OTHER.

The diameter of the cosmic structure holds the value of r and singularity holds the dimensional value of Π meaning that the radius or diameter (r) extends to become the diameter multiplying the value of singularity. But since r already consists of the square of space holding a definite positional relation with the value of singularity being Π the diameter comes into effect.

Π extends each to an individual value to a point where the singularity on each side meets, bringing about a mutual Π^2 to the value dominance of the larger singularity control.

At this point the equality of the straight-line dimension to the triangle and the half circle holds prominence as a straight line, a half circle and a triangle is dimensionally equal. The common denominator will bolster all factors to an equivalent ratio.

 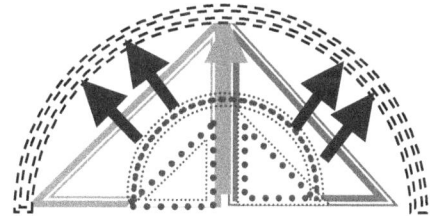

When singularity by the straight line increases the singularity by the triangle it will also bolster giving equal potency in singularity by the half circle. As the singularity of the major component revives the lesser singularity to equality, the **triangle in singularity** will match the performance and so would the half circle respond in precise ratio setting equilibrium in order. The major partner's singularity in the straight line excites the minor partner's singularity in the straight line affecting all other aspects holding singularity in both objects to match equilibriums in all aspects of singularity. That is the Roche lobe.

From this the lesser partner will fill by the extent of the larger partner and as soon as equilibrium sets in the growth will duplex to matching in both accounts, normally to the fatality of the lesser partner, as the lesser partner will be capitulating under the strain of the dual. In that way the inner planets came in place as I explain in part 7 of the theses.

The Titius Bode configuration in accordance to orbiting formation holds a slightly different explanation to the explanation that applies to cosmic structure surrounded by space. It is moreover the individual singularity in maintaining the major singularity, which sustains the governing singularity providing equilibrium in space-time. Not only does atomic individual singularity maintain self preservation, but in doing that it also sustain a governing singularity holding structural composition and form within a cluster of matter for example a star. As there is between stars so there are in the same manner a mutual or bonding singularity between atoms in stars, which we see as fusion. From this one may freely deduct that gravity is not forcing material closer but is destroying space whereby it converts the space to a density the senior partner has in the atmosphere of the senior partner.

It started with a dot, because that is the only form, size and dimension mathematical logic will allow our brain to accept. From the one dot had to come a second dot and a third dot. The dynamics of such a dot is smaller than we can understand because such a dot is in negative relation to what we see Π to be, and the deeper we delve in finding the smallest fragment where space started, in the spot where time is still eternal as much as we can accept eternity to be. This we find in the aligning of planets where the one dot from which the aligner stem becomes the reference too the distance applied between the aligner and the original dot, or governing singularity or structure in charge of holding position to all orbits following. The reason why we should first locate the spot is because we can only work from that point forward. By working forward we have to work backwards to locate where we are heading. The cosmos started at a point and where such a point is, we will find the universe. Every one knows where the universe is, because we can see where the universe is, but if we can see where the universe is, then we should find the centre of the universe in that spot. Einstein theoretically positioned the point of beginning at a place he indicated where singularity should be. With the cosmos the size it is and space so large compared to our smallness we have no chance in finding the centre of the universe. The universe started where singularity is and singularity is the sure indicator of the universe. With all spinning objects holding singularity we then have located singularity in as much as finding the centre of the universe. The universe started with a dot forming. That answer arrive from taking mathematics back to a point of being the smallest possible position, far smaller than we may be able to calculate form.

My approach might seem unconventional but through the abandoning of the accepted, it enabled me in locating the precise location of a universal singularity forming a connecting basis of the universe (this I say with some degree of confidence). The smallest figure there can be must be a dot. The dot is the only form that leaves all the options open to extend in any and in all directions should the opportunity arise. The only mathematically sensible option about extending a line from the dot will be non-bias progress in all directions equally in order to give a meaningful flow of mathematical equilibrium.

The Pythagoras mathematical principle is the proof and that I explain. The obtaining of singularity is in my rejecting of nothing by replacing it with something being the dot. With the clepsydra or "water thief" Empedocles deducted that air was composed of innumerable fine particles, braking the thought that what we now know is air, was also believed to contain nothing being altogether a space filled with nothing until proven to be wrong so many years ago. Never did science take the lesson learnt back then to the future and out onto outer space. If there is space, there cannot be "nothing" as space is something. The claim becomes obvious when observing the connection between the half circle, the straight line and the triangle, which could also promote all the qualities lurking behind the pyramid. Consider the connection between 180^0 sharing and then one may realise much of the pyramid mystique becomes less spectacular in considering the very basic in mathematics being the Law of Pythagoras on which all mathematics are focused. Once the water thief was eliminated by some human intelligence the matter was left at that. Nothing shifted out to an area we think of as outer space. In outer space we now find nothing. There is nothing but an atom here and there and even the atom is covered in nothing.

I wonder why the nothing landed there? Could it be that the reverse came about and because there was no visible "water thief" the very limit of man's suspicions came into practice. Man has always been extremely good in flying from one outer edge to another and if the water thief proved something was present, then the mere absence of a water thief must therefore prove that nothing must be in outer space. But what is space as such. What can space be, because with explosions we can clearly witness space created from heat. Our culture prevents us from admitting our vision, but the release of heat produces a *"shock wave"*. That *"shock wave"* is nothing less than space created from heat released. We have to brake free from culture of the past and a rigged mind set narrowing our vision.

<u>Einstein's Critical Density</u> lacks the accepted matching facts we need in proving the critical mass factor. But our inability in securing such required evidence defies the most basic logic. It seems all new evidence we receive from outer space is disputing all Newton laws findings that disprove <u>Einstein's Critical Density</u> as the answer. The universe will not reach a point of contracting, not withstanding whatever dark matter astronomers try to locate in the vast space.

Why would the expansion turnaround and do a reverse by going back to where it came from. Consider the momentum alternation such a change will bring about.

The sun is not a gas-filled sphere holding hydrogen in its "natural gas" form, but it is all fluid and is in a liquid form where singularity is liquid- freezing hydrogen at 6500^0 C while outer space is boiling over at -276^0 C. This book explains the Roche limit in the practical sense… when applying cosmic laws instead of improvising cosmic laws uncovers that reality then becomes awesome. It becomes clear the universe is as much expanding as it is contracting and contracting by expanding. As there is no hot or cold, no big or small, no grand opposing but relevancies in ratio to one another. If you do not believe me, then believe your eyes when looking at the picture. What ever the sun is it is fluid falling into fluid.

Consider the time it took from 10^{-43} to 10^{-5} seconds to create a cosmos the size of a neutron. Compare that to what is happening now and see how many events took place by the creation of every lepton and every hadron and it is true that that period took longer too complete than it took the universe to create the solar system. The flow of light through the density that space produce heat gives the speed of light the relevancy of time in space. The thicker the "soup" of heat is that space forms, the longer it will take light to over a distance. It is very important to note that the speed of light is a relevancy between time (seconds) and space (kilometres). The speed relies completely on the value **k** holds on space –time. The speed of light is forever a constant but the constant is part of the relevancy of space-time

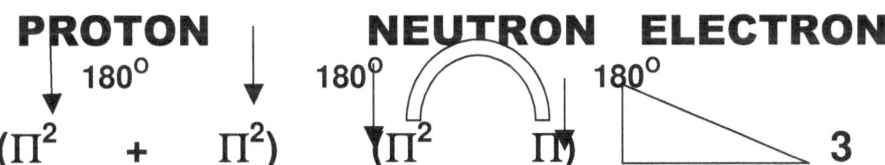

If one looks at the transmission of sound, it too depends on the relocation of matter, but to a very small degree, and in this process lies the transmitting of sound. To make the error of judgment in confusing the process with the breaking of the Doppler rings are quite understandable.

It is about confirming space **conforming space and** converting space.

ELECTRON is about confirming space

NEUTRON conforming space

PROTON converting space

The Universe connects in a way Kepler established through his relevancy theory. Those not convinced answer this: where would the Planets be if not for the sun securing planet positions. The relation proves the ratio of one in all cases to be valid. It proves much more than merely connections at liberty of holding positions where ever the randomly opportunity placed the structure. The structure does not come closer by a pulling and tugging. Kepler's figure must still be around and by repeating the task but this time made much easier with the help of computers and telescopes of magnificence compared to the which Tycho Brahe felts exited about.

The role of the electron the neutron and the proton is very commonly accepted the role each sub atomic particle plays. But galactica and stars are just as much just more cosmic atoms playing their part in the very same way as does the sub atomic particles.

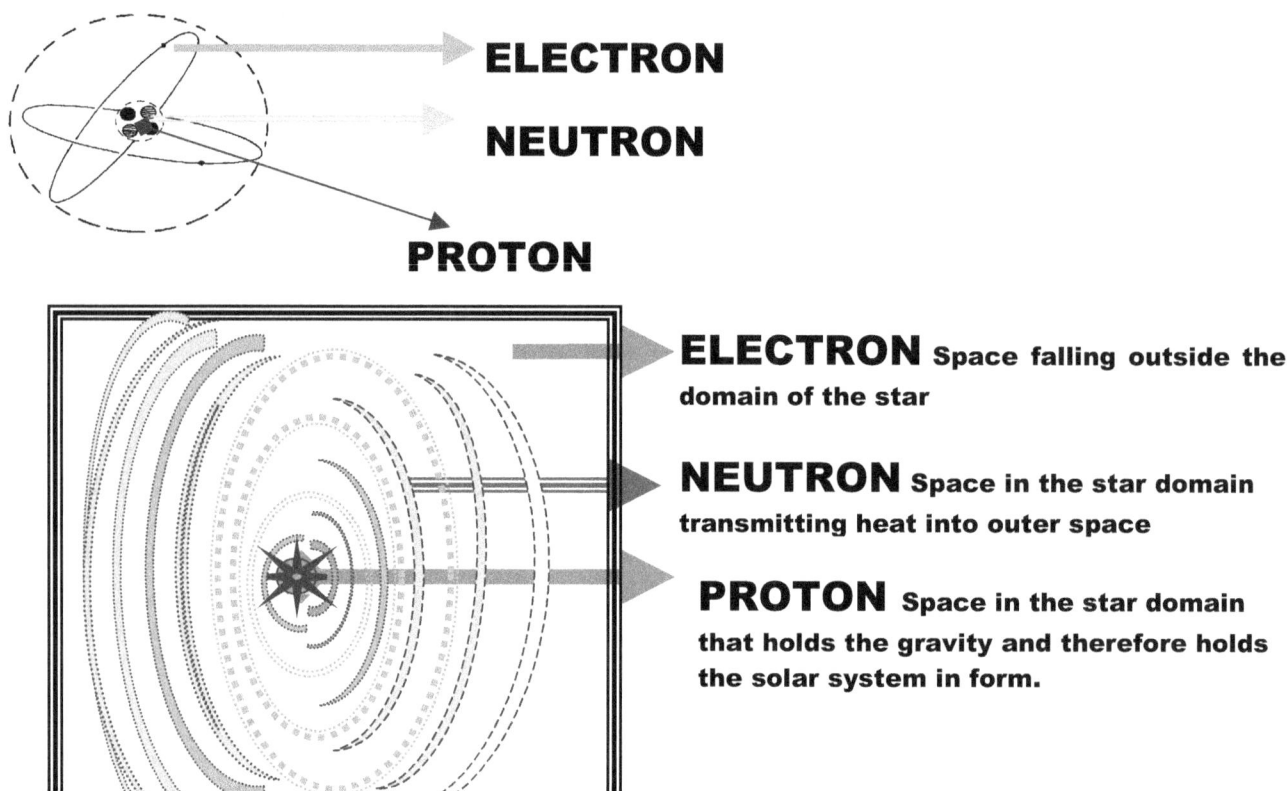

Science should become serious about science and not about self-protection and self-preservation. I found on all and every campus I went that any remark about Jesus Christ supposedly making a mistake generated immediate interest with even the most adhering Christians coming to hear the argument. Making a remark about Newton making an error gets you marched off the campus by security. Why not test Newton's $F = G(M.m)/r^2$ from figures Kepler left us and see how far did planets shift closer. I guess this will again make this book as successful as the others with my openly criticising Newton and Newtonians but Universities are not about knowledge but t about protectionism. Universities protect their own without any willingness to test that which it protects. It should be clear in confirming that the basis on which the entire world science union is founding all their policies and beliefs are correct and not only that, how far did the structures move closer. From that we then can see what we are waiting for and how long before the big solar clashing will begin. The absence in they're just mentioning such possibility confirm to me they know as well as I do there is no tugging and the Universe is in synchrony more that any person may ever be able to prove.

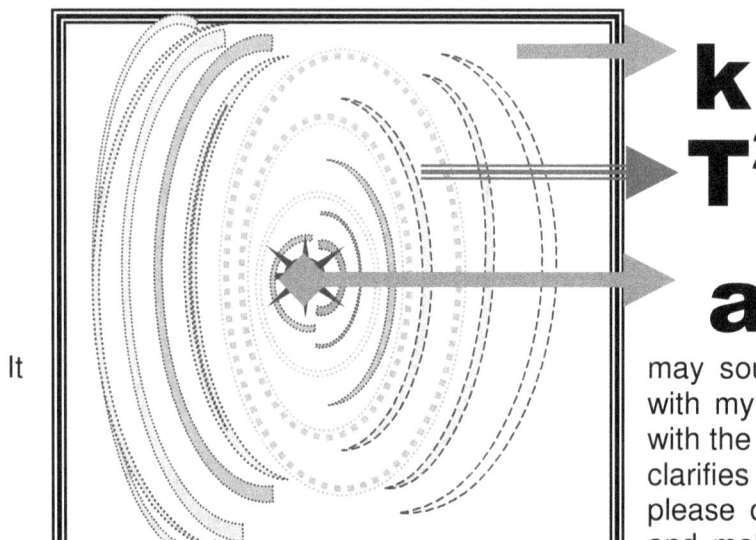

It may sound incorrect and **unscientific madness** but with my applying of Kepler's formula in alignment with the position I located and valuated singularity it clarifies the possibility of the above statement... but please do not take my word for it, use your eyes and make sure you look past the culture bias of past incorrectness. See the fluid push out of a bowl of liquid, from within stars, spilling both sides as it falls back into liquid pool forming the sun. The inside of the sun is not gas but it is fluid. In all of nature there is no NATURAL **gas** as much as there is no **natural solid**. Hydrogen is as much a liquid as iron is a gas and neon is a solid. It depends on the element relating to the space/heat in the circumstances surrounding the substance at that very precise instant in time. We have to stop telling the cosmos to show us what we wish to find and start accepting what the cosmos is telling us what is out there that we should look for and find. I hope by you're reading this book and find out that the Universe is already contracting as much as it is expanding and it is contracting by expanding because it is through the contracting that it is expanding; the answer comes about from $a^3=T^2k$.

Everything in the cosmos is moving, either by own individual accord, or under the influence of some other singularity dominance. In explaining we return to the top.
When the top is in a state of motionlessness on own accord it is everything but motionless. The motion it adapts are synchronised with the earth in harmony with the solar system and according to the greater picture of the cosmos. When an energy source not related to the cosmos called life intervenes and energises the tops motion, the singularity in that top suddenly jumps to life. By adopting a rotation energised to an unnatural state of energising because of life's intervention, the singularity of the top is not in charge but as it applies more and more energy, it will begin to find a means whereby it can escape and apply individual singularity as the top starts to separate from the singularity the earth holds. The singularity holding the earth would then allow the singularity of the top to rotate within a specific band where that a specific band of being active before the earth's singularity will start to destroy the singularity in rebellion. The top on the other hand will try its outmost, when the singularity it holds gets by individual spin is too strong to remain be in domination of the earth's singularity. The motion of the top is an attempt to begin applying an individual singularity space-time defying and standing apart from the earth's gravity. That action we see as the top starts rotating in a manner where the top does not align with the earth's singularity. With the adding of spin, the time the top holds becomes unrelated to the time the earth holds and the top will start a campaign too escape from the singularity domination the earth has on the top. When the time or spin of the top exceeds the limits the earth places on the top, the top would emerge by trying to escape from constrains placed by the earth. The view I represent at this point is known to science for almost as long as science knows mathematics.

If we wish to find the future we should locate the past. If the cosmos is contracting, where to is it contracting? The direction of contracting must be in the opposing direction the direction of expanding. If we wish to locate the past from where the cosmos came and through that in what direction the cosmos came, it must take an effort to backtrack the direction it came. Should the argument come about that all came from nothing, then everything either still has to be at nothing, or our understanding of nothing leaves much to desire. Nothing means not existing, not being, never found and unable to produce any multiplication of any growth.

The above questions, but mostly the fact of what is more nothing and what is less nothing draw me to the realisation there can be no such a quantity in space as nothing because even space has to be something. Clearly as it is for any one to see one create space by nuclear explosions. The wind is shock waves, but what is the shock wave other than new space coming into prominence. In that way it is clear that releasing heat brings about the expanding of r as part of the sphere forming space. Hubble proved the universe is expanding. Then by backtracking we have to set about reducing the sphere constituting the expanding universe. If r in the circle is growing we have to reduce r to backtrack.

When the circle reduces, the value located to r will become implicated because r determines specific size. Not so in the case of Π, because Π in the true sense only indicate that the circle is a square without corners and therefore Π dictates form and not size. By reducing size only r comes into contest and will point to such reduction. By reducing the circle radius r by half continuously will lead to an infinite small circle but Π will remain because the circle as a form remains even being infinitely small.

In the past, and even in some quarters today, science is on the search for the 100 % efficiency machine. That theory runs on the surmising that a machine can drive as an output delivery without receiving input of energy. A few hundred years ago many Kings were fooled by such notion and some scientists truly spent a life in honest search of just such a device. Mostly the accomplishment came from cheats that very well new their machines were not up to the task, but in fooling a rich investor, brought about wealth to the inventor. As science progressed the no input giving all output machine became less and lesser a feature of the honest inventor. But the idea does not exclusively come from crooks finding a way to cheat the world. The practise of receiving without giving comes from science in the form of physics. It is physics taking the world on a wild goose chase in the way physics present the cosmic motion. Physics propagates that the cosmos is all about running without input driving energy. The cosmos is all about wasting matter to a supply of motion. This idea prevails even after the world of science saw clearly in the past that there could be no such machine anywhere. Even the cosmos must be a machine driven by an input and an output. It is the input / output driving energy that must be located and the driving ability we have to locate. Science hold the mass drawing power to prominence, but what if it is not the drawing power of mass that holds prominence, but it is the reducing or contracting of space that is the driving motor behind the cosmos. All energy we humans at present use to accomplish matter motion, holds some form of heat redistribution. Even electricity is a form of pure heat. I say that in mind of what apply when the energy of electricity becomes over abundant and the machine overheats. By overheating it means that the motion the machine creates comes about from heat control and precisely planned heat distributing.

When I realised that it is not me that is drawn towards the earth, it is the space in which I find myself that reduces, and that produces the effort bringing me closer to the earth. The formula $F = G (M_1.m_2)/ r^2$ suggests driving, moving in a direction and contracting. It suggests the reducing of space and not merely drawing or moving closer. When looking at any machine in practice, the machine draws power from space reducing whereby heat increases. Not releasing the heat to form space will lead to the destruction of the composition forming the machine. There is no form of matter, or element strong enough to resist matter deformation brought about by overheating. Having this in mind that matter does not resist heat, it is of importance to recognise that it is heat that is allowing space to give matter form. Looking at the manner in which energy is utilised it is space and heat forming matter allowing motion that allows work to achieve value.

At this moment science is all about a body falling where the two bodies are producing a force whereby the bodies draw one another closer. The bigger the mass, the bigger the drawing that comes about from the force unleashed by the mass of matter. The idea about this practise was phenomenal in 1602, it was impressive in 1802, but it is really ridiculous in 2002. Why would Boron form a solid having 5 protons weighing 10.811 g / mol and Argon a gas having 18 protons weighing 39.9 g / mol. But the "heavy" element with the biggest drawing power is a gas and the lightest element is a solid. That denounces the contracting force theory. The way we compile and use energy must be in a similar manner to the way the cosmos uses energy distribution. **We humans can create nothing, but nothing is all that we humans can create**. The rest of our achieving is by duplicating whatever nature provides. To establish what drives the universe except for blaming some medieval magical force coming from nowhere going nowhere we have to find what drives us. The energy we use in all forms is producing heat in space by either converting space to heat or heat to space. Explosions are about converting heat to space. Compressing is about reducing space to heat. That is all energy composing work and is the only method of producing energy notwithstanding the immeasurable many names we use to express the same function in different forms.

Arriving at the question about locating the space and time forming the centre the centre of the universe one has to realise the centre of the universe are in every singularity forming matter weather it is big or small, size carries no significance. It is the impartiality of singularity that is claiming the value and not the differentiation of matter. One must realise there are no big / small or hot /cold or near / far. It is all relevancies between matter claiming space and space is heat in a turnabout manner. Every aspect in the cosmos are locked-in universes, sealed off from other universes and inclusive or exclusive depending on singularity holding relevancies relating to one another. The relevancies rely on inter dependence and inter linking, but there are no differences according to human sizes or standards. Accepting that principle unlocks the "so called mysteries" of the universe and brings about clear understanding. It is all about accepting, acknowledging and interpreting the role singularity maintains on matter.

One should not try to focus on an image of such a spot or dot because there is no image. The line dividing the cosmos and that run through every particle, no matter how large or small is beyond our vision. Such a small line, so small it is not even noticeable is large enough to part the cosmos into sectors. It splits the biggest there is into particles and we are not even able to notice the precise location of such a split. In truth there is no top or bottom that we living in 3D can see. We shall have to use a general conception brought about by intelligence. Your intellect tells you about such a spot, but that is all because that spot is on the other side of the universe (quite latterly). From the centre of the dot there is a top and a bottom spot. From those points there is connection with four quarters. That produces six connecting points that are all aligning to the centre. Because it serves big and small, hot and cold equal and alike, and it is the smallest cutting the biggest into equality, size is of no issue. Size is what man makes of it. In the universe there is no size in hot and cold, large and small. For the smallest there is, it is serving the largest there is equally.

Our instincts, our logic and our calculating process all indicate that the sphere holds a centre point from where six evenly positioned point's position matter to be. Using The formula **$F = G (M_1 \cdot m_2) / r^2$** it indicates to a force pulling objects closer, where each force is coming from each centre point the body in question has. The contraction must commit the two bodies towards a point in each case being spot on in the middle, not withstanding what direction the force is applying, the body will draw to the centre.

If the universe spins around a centre point holding singularity, and singularity confirms the centre of the universe, then every particle holds the centre of the universe making the number of universal centres immeasurable many, and every atom and sub atom particle presented outside the atom in smaller bits, are all not pieces of the universe but they are a universe surrounded by many universes. If every atomic particle no matter how small is holding the centre of the universe, then the gravity is coming about from that point because that is where the gravity applying in the universe are applying contraction.

It then is the atom in the most centre part where space and time meets singularity, that Einstein found a universe collapsing to a single dimension, and every atom at a point post of the proton where gravity initiates in according with the proton dimensional colas of $(\Pi^2 + \Pi^2)(\Pi^2 \times \Pi \times 3) =$ 1836

The formula of $F = G (M_1 \times m_2) / r^2$ only apply in a very specific range, and at a very determinable point the formula does not effect objects in the air. After such a point one will find satellites able to orbit, be it art a definite pace that matches the rotation of the earth. Still…below such a point (B) orbiting objects will come crushing down to the earth.

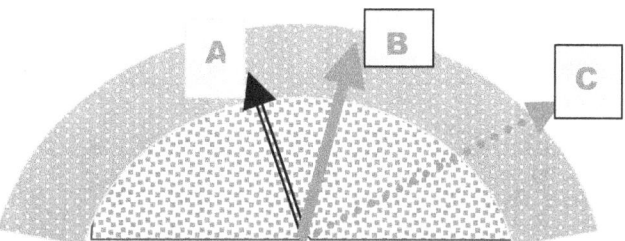

From point (B) to the earth Newton's formula apply and from point (B) upward Kepler's formula apply, but my pointing this out brings about all sorts of annoyance concerning academics. It must be clear to all persons that there are a big difference between the applying of Newton's $F = G (M_1 \times m_2) / r^2$ and Kepler's $a^3 = T^2 k$. When the objects reach some point they will drop to the earth and when that happens, mass do not play a part in the speeds they come to reach.

When examining the case where two balls drop vertically, gravity, as a force does not apply and therefore gravity does not come into effect because there is no difference in speed or duration.

With out any apparent reason the formula is substituted with the following formula:

$g = G(M \cdot m)/r^2$ where:

G = the gravitational constant,

M = the mass of the body,

M = the mass of the lesser body

r^2 = the radius between the two bodies.

Let us take this formula back to the accepting of the Big Bang and find sensibility amongst a lot of confusion that I can see.

There was a beginning that saw a radius between objects so small the size will never again repeat. The diameter of the particles were also next to nothing but that should not be a contributing factor surely…the main focus point is that particles were as cramped as it shall never again be repeated.

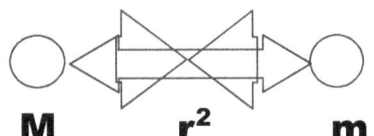

With the radius in the square dividing the shared and combined mass of the particles the relevant mass of the particles rises by the square as the radius reduces. If the radius becomes infinite, the relevant mass that the particles will produce goes up eternal. No force in the world would keep particles apart drawing on each other with an applying force but such a force is divided by an infinitely small separating radius. This is a recipe for joining and not dividing. Still according to the Universe I am able to witness the dividing became enormous and the joining practically irrelevant. The gravity was more than words can describe, the heat was able to melt it all in one structure, but that did not happen. It split into billions of individual atoms.

See the fluid push out of a bowl of liquid, spilling both sides as it falls into liquid. The inside of the sun is not gas but it is fluid.
In all of nature there is no NATURAL GAS as much as there is no NATURAL SOLID.
 No element is either a gas or is a fluid or is a solid. We arrange the elements in such a manner, but that is only applying to the situation the earth grants the elements.
When an element freezes it is solid notwithstanding…
When an element melts it becomes a liquid
When an element boils it is a gas again notwithstanding.

Hydrogen 1	melts at –259° C,	boils at –252° C,
Helium 2	melts at –269 ° C	boils at -268,9° C
LITHIUM 3	melts 180° C	boils at 1300°
BERYLLIUM 4	melts at 1287°C	boils at 2770°C
BORON 5	melts at 2030° C	boils 2550° C
Carbon 6	melts at 804 °C	boils at 3470° C
Nitrogen 7	melts at -210°C	boils at –195.8° C
Oxygen 8	melts at –218.8 °C	boils at -183° C
Fluorine 9	melts at –219.6° C	boils at –188.2° C
Neon 10	melts at –248.59° C	boils at –246° C
Sodium 11	melts at 97.85° C	boils at 892° C
Magnesium 12	melts at 650° C	boils at 1107°
Aluminum 13	melts at 660° C	boils at 2450°

Hydrogen is as much a liquid as iron is a gas and neon is a solid. It depends on the element relating to the space/heat in the circumstances surrounding the substance at that very precise instant in time. We have to stop telling the cosmos to show us what we wish to find and start accepting what the cosmos is telling us to find. The culture that I am referring to is all about **nothing.** At present we find that there is something we think of as nothing in outer space. Because nothing is what we wish to find and nothing is precisely what we are getting because we think of outer space as nothing. If you accept the cosmos to be nothing, then please define nothing to yourself and find the definition in the cosmos.

Another point I question about the Official Policy is that they as I am are in agreement that the heat melted particles onto particles and in those joining better combinations of particles came about. How it happened is another bone of contention but more about that a little later on. There was heat on the outside and there was matter on the inside. The heat was liquid because the sun and other stars still indicate masses of liquid fluid inside. I can only imagine that that liquid inside the sun holding temperatures as low as 6500^0 K and up to 1.8×10^6 K the heat already is in a molten form. What about the heat then when the frozen outer space was 10^{34} K and such temperatures were the general order of the day back then. If the sun is liquid now then those temperatures raging back then must put the heat in form available in outer space at the time as thick as mud.

From the outside drawn onto the particle inside the blanket of heat came a flow of soup that became matter. That much I do understand. This carried on until…when? When did this stop. When did the universe run out of heat. When could on consider outer space as the coldest all around? Where to did the Universe dismiss the heat that was once there but now is empty. How did the process stop of bringing from space intense heat and from that particles grew. When did it stop affecting the growth of particle, the growth of space, in fact the growth of everything that grew came from this first growth. What you see or do not see grew since it was part of the Big Bang and everything in the cosmos at present was part of the cosmos during the Big Bang. I say this process of collecting heat from outer space never stopped but is an on going process we now gave a nice name calling it gravity. Outer space never became empty and void but relevancies changed concepts where centres form that should not be as it then interferes with concepts about relevancies Gravity is not and never was about particles pulling each other closer. If it was, no Big Bang was possible. Gravity is about turning space, which is released heat back to heat and concentrate the heat where gravity is the strongest and heat is the least. Space is the transverse form of heat and visa versa is also true. Should any one not believe me try a bicycle pump by compressing the plunger while blocking the valve but. The heat will burn your finger to blisters if the force on the plunger is strong enough, the plunger seals enough and your ability to withstand pain can last that long. Then answer your own question about where the heat came from because sure as hell is hot, it did not come from friction with air particles such as oxygen and nitrogen escaping through the valve bit. Heat is unleashed space and space is concentrated heat. Reducing space to heat is gravity and antigravity is expanding from overheating blowing into space accumulation.

When looking at a sphere the inside has always (in a cosmic relevancy) the location with strongest heat also always has the strongest gravity in any given Cosmic sphere.. The centre of the sphere clusters the combination of particles forming the sphere into unity. By holding a specific centre the sphere becomes the strongest form any object can be. The sphere is without any doubt the favourite choice in form of gravity. Where gravity has the last say without other influences changing possibilities as collisions leaving debris in space or natural out burst like Super Nova explosions, gravity will enforce the sphere to be the form taken by the particle. But there is no evidence of particles of similar size joining in matrimony through gravity being the shotgun at the wedding. In cases where there is a mismatch of size outside any proportions of equality then there is a contracting of the lesser by the greater. In such cases the lesser is not qualifying as material (and that I prove later on) but the greater consider all the lesser to be heat. It is humans bringing distinction to matter in form.

The two objects should have their own value of gravity and gravitons and in comparison with the gravitons of the earth; their value is insignificant. However, these two balls are in their own individual deuce to see who reaches the earth first, and the iron ball's gravitons should give it a superior advantage. This comes about because the two objects are in a position where they compare in relation to one another and share a common second factor, which is the earth. In relation to the earth, the gravitons of the two balls do not come into consideration, but this do not play a part since the earth is a common factor. The balls, however, is put in a situation where they stand in relation to each other. When compared to one another, the gravitons should give the heavier ball a sizable advantage. The sensible example one can show to prove that where some matching structures in size come into conflict about occupied space sharing one of the structures are turned to heat in space by the other and larger structure. If the structure proves to large the superior structure turns the lesser compatriot into heat. Then being heat it will apply gravity and admit such heat into the ranks of its atmosphere, but not before it turned it into fragments good enough to be heat. The Roche limit is:

The Roche limit is:

The region surrounding each star in a binary system, within which any material is gravitationally bound to that particular star. The boundary of the Roche lobes is an equipotential surface, and the lobes touch at the inner Lagrangian point, L_1, through which mass transfer may occur if one of the components expands to fill its lobe. It names after the French mathematician Edouard Albert Roche (1820-83).

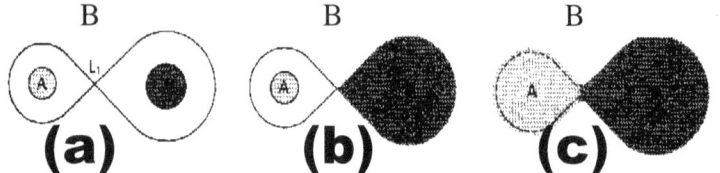

(a) **(b)** **(c)**

THE ROCHE LOBE: In a binary system, the Roche lobes of components A and B meet at the L_1 Lagrangian point. (a) In a detached system, neither star fills its Roche lobe. (b) In a semidetached system, one massive component, B, fills its Roche lobe. (c) In a contact binary, both components overfill their Roche lobes and share a common envelope. As with the graph I can see the two sides forming a connection therefore relevancy has to apply, all contradicting Newtonian claims of no connection but through mass attractions. The mass does not attract but one interferes with the other total influencing the space surroundings.

 Considering Official Science policy the collision must be devastating and total destructive to one or both. Where r that is the radius between the two colliding structures disappear from the equation since the collision is already in progress the structure would be unable to maintain any viable distance

$$F = (M \times m) / r^2$$

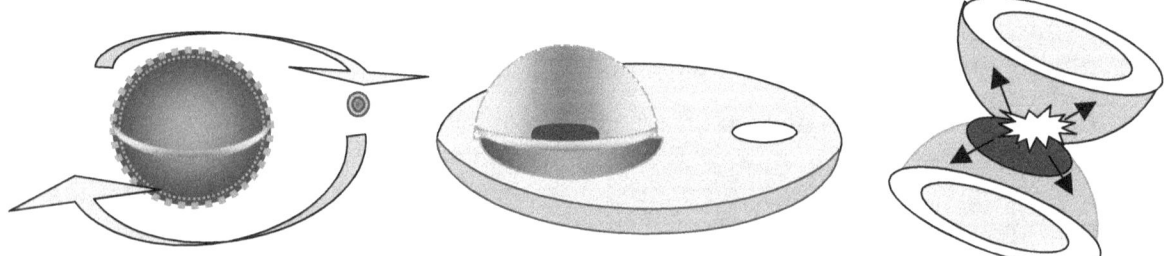

Even more astonishing is facts about the Binary star system that is seldom to never mentioned.

The Official Policy Protectors never tries to explain the relation between Newton's laws as mentioned above, and the binary star system forming the principle we Know as the Roche limit.. The binary stars are systems where two stars spin around each other and never collide. These stars are many times over the size of our sun. When one applies the same Newtonian formula as given above, these massive giants must crash into each other, destroying themselves in the process. The enormous mystery is not in the apparent misbehaviour of these giants, but the fact that this is known to science since the previous century. Relate the binary once again to the comet/ sun relation and there is a distinct similarity.

With the comet, the Newtonians regards a force to attach to the sun in some way where this force pulls the comet towards the sun. At the same time another force join in that pulls the sun closer to the comet, but such is the mass difference between the sun and the comet, the force the comet applies never realizes. In view of this, only the force the sun applies, comes into effect. The comet proves this force by speeding up its movement as it comes closer to the sun. If the force did not become greater, why would the comet gain momentum?

With the arrival of the comet in the sun's domain, the Newtonians leave the argument to be. The sun applying the force should remain applying the force and the force should increase all the time, accelerating the comet to the point of splash down. We must all argue that gravity is a force, which pulls an object to the centre of the larger object wherever that centre may be. The very same force that pulls the Chinese down, is pulling the Americans, and if not for the surface of the earth's intervention, the next world war would be between the Chinese and Americans for King and country, honour and glory and to find who has the most powerful gravity force that will provide space to live in. If not for the earth stopping matter falling right through the earth because of conflicting forces on both sides of the earth the Chinese and Americans will then have to establish border checkpoints in the centre of the earth. The checkpoints will indicate where the Chinese gravity meets the American gravity and by allowing the force of gravity to find borders, we will finally have world peace. The only problem is to find the position where the Chinese gravity meets the American gravity and the two forces nullify each other. Just think if the forces of gravity, and not man, will intervene to set border standards: that must be the answer we were always finding a question for. This is a study far to complex to bother the United Nations, so we can find a more suitable group to investigate this fact to bring about world peace.

I am personally part of Africa, born and bred in Africa as an Afrikaner. I know the African solution to such a problem. In Africa, appoint a committee to investigate and then wait for everyone to forget about the problem in investigation. Therefore, such a problem is far better solved in Africa, because those in government aim to receive maximum western aid but never aim to solve problems, you make it go away by postponing the solution to the unsolvable. The African way is to ask the west for aid in order to create another useless committee to become over paid and under worked, quite capable of dealing with any non-existing issue of any magnitude that will never find an answer. Then sit around at leisure and wait each month for pay day to come for many years while the west is paying the committee to be bored until their pension dates arrive. By then no one would now the name of the committee and much less remembers the problem investigated. On the other hand, the Newtonians are doing quite fine by their method on their own using a technique they apply for three and a half centuries. To solve such problem, the Newtonians will apply a very different solution: Blame gravity's boundaries on a non-existing force, brainwash all future students in accepting it to be a force by telling them they will accept the force and forget the problem or fail the examinations and be chucked from campus, because that solved the problem so far. By the time, the student reaches a senior position he (or she) will no longer bother their mighty brainpower with the little aspects. They will advance to a point where they can move Black Holes around, travel at the speed of light, and divert time back to the past while others calculate all the mass seen and unseen in the universe and any other ridiculous notion they may find to test their personal brilliance. If you for one second think everything about this last paragraph was silly, the silly ness started with GRAVITY ON BOTH SIDES OF THE WORLD, opposing each other, and that idea is not mine!

This is where the century, old trick of the Newtonians work best; do not think any further and no further problem will arise. Leave it at a force because with a force and thoughtlessness applying even-handedly, the problem never surfaced yet and that continued for the past four hundred years or so. So why bother with a problem that bothers no one. When a fellow like Hubble proves quite the opposite to Newton's claim of attraction, get a man who has a bigger ego than a brain and tell him to measure the universe. It will keep every one involved occupied with something senseless while the problem vanishes through the many centuries to come. It is a force, and the way of all forces is mysterious, but never admits in believing magic. Those that do not accept forces to be of a mysterious nature should just contact astrologists and come to their senses about forces being not understandable. With everyone in agreement about forces, their nature and unpredictability, who then needs more real problems to solve?

No big-brain should bother about little issues like comets when there are so many galactica to conquer. Apparently the comet-problem just will not disappear. Something broke the force, something interrupted gravity. Let us see what happens. A force means it acts the same way as tying a rope on one object and start hauling the tied object in. The longer the rope is, the less control will be on the lesser star. As the rope shortens, the better the control will become. By implying that gravity is the force, our Newtonians tell us that we have to regard gravity in the same way the rope is hauling in a comet. It is something like fishing where the comet pulls, and the sun pulls and eventually the angler gets his fish. One may argue that the rope is not the force because the force is actually the hauling, or shortening of the rope. I have had Newtonians trying to avert the problem they refuse to see by bringing in this argument. This manner of reasoning has the same value as introducing the African committee of investigation that will never uncover an answer. The rope is the extension of the force in a way being the sole representative of the force and the instigation acting out the force. The rope therefore is the force, extended somewhat, but still acting out the application of the force.

Weather the rope eventually broke, or hauling stopped, the effect as far as gravity applying its force, the process came to discontinuing ... and we know that gravity is a force that pulls something to the centre of the body in control of the gravity. What made the force act in defiance of its nature? Why did gravity change its mind? What stopped the sun applying its veracious onslaught of the body holding the poor defenceless comet? Non – Newtonians will blame me for exaggerating, but I know that there is no Newtonian that can understand my argument. In that light, I ask non-Newtonians to show patience, because there may be a few Newtonians that will also read the book and to them everything said this far does not make sense.

This is where tutoring comes in best. Should a student bring up such un academic and spiteful thinking about the mysteries of a force, then the lecturer set a date on testing all the students' reaction about how much they accept the force. When any student show signs of defying the force the lecturer can fail him out rite and have a good reason to drive the silly youngster from campus.

By ignoring the problem as to why this comet brakes free from the gravity of the sun, and continue in its freedom until gravity is at a point where it is most weak, may not bring answers, but it surely avoids nutty questions! Questions are not there to interrupt Newton's laws! Ask any Newtonian High Priest and he will either tell you that in a very roundabout way or he will simply ignore you by telling you to your face that you are incapable of understanding Newton. The best way to get out of the answer of course is to tell the sod with all the questions he has not the qualifications or the mental capacity to understand Newton. That will make the pest retract to some ditch he should be in, in the first place without bothering the greater minds with some stupid minor issue. How do I know this you may ask? I have been down that alley many times and treated with that precise treatment on occasions more than I care to remember.

Still, the comet defies the force of gravity and my questions remain unanswered.

Dear reader, if you wish to read the funnies, jokes and laughs -a- minute, treat yourself to some real good clean jokes. Read the Newtonians explanation about how comets came about; how they get to the sun and where they came from. It is going from the ridiculous to the thoughtless and ending in the realm of the mindless. However, be warned! Only do this on occasions where you feel very depressed. The jokes will otherwise drive you in a state of laughing hysteria. Poor old Newton was considered a very dry humourless chap in his day. To think what silly ideas can come from his forces.

Hauling in and releasing something caught on a line is called playing with fish. We might say that Newtonians love fishing and confuse planets, comets and fishes when they regard the interaction of comets with the sun. There is only one small problem with that argument and that is that fisherman and fishes form part of a second natural force named life. Life stands apart from the cosmos. Life and the cosmos only share time in space, not a joining of forces. Beside that, comets were part of the cosmos long before life had any role to play, so blaming it on someway life interacts with life does not cover the solution.

Why would the comet brake free from the sun's gravity? That is defying the law of gravity. Far worse than that still, is the fact that the comet's actions have the nerve to defy Newton. No one alive can defy Newton and remain alive. Does the comet not realize his actions contradict the all-important Newton and the gospel of the Newtonian - Priesthood. The best way the Priesthood of Newtonian gospel can deal with such defiance is to ignore it and no one will notice the actions of the comet. That is the scientific approach. Ignore and forget the problem. It is as simple as that.

With that let us conclude comets and really enter the world of forces at work! Let us now apply our attention to the forces of planets.

The next formula is very simple to understand. It is the fight of understanding the applying that becomes not applying it that is troublesome. If you understand the applying of the working and never spotted it not working, then forget it. You are a brainwashed Newtonian and if not…, well there is still hope that you have a clear mind left. The resentment you carry with you from childhood about the formula is in, not understanding it, but accepting the outcome of the formula you never could understand. Newton said that the force between two objects depend on the mass of both objects multiplied with each other and with the gravitational constant and the derived product you divide by the radial distance square that separate the objects. I shall put this in a mathematical language for your enjoyment that will explain the life-long not understanding to better effect.

$F = G (M_1 \times M_2) / r^2$. What does this say?

The greatness of the force depends on the masses of the two orbiting objects, aligning that product with the contribution of the gravitational constant. This then, you divide by the square of the distance between them at any given point.

Please, in all fairness to you, the reader, I have to warn you that quite a number of professors in physics told me that by reasoning in the manner I do, I only prove that I know nothing about Newton and understood even less about his work. Considering such allegations, I shall explain to you what I understand in as much as telling you what I know.

The Newtonians formula states at the force between the planet and the sun will improve as the mass of the planet increase (becomes bigger) and by multiplying that with the universal gravity constant you will get a value that will become lesser, the larger the distance are between the sun and the revolving planet. With the reducing of the distance the mass on either side must therefore be on the increase because it holds an inverted relevancy. This means the sun is pulling according to its mass. The planet is pulling according to its mass. The gravitational constant is influencing the pull evenly at both ends and the distance between the objects will reduce to the square value of the force's total application. I could never see what part I do not know and what I did not understand. No professor ever explained to me what it was that I did not understand either. That left me in a place where I did not understand what I did not understand and I never could see what I never could see. I shall try and make sense of my not understanding my not understanding as follows:

This is like having two balls attached by a rope on a floor that holds the same drag on both balls. When I reduce the length of the string, the bigger ball will show a greater resistance than the smaller ball, therefore the larger ball will apply a larger tug than that of the smaller ball. The rope will reduce (become shorter) at the end where the larger ball is than at the point of attachment where the smaller ball is. What is wrong with my argument? When the two balls are so miss- matched in mass as is the case with the sun and the comet the one ball will do all the moving, laving the larger ball stationary.

Surely the tugging at the larger end must bring the smaller object closer. By comparing the mass differences, you will find there is no comparison. The smaller object just has to come closer with the application of such a force as gravity. We know that gravity can really pull. By standing on a tall building you will find proof of this. Drop a tennis ball down from the buildings roof and see for yourself how it falls. The distance between the earth and the ball's reduces by some speed. With that being obvious, the distance between the sun and any planet have to reduce as the planet orbits the sun each year. Even if it is small, there has to be a visible reduction after four and a half billion years of pulling and tugging! Today after wrestling this problem for the duration of twenty-five years I can say (with a clear mind) I finally know how it works. It does not work!

I was always looking for mistakes on my part. At first, I thought there are a fifth force that I am unaware of because of my slender education, a force the academics can obviously see, but I cannot through obvious lack of education. I thought that my personal ill literacy gave me a blind spot that every non-educated have and was born with. The blind spot cleared only thorough education as education removes it in the way only education in science brings knowledge. I thought the removing process similar to the way washing removes stains and spots from whites; education can remove blind spots through the process of intensive tutoring. All I wished for was some academic to help me remove my blind spot about comets and their behaviour. The comet's behaviour, I could see, was an exaggeration of orbiting patterns applied from our planets orbiting around the sun; in the way, we observe galactica in the sky.

Then finally I came to the point of accepting defeat. It was not I, with the blind spot; it was all the academics brainwashed into a state of having such a blind spot. Science insists on repeatedly ignoring mathematical principles, because Newton had his claim to fame with one single calculation, THAT HE, IN FACT, DISCARDED, BY THROWING IT AWAY.

He made a brief calculation as a young man that saw an apple fall from a tree. Seeing this he jotted down a formula and the chucked it away. His piers and elders picked up the trashed paper with the calculation, and got all excited by the logic implication it had. $F = r^2 / (M_1 M_2)$. The mass of the two objects destroys the radius between the objects. Everyone went ballistic, proclaiming him as an instant genius, the one the world was waiting for after the crucifixion event.

I do not, for one second, deny or dispute the revelation. What I do encourage is place the event into its correct context. It was merely, and simply an apple that fell from its branch to its roots. The apple did not pretend to be a meteorite that fell from the heavens. If it were a meteorite, I am sure, with the man's genius, science would be somewhat different at this stage. However, as a young man, being very impressionable, as all young men are, and with the attention this brought about in the world of science, the matter overshadowed the fact.

I am not disputing Newton; I am disputing the relevance of Newton's scientific breakthrough. It was not two objects of cosmic proportions, colliding in a show of spectacular. It was, after all, only an apple falling from a tree. With this miracle revealed Newton found he was competent to improve on the work of Kepler and if I may dare say this, there must have been some political agenda behind this act and the accepting of it for Kepler was a German and what German can ever teach any Brit. The very same politics are still the order of the day forming international rivalry on all fronts.

Newton, and science, made one enormous blunder, from this stance. They took the radius of a wheel not to have any influence on the wheel. In doing that, they removed the very fact that keeps the universal attachment together. They put two objects in a attaching relevancy and then announced no relevancy. Doing that is breaking the most fundamental mathematical principle.

$$\frac{dJ}{dt} = 0$$

This disputes mathematics. DJ / dt can have any number from eternity to infinity, only excluding one; it cannot be 0. By placing the one in division of the other, you bring in relevance. You cannot then say there is no relevance. By doing such, you proclaim that one of the factors is non-existent.

$$\frac{dJ}{0} = dt \text{ or } \frac{0}{dt} = dJ$$

In both cases, one of the factors then does not exist. Such a claim is incoherent, because you proclaim that a circle has no radius, or a radius has no circle. When calculating a circle, you multiply either the square of the radius by Π, or the quarter of the diameter at a square by Π.

$\frac{dJ}{dt} = 0$ constitutes a circle and is also therefore Π X r² = CIRCLE

If you remove r it then is Π x r² / r² = CIRCLE.

You cannot then say r²/r² = 0 and therefore Π x 0 = 0. That is nonsense. Πr²/r² will always be Π x 1, and that is the eternal circle.

When looking at any rotating object, there has to be a point of no rotation and no rotation means "no rotation", not no existence. No rotation means a factor of 1, not zero. That then is singularity. The eternal Π, the Π that may not have significance but still it is a Π of value.

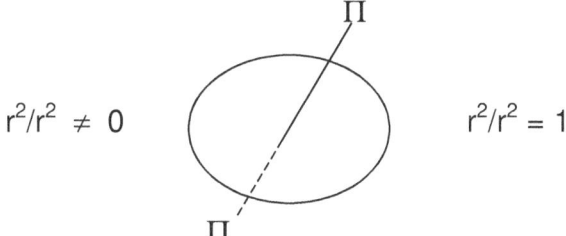

The relativity remains one, eternally one, but it cannot be zero. Therefore, dJ/dt cannot be zero. dJ/dt can be eternal or infinitive or at the worst it can be dJ/dt =1 but dJ/dt ≠ 0

When explaining this to any child, they can immediately see that. Explain this to any Newtonian High Priest and he may have you removed forcefully from campus. I cannot find one Newtonian, large or small to accept that. By not having a wheel rotate the rotation seize, not the wheel. When the wheel begins to rotate, you cannot state that all things remained as it was. With the wheel in non-rotation the rotation still exists forming the infinite possibility of rotation. Then afterwards the wheel starts to rotate and by the start of rotating the circumstances surrounding the wheel changes. A wheel in rotation is very different from a wheel not rotating and therefore cannot be the same thing. By establishing non-rotation, the wheel becomes the factor of one, and the rotating action becomes zero. The wheel does not disappear. But in the same manner does a wheel in rotation not remain still.

In the cosmos, everything is rotating because nothing ever stands still. Therefore the mean equilibrium, the common factor there is to share, has to be one, eternity, the eternal Π, because all rotating objects has Π in singularity, and sharing singularity, gives every object in space a relation with all other objects in space. After trying for many years to bring our Brainy Bunch the candle, I concluded that Newtonians are incapable of realizing that mathematical principle as a reality. They maintain they know mathematical principles far better than an ill literate such as I and yet ….

The comet rotates the sun, and the sun by itself has a point of singularity where Π remains without r. The comet, holding the orbit, also has a point of singularity, but since there is space separating the two objects, they cannot share a mean point of singularity, the very point of existing. Since singularity means just that, being single, there cannot be two. The comet and the sun have a mean point of singularity but the space they occupy divides their common singularity. That is why they orbit in an oval path, a path where the one structure holds on to more space from its point of singularity towards the space it claims. Since they do not claim equal space, BY THE DENSITY they hold, the space will not be in proportion. They do share in the common fact of singularity and singularity cannot be two, because then it will be "dualarity" or (in case there is no such a word) duplicity where both find the space they occupy, with the space they hold, will be their individual eccentricity from singularity. The two objects are holding eccentric space around their individual but common singularity forming a point of mutual singularity in accordance with the individual singularity both claim space from. That point of singularity is Π the circle without the radius because the singularity removes all forms or values of r, leaving Π to be singularity.

That is why Newton is bullshit, and his F = G(M₁M₂)/r² is utter nonsense. The moment you say Newton or any of Newton's laws, the Newtonian brain stun. For all the life in me, I could not once find one single Newtonian to see this. If you say Newton is wrong, they spiral down to frenzy, and just mention gravity and they all fall on their knees, cover their eyes in the ground, start praying and you cannot make them say anything other than Newton is correct. Dare say there is no such a thing as gravity and Newton is wrong, they have you in an armed escort patrol, straight to the department of mental disabilities and psycho diseases in preventing you committing acts of extremely dangerous life threatening behaviour to yourself and others.

What is it the Newtonians fail to see? If an electron is orbiting around an atom, the inside of the atom must be a circle. If the atom was not a circle, it then had to be a cube. The electron cannot rotate around a cube; therefore, the inside of the atom is a circle.

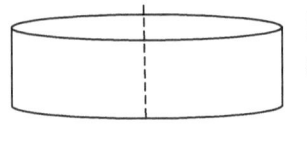
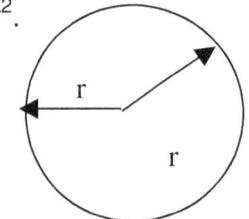

In a circle, there is a radius that initiates the circle. The calculation of such a circle is Π X r².

The radius r runs from the circle outwards, from a circle centre point towards Π, the value of the circle. In the centre of the circle, there is a point where the radius starts. It runs outwards from that point in all directions towards the circle Π. Technically, there then has to be a point where r is infinite and not zero, an absolute infinite. However, the circle therefore remains Π. The circle does not disappear; it remains there for all to see. It is only the radius that almost disappears into the infinite, but it does never become zero!

$$\frac{\Pi r^2}{r^2} = \Pi$$

If one removes the radius from the circle, the circle remains, only holding the value of Π. By removing the value of r, Π becomes singularity with no place to be. Singularity is the place where there is no space to be in place. However, Π remains because once r receives the slightest of space Π will find space. Then the circle will grow to Πr² and r would determine the space. Without space, there is no r but there is a circle with the value of Π. Singularity is in every single rotating object, be it the proton or the combining effort of all particles in the universe. That is what light and the photon is. It is concentrated heat that the sun (or any other generator of electricity) connects heat to singularity where the heat receives either temporary connection to singularity or a small piece of individual singularity.

At first you as the reader may think I am trying to create a mountain from an ant heap, but in scientific terms the human race is preparing for the start of the cosmic journey. By completion of this book you will realize how Xepted science believe they built science on a solid foundation, and, boy are everybody in for a rude awakening. Compared to the leaning tower of Pizza, science is about to start with the next section of a much bigger building adding many levels and already the view at the bottom where I am looks far worst than the leaning tower does.

If I contacted and argued with one Physics Lector or Professor about Newton, I have been in correspondence with at least a couple of hundred. What prompts me was the comet's orbit. The commit truly fascinate me from my childhood days, in the way it defies all the laws of gravity. Since my very young days, I was in search of what I at first believed to be a fifth force. I have raised the argument with just as many people not schooled in the art of physics and received a very different response. The most amazing aspect was the fact that the two groups were that far polarized. The non-physics group reacted astonished, amazed, disbelieving and reserved about my view about comets at first, but with their distrust not withstanding, everyone saw my point. The non-Educated responded in the same manner that I did at first. They argued that I was missing something of vital importance because "why do the wise not see it", was their argument. They always were of the opinion that I was too little educated to understand, while the educated was of the opinion that I was too little educated to understand. Neither party had the same view about my not understanding. The non-Educated understood my argument, but dismissed it on the fact that it was so obvious, I missed the rest of the knowledge behind the facts that makes my arguments too difficult too understand while the educated dismissed my argument that I could not see anything they could not see. Education brings the ability, which then made me unable.

In short, they thought I was too stupid too know the rest of the story. Polarized to the non-Academic view was Official Policy Protectors where not one academic could understand my argument. The academic response was as much defending the Newtonian view as it was drawing a blank about my questions. They all seemed as if their ability understanding my view completely lock behind some wall. The non-Educated, of which I am a member, at least understood what I was saying, but dismissed the simplicity about the argument. In the corner of the Official Policy Protectors, was no response of any kind, but to feverishly defend Newton by raising the dumbest arguments I have heard. The arguments, even the most highly educated brought about, seemed motorized and non-responsive. It seemed when their accepting the points I raise with my questions will demise their senses and in defence they put up a block. There is a peculiar sense of numbness in the way they could not understand what I did not understand. The academics showed no signs to indicate that they could even argue my point of view, by responding that I have an argument, and from that launched a responding argument to explain how or where I made my mistake. Their abilities in even understanding always seemed to hide behind a wall of not understanding that someone may not approve of Newton's arguments.

Newton says two pieces of rock will draw each other closer by reducing the distance keeping them apart. That we all can see by merely jumping in the air. No sooner have you lift off than you are back on the ground. That is what Newton said about three hundred and fifty years ago. Even Trying to tell the Official Policy Protectors that Galileo said mass of an object has nothing to do with the falling, seemed to pass the Official Policy Protectors sense of comprehension by miles. I was told on so many occasions that I did not understand Newton, but there it stopped. No one could explain to me what it was I did not understand about the commit missing the sun by miles, where it was supposedly to hit the sun with a dazzling impact. To this point, I cannot get through to them as much as they cannot get through to me. Our understanding is so far apart, we do not share the same planet, and yet after all my arguments and investigation no one, and I repeat: not one could once clearly tell me what it is that I do not understand.

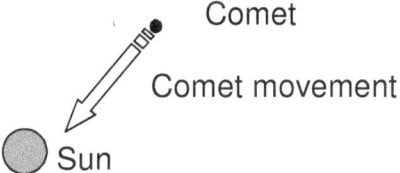

You have the sun and you have a tiny piece of rock covered by water also better known as the comet. There are thousands of them flying around, but never aimless. At first Newton's formula makes pretty much sense. The sun draws the comet towards the sun, as Newton said it does. The comet responds by speeding towards the sun, also as Newton predicted. Anyone can see a collision coming ten miles away. The sun applied gravity, the comet applied gravity, the sun is far too massive to fly to the comet, so the comet with much less mass does the flying on behalf of both objects. Every person with even the least of knowledge about science knows how the gravity application works.

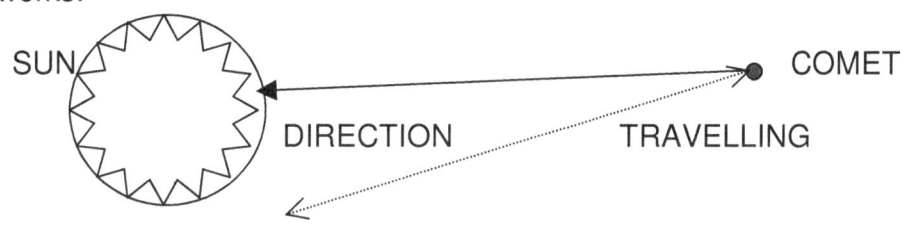

The gravity of the sun collected the comet from no-one knows where, pulled it through billions of kilometres to the area the sun produces the gravity with which it pulls the comet where the comet is to find its last resting place. The mass of the sun is obviously so large, it could produce gravity that can locate any comet hiding anywhere and collect it as a souvenir. What is there to understand?

The gravity of an object always points directly towards the centre of the object, the very, very middle point. Concluding from the fact that the comet is heading towards the centre of the sun, just as much as the sun is heading towards the centre of the comet, would not be out of line. The two centre points are heading for a direct collision, the collision becomes more and more unavoidable as the radius reduces by the value of the gravity that the mass-produce in accordance with the gravitation constant. The comet is heading towards the sun, and by not even moving the sun is moving towards the comet by attaching the movement the sun were suppose to have, on the comet. Newton's law proves to be exceptionally correct.

As the sun/comet, radius reduces, the radius separating the mass of the sun and comet effectively increases the relativity of the mass influence on each other in the form of gravity. The mass of the sun and the comet increases by the factor of reduction of the radius separating the two objects. That will produce a growing gravity force as the comet / sun radius becomes smaller. By the time, the radius becomes one the mass will grow on either side by a relevancy of 100, and when the radius becomes infinitely small, the relevance to the mass of both structures will raise a force with eternal power.

At a point, where the comet / sun apply a force of immeasurable strength, the comet brakes this immeasurable force. Remember the direction of gravity always point to the centre of the object, and that is where the collision is heading. As the objects draw closer, the distance reduces, but in accordance to the relevance the objects also become that much bigger in drawing power. It depends how one consider the relevancy to grow by the approaching nearness diminishing the distance between the objects.

$$\frac{M_s \times M_c}{100} = 1 \times F \quad (r^2 = 100)$$

$$\frac{M_s \times M_c}{50} = 2 \times F \quad (r^2 = 50)$$

$$\frac{M_s \times M_c}{25} = 4 \times F \quad (r^2 = 25)$$

$$\frac{M_s \times M_c}{5} = 20 \times F \quad (r^2 = 5)$$

Then out of the blue, the comet finds the ability to eliminate the eternal powerful force of gravity, and keep at a safe distance around the sun. At this point, Newton goes sour. Nothing Newton predicted is happening. The comet and sun not only stabilized the force, the force begins to decrease as the radius between the comet and the sun is on the increase AT THE POINT WHERE THE FORCE IS THE STRONGEST, THE COMET BRAKES FREE AND SLIP AROUND THE SUN, UNSCATHED.

UP TO THIS POINT I STILL SEE WHAT THE BRAINY BUNCH AND NEWTON SEE, BUT IT IS FROM THIS POINT ONWARDS THAT THERE COMES THE POINT THAT OUR MUTUAL POINT OF CONCENT DIVERTS POINTING OUR MUTUAL POINT ABOUT THE POINT OF AGREEMENT TO THAT OF OPOSING POINTS WHERE OUR VIEW SEPARATE BOTH HEADING IN OPPOSING DIRECTION ON AN ETERNAL DIVERTING PATH.

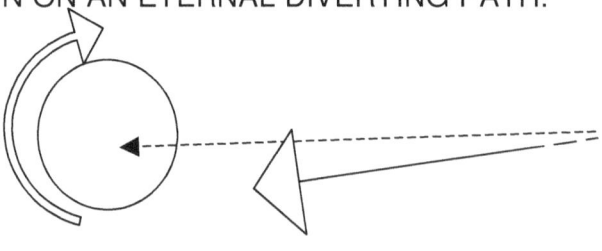

Then, in complete defiance of the Newton Law on gravity, quite the opposite applies. At the point where the radius that is separating the two cosmic objects is at its strongest, it will also bring about that the gravity force is at its weakest. At the point of almost no ability the gravity force suddenly releases enough strength to break resulting in the parting of the two structures. The force now curbs the rebel comet on its way escaping the sun gravity for the very last time.

At the point where the force was the greatest, the comet overcame the force, but where the force was the weakest, the force overcame the comet's rebellion.

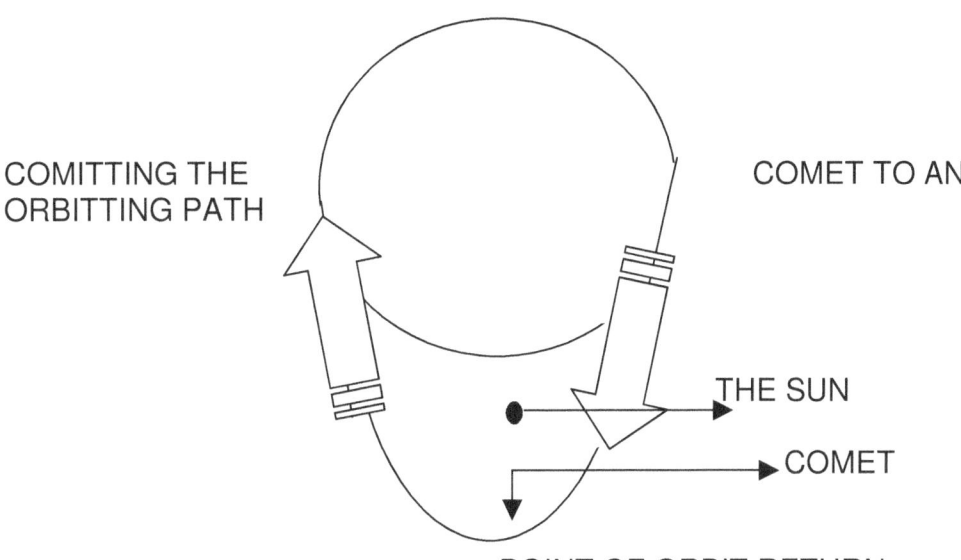

The correctness of my argument is no longer the issue. It was twenty-five years ago, when I still held the impression that I was missing some point here. I do not state this phenomenon any longer in the hope of bringing across some flaw in my understanding. The flaw in my argument is not there because the flaw is science as a whole.

I could never understand the reason why "the ordinary", like me and others with my development level, can see what I can see, yet academics that has more brainpower in their heads, than I have life in my body, were unable to see such an obvious conclusion. You; those Official Policy Protectors are my superior in every sense a human can have, with the brainpower to break a wall, and yet you cannot see how far the tower of Pizza is leaning over.

I make the point to help you the reader to judge yourself. If you are able to see the validity in my argument, you are not brain dead. Education has not yet bashed your thinking ability out of your scull. However, if a cloak of not understanding role over your brain, and a numbness sets in on your ability to reason about this phenomenon, beware, you are a Newtonian. Newtonians should read this book very slowly because the effort you are about to launch, may be the most painful you shall ever experience throughout your academic career. You are going to suffer from reconditioning and Newtonian withdrawal, not that dissimilar to that of an addict in rehabilitation. You are going to reject me, hate me, despise me, loath me as you never felt about anybody else. If you think I am sarcastic, I am not. You will reach a point where you will abandon the reading of the book. You have my sincere sympathy and with all the soothing it may bring, know that you are not the first I saw getting such painful Newtonian rejecting.

Once more, this phenomenon should not occur with Newton's presumptions about gravity. These bodies will and must collide and destruct, without a doubt. When the formula $F = \dfrac{M_1 M_2}{r^2} G$ apply, there should not be any force which is able to keep them apart especially when r reduces to almost infinity compared to what it is at maximum. However, they do exist and what is more, they maintain a certain distance apart.

A Road Leading to A ROAD LEADING TO THE CENTRE OF THE UIVERSE.

With the "force" of "gravity" "pulling" the stars closer using the accumulative mass of the stars and multiplying that value with both objects by the mass component, this will reduce the radius r^2 progressively until r^2 reduces to zero. Seen from this view, it is little wonder that the significance of this was lost in the notion that this is yet another "mystery" of the universe. The scientists of the day (and the past) lost the importance, which this holds for us as earthly dwellers.

A most surprising aspect of this is that it is not that an unfamiliar or rear phenomenon. However, any answer to this would clash with Newton's presumptions, and before the scientists allow that to happen, they would much rather ignore what is obvious. However, what is the obvious?

The cosmos work in relevancies to singularity. The earth has the value of $4\Pi^2$.

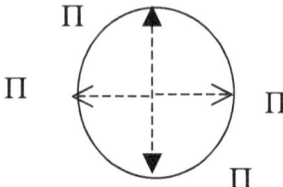

From the earth's perspective matter is 7 and time is 4 therefore
$4 \times (\Pi = 7) = 28$.

The moon holds its singularity in relation to the earth in singularity. Because the moon is in a Roche limit (near enough) the proton value the moon accepts is $\Pi^2/4$. The neutron position the moon holds, relating to the earth is 10 and the moon has an own point of singularity, forming the electron edge of the earth.

The earth is $4\Pi^2 = 28$ days to rotate to one moon cycle of 1. The relevancy of the earth, taken from its point of singularity is 7 (matter in relation to space will always be 7 to the space value of 10).

Therefore Π in factor of Π^2. 7

In the relation from

singularity accepts 7 as the matter

singularity Π holds 4 positions of 7 = 28

OBVIOUSLY ALL RELAVENCIES HOLD TWO SIDES TO EQUAL PROMINENCE. THE RELEVENCY THE MOON HOLDS TO THE EARTH WILL BE DIFFERENT TO THAT WHICH THE EARTH PLACES ON THE MOON.

The earth holds singularity and the earth's singularity holds the moon's position in singularity. To the moon's relativity, it holds value to space from the earth's singularity of $1 \times (\Pi/2)^2$ (the relation between the earth and the moon) x 10 (the fact that the moon is within the space of the earth, as the moon has an individual point holding singularity (Π). Therefore the moon holds $(1 \times (\Pi/2)^2 \times 10) + \Pi = 27,8$ days as one and the earth holds to moons single day of individual singularity at one to 28 days.

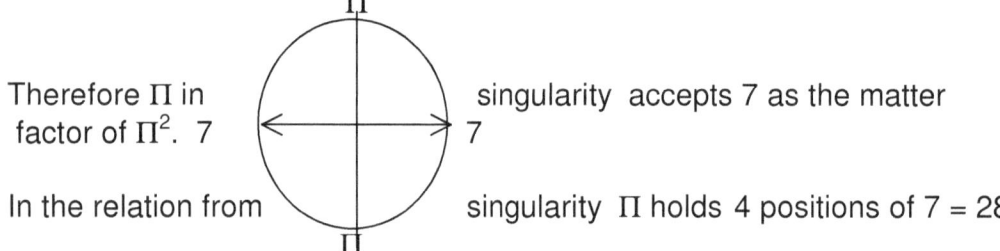

Behind this is the principle is the sound barrier and the reason why aircrafts break the sound barrier. It is all a relation in different positions of singularity. From the time, I wrote the first few pages I was on a quest to find a more suitable person to take over the work. I, more then any one else, know my limitations and limitations they are. I knew that anyone of the more than one thousand five hundred academics I eventually contacted held more knowledge in the tip of his (or her) little finger, than I have in my entire body. I tried in desperation as I tried in vain, to convey my message to the right person that could see what I could see. It all came to nothing because the academics all sat on the same mighty Sear Tower far too high to even notice me pointing at the cracks down below. From where the Official Policy Protectors sat, they did not notice me. Those to whom I drew some attention be it personal, by mail or on the Internet saw me as a nonsense proclaiming nonsense.

From where I stood, I could see the mighty tower they sat on. I could also see what construction held that mighty tower together. I could see how much that mighty tower was leaning over like the tower of Pizza. I could see how the tower will fall one day if not soon then later, because I was at the foundation of the tower. From where I stand, at the very bottom, I see the foundation of this enormous Petranos Towers collapsing from the misconceptions it holds as a base. Down at the very bottom where I am, I could see whatever other one holding a position in this tower could not see. Those that are at the very top, is so high, so very secure, they would not even hear me or take notice, and yet they are the only ones that can do something about the inclining tower.

After attempting for five years to make myself heard to get the High and Mighty to notice the insignificant me down below, so small in relation to their greatness, I decided to show the world what holds them that high. I decided to show everyone how great "they" are, especially about the greatness of their misconceptions that put them at such a dizzy height.

To every one of the Official Policy Protectors, I say this: Every opportunity you had, you thoroughly rubbed my nose in the knowledge that we are not in the same class. I accept that fact as I accept my academic qualifications being so very poor. I shall never be your next-door neighbour or the bloke living down the road from your house, because I am not in your league and I shall never be. I do not begrudge you your academic position, your mighty achievements or the height you have reached in you sphere. I shall never enjoy your company as an equal, because I can never be your equal. Your brainpower puts you light years ahead of me, and for that reason I do not even wish to have the honour of your company.

All I ask, is listen to a mere mortal, a mindless illiterate compared to you, one sod down here at the bottom where you are at the very top and that may just see things you cannot see from the height you hold. I do not wish to join your company for I shall not fit. I never had or have any ambition to fit either, because I am quite happy being in the sub-minor league. All I ask is to be heard. So many times you, honoured members of the clan of Newtonian High Priests did not even attempt reading my book that I sent to you. You did not even try to pretend I had a point therefore you merely through my book away in disgust. To you my illiterate arrogant views about science being wrong for the past few hundred years are the epitome of a mindless. You saw me as a totally mentally underdeveloped excuse for breathing and you could not bare my company because for me to have my view such as my view is in rejecting the view of an establishment centuries old. I can understand your disgust, but that does not change my point and that does not change the incorrectness of Newtonian science!

Before you throw the book down in total disgust, first answer the following argument and if you can answer it truthfully then throw the book down. If you cannot answer it, go on reading the book and you may just set your thinking mind in motion. Hear this from a mindless: you may have the ability to learn and afterwards reflect on that which you learn, but you cannot think, and I do not know who is the most mindless, me without education or you with education and without reason.

Newtonians declare that a force brought about by the content of the mass of a body will therefore pull objects closer where the pressure coming from the weight brought about by the mass of the bodies, will lead to heat.

If a cylinder is pumped with air, the pumping is a force. The force comes about because the force is that of the intentional action of the only force in the universe, the force of life. The more pumping there is and the longer the pumping will last, the hotter the air will become, and therefore the hotter the walls of the cylinder will get, as it transmits the heat from a point where the heat is most abundant to a point where the heat is least abundant. Heat flows from hot to cold.
After pumping stops, the heat on the inside will reduce in value up to a point there is equilibrium between the heat on the outside of the cylinder and the heat on the inside of the cylinder. The heat

reached equilibrium with the event of time. Please for the sake of sanity, do not reply that it is the molecules in the air tank is bumping each other and through that collision, friction causes the heat. That is as much Newtonian rubbish as one can ever find. Should you insist on that being your answer, then ask yourself what will calm the molecules down afterwards where they get so calm, there is no more heat in un-equilibrium. Did the molecules take drugs, or did the force calm them by telling them gentle nighttimes stories. I had so much bullshit thrown at me wherever Newtonians defended their Master it sickens me. I may be uneducated, but I am far from mindless!

No molecule can ever, ever touch another molecule because the electrons guarding the outside are equally negatively charged, and will therefore reject any contact or coming closer to one another or with another molecule.

When we look at the earth, we find the coldest region on the outside of the atmosphere, where the least mass, weight and gravity is. As the circle grows smaller, the molecules become more, the mass becomes more, and this will increase the weight that brings about heat from pressure. This is not my saying this is Newtonian science.

The increase in mass, bringing about an increase in weight, puts most pressure at the centre to be the hottest. There is this force in matter that pushes and pulls, until everything is boiling hot, and that is gravity.

There must be a point, where all the matter has finally found a position to bring about the overheating that occurs in the centre. The heat is energy and cannot be manufactured, but has to come from somewhere. The heat cannot be lost, because heat will transfer to a colder region to bring about equilibrium. The heat cannot continuously come from nowhere because at a point, the molecules will settle in their individual positions, find equilibrium and maintain the heat balance in that spot. It cannot produce heat from nowhere, on a continuous basis, because heat is energy and being energy it must come from somewhere as much as it is going somewhere. Either the atoms lose their mass to heat and the mass becomes heat in order to generate heat on a continuous basis, or the heat must stop, decrease and become as cold as outer space because no further heat comes about because the heat is exhausted.

If the matter as much as mass becoming weight established heat by applying continuous pressure, the earth should decrease in size as mass turns to heat with the consequential loss of mass and gain in heat. The earth must then deflate and be a pretty small place by now.

You may say there is a lot of mass-producing a lot of pressure becoming a constant flow of heat, but that will mean some of the mass must have disappeared to heat because four thousand five hundred million years is a pretty long time. In four thousand five hundred million years, some size diminishing should show, or the earth should have reached a point of equilibrium by now where the heat supply is completely exhausted because again four thousand five hundred million years is pretty much enough time I would say for matter to have cooled down by now.

Yet, the flow of heat maintains in the earth, very much uninterrupted and in no way showing signs of decrease. If the force of gravity is manufacturing heat, from where does it get its raw material and where is the manufacturing plant? If weight and pressure leads to heat, then the heat should have transferred altogether to outer space, the coldest place we know at minus 276°C. Four thousand five hundred million years have come and gone since the earth became the form it has now.

On the other hand, if the force continued its pushing and pulling it applied when it formed the earth, the earth should by now have incinerated. Getting enough pressure to push hydrogen and dust into iron and release as solid as the earth is, required a lot of pushing and compressing, and if that pushing and pressure continued as much as it had to increase with the demise of particle space separating the molecules. Keeping that in mind as a natural law, everything on earth should be covered in flames by now. When you go with the argument that there is not enough matter to produce such pressure, then there was not enough matter from the start to get the place as dense as it currently is. The matter did not decrease, the force therefore had to become weaker with time and the force is so weak now, it does not collapse the earth into a smaller space than it was say three thousand five hundred million years ago. However there are very little signs of that, in fact it seems the whole thing is getting bigger, with all the lakes losing their depth, and the mountains rising, including the volcanic activities establishing new islands.

If you cannot state why the heat has not reached a point of saturation or has decreased to equilibrium with outer space, it proves you are not thinking. In that case read on … it may arrows your thinking ability once again. This I say, not in arrogance because I, of all people should know how poor my abilities are. I sat day after night, night after day, breaking my brain to find answers, or only clues to the questions in hand. I knew at each point I arrived, that the answer is right in front of me, but through the darkness of my personal ignorance, I could not see what I knew there was to see. From that feeling of incompetence I tried once more on every occasion to contact any person with more brains than I but every time I was luckless to energise interest and had to continue with my personal incapacity.

Every time I contacted an Official Policy Protectors in desperation for help, they ignored me flat. Every message I sent telling whomever I contacted, that there is no such a thing as gravity, it is all a medieval hoax, Newton is altogether completely wrong with his gravitational laws and the Bible is one hundred percent correct about creation, they would not even reply or at least respond. Well… to them I say this: I still maintain that there is no such a thing as gravity: it is a hoax.

Obviously, the obvious is that the Newtonian Order of High Priests would lead every body to believe that Newton and Einstein's findings are flawless. All these mentioned discrepancies are known to "Xepted science", yet they keep the charade going about other planets they are about to find, only to mislead the public and milk their tax money, in the name of research. If it is not for funding and an effort to provoke general interest in skimming tax money, then why would they deliberately spread such malice.

Scientists know about this discrepancy in the Newtonian laws, yet there is never any mention about it.

The so-called "evidence of the existence of planets" is based on just as laughable principle. Allow me to explain:

The findings which science base their proof on about the location of other planets are the gravitational pull.

At first the way in which the facts are presented does not sound that unfamiliar in an argument, and one tend to except it without a second thought. When given the second thought, the blatancy leaves one breathless.

In the evidence, the stars and "planets" are presented to be in a tug of war. This one can see in the sketch above. How this supposedly works is that the one star first pulls the other system closer with the force of gravity, and then it is the other systems turn to apply gravity and jerk the first system to its side. What they do explain in explicate detail is that they do not understand the first thing about the matters they pretend to understand.

The Official Policy Protectors knows very well that all children play this game, and therefore every one will associate this explanation with familiar events in their past, and no further questions will be asked. Let us examine this principle with obvious general knowledge about space flight and how this applies in outer space. An Astronaut is capable of lifting fore tons of equipment by himself. He (or she) can perform this action effortlessly. However, what is impossible to perform in outer space is to correct his position when he is not secured to a stabilized object. Every person knows it can be life threatening when an astronaut loses his grip or connection to the spacecraft. In this, the question is; how can two heavenly bodies have a tug of war under such conditions? There is obviously some stabilizing factor on earth one do not find in outer space, and that is not gravity because such an answer is avoiding the issue

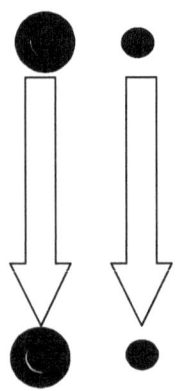
In the past so many Official Policy Protectors dismissed my rejecting the Newtonian claim about mass being a factor in falling objects. Some even went as far as refusing to read my book beyond that statement, that being on page four of a one thousand seven hundred page document, claiming I do not understand Newton, but no one can explain why oxygen is a gas, and yet oxygen is more massive than boron and carbon. If mass were a factor of the essential being the claim of Newton, then oxygen would fall to the earth faster than boron and not float in the air as air. Yet boron is a solid and oxygen is the gas.

The reason why the **mass of the smaller object do not apply**, is because **it is not the object** that is **drawn** to the earth, but **the space-time in which the mass of the object finds itself in,** that is being **drawn towards the earth**. The most obvious proof that something applies movement to something directing the movement towards earth is the simple pendulum. If gravity was a force, one should get the same result by applying a pull spring to the pendulum. With the spring connected to the bottom of the pendulum, both time as well as space would compromise. However, we know that the result proves the opposite, which proves that there is no force applied of the pendulum. On one of my many crusades I met one of the most influential academics on astro-physics in Africa.

I tried to explain the pendulum swing to a high ranking man of much academic importance, doing great work on behalf of NASA in South Africa but with me not knowing his superiority on the matter of the pendulum, I took this man on, on one of his specialties. This man was apparently an expert on research or lecturing or what ever on matters about the pendulum. Off course, as usual, the very first thing he asked (as all Newtonians do) before even asking a persons name, was at which university did I study and what my academic qualifications was. By replying that I have never been at any university for longer than a few hours in my life, and therefore my academic qualifications was less than zero, this highly rated person of high standings was less than impressed to spend any of his valuable NASA paid time with me. To complicate the whole aspect of my un- welcome visit was that I tried to explain to him being the expert on the pendulum that he is, about the pendulum. Boy, was the man annoyed with me.

He was very polite and very civilized about the whole issue, but his annoyance with MY EFFORT ABOUT EXPLAINING THE PENDULUM TO THE EXPERT ON EXPLAINING THE PENDULUN WAS MORE THAN HE COULD BE CIVILISED ABOUT. There is a point where a person gets a little too civilized to be true and at that point where a person gets a little too polite to be civilized about a topic. Any person can sense such a point, the point when one realizes that is the point of limits. I also realized that what ever I had to say on any matter concerning all relevant matters was as good as never said as far as our Professor Doctor, was concerned. As a matter of fact that Professor from the University of Potchefstroom is one of the most exceptional men to walk this planet. I gave him a copy of The Thesis and after reading only four pages from a book containing over two thousand pages he was able to draw a conclusion and condemn my work. Off course the reason was the usual: I did not understand Newton obviously because of the lack of education on my part.

You may ask yourself: *"What was me (the un-welcome person) trying to say?"* This is what I am saying:

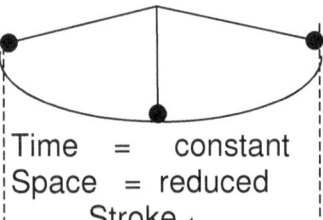

Time = constant
Space = reduced
Stroke $_1$

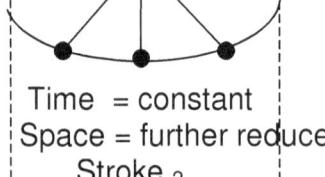

Time = constant
Space = further reduced
Stroke $_2$

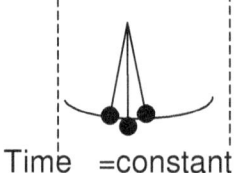

Time = constant
Space = Reduced even more
Stroke $_3$

Time remains the same. The swing distance tarnishes. Period $_1$ = Period $_2$ = Period $_3$
Swing distance $_1$ ≠ Swing distance $_2$ ≠ Swing distance $_3$

In the pendulum principle that brought Galileo his everlasting fame, the pendulum swings at an even interval. As the **pendulum swings**, the **space tarnishes** while the **time (period) remains** the same. This is the principle on which all clocks work. What is it that Newtonians are missing for three hundred and something years about the pendulum? Newtonians are not seeing the very best example there is to indicate singularity outside singularity

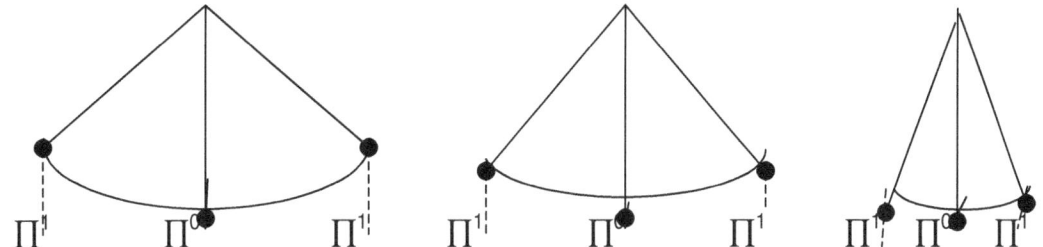

The pendulum indicates the very point of singularity the earth holds and the marks on both sides where singularity deviates in space, giving time to that singularity diverting.

If gravity, which is a force, did apply, then it would be as if a spring was fixed to the bottom of the pendulum and to an unmovable object below the pendulum. With the applying of the springtime will not remain at equilibrium but will tarnish in a vector with the declining of space. Something is holding time steady to the demise of space.

When a spring of 9,81 Nm is mounted to a pendulum, which is an equal force to that of gravity, both the time period and the swing distance would equally be affected, but to a lesser degree as the swing distance declines.

ALL SCIENTISTS GO INTO FRENZY BECAUSE GALILEO WAS PUT IN HOUSE ARREST FOR TEN YEARS! HOWEVER, THESE VERY SAME SCIENTISTS ARE STILL HAVING GALILEO'S WORK KEPT IN HOUSE ARREST AFTER ALMOST 350 odd years. HOW DO THEY EXPLAIN THAT? Galileo introduced the best devise indicating space-time and half a millennium onwards, nobody but me can see it.

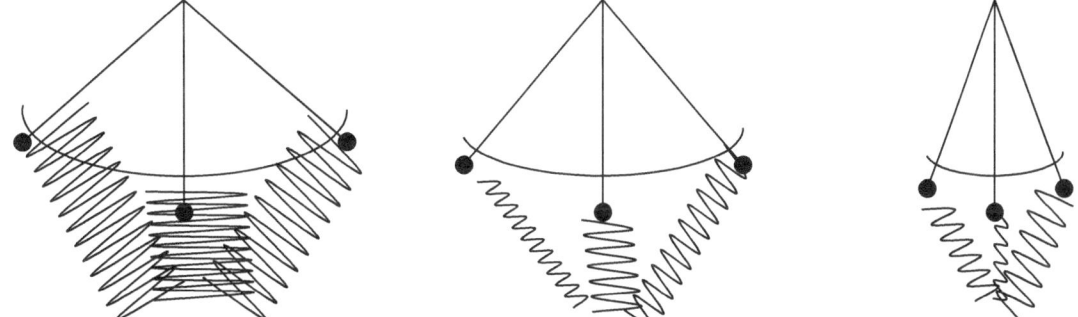

This means there is **NO FORCE** and therefore there is **NO GRAVITY**. There is only **space** (stroke) **time** (period) = **space-time.**

What is space-time exactly? Einstein was the first to explain the existence of space-time, but Galileo was the first to indicate space-time and Kepler was the first to pin point the position of space-time. Einstein made one big error in judgment. In his all to well-known formula $E=MC^2$, he relates to space-time as if space had a factor of one and time was the altering factor. According to the Einstein / Newtonian Order of High Priests, we live in a total dark, totally flat, and single dimension universe.

Why would it be a total dark universe? According to Einstein, the speed of light is the same as the speed of time. Should that be true, photons have to freeze in time, and must be unable to move through space in time! (I shall elaborate on this in due time.) This proved how far the greatest Newtonian outside Newton really were off the mark

Look around you and see all structures (**space**) are different in size. Therefore, **space** cannot have a factor of none converting to one and back to none, but relate to the size of the object, whether it is an atom or the cosmic universe.

Time (C^2) is at an even factor as all things in the universe relate to the same time, (although not the same duration of time). By implying that $E = MC^2$, he puts R^3 at a relative value of one throughout the universe. Space can hardly disappear, but can compromise under abnormal star growth.

Every round object has a point establishing a very centre, a middle dividing one side from the other. That division determines the space from one side away from the other side. At one point there must be a point that does not fall on either side of the divide. Such a point will still be a circle, because from that side the circle divides into two sectors.

Every solar structure is spinning around an individual axis while the whole lot is spinning around a mutual axis the sun provide The spin that shows on the different planets is the most crucial aspect of their orbiting the sun Calculating a circle involves two aspects where the one is either the radius or the diameter that is double the radius. The other is the factor Π
$\Pi \times D^2 / 4$ = circle and $\Pi \times r^2$ = circle

The point of singularity cannot be in space at large because space is not there and secondly what ever is there spin to slowly to have a connection with singularity directly.

The pendulum indicate the very point where all the universe conjuncts placing space in relation to the time-Zero singularity as indicated through the position the earth maintains individual singularity parting from cosmic singularity

NEWTON ON THE OTHER HAND HAD A MUCH DIFFERENT IDEA AND FOR THE PAST THREE AND A HALF CENTURIES THE BRAINY BUNCH ECHOED HIS SENTIMENTS.
Take the pendulum.

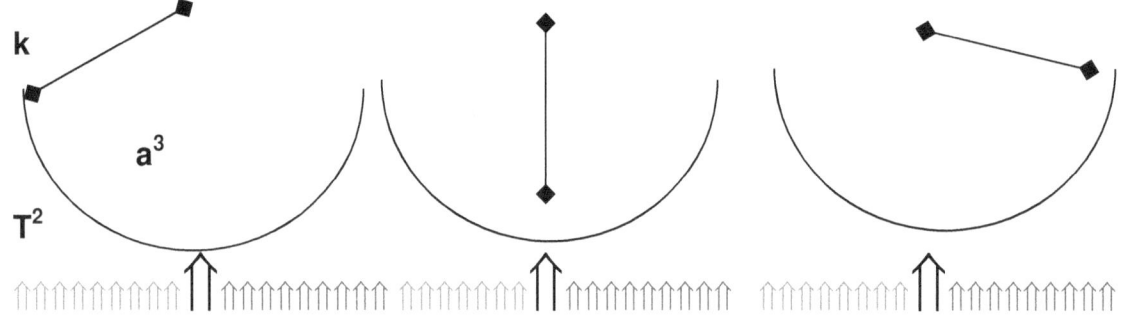

Every time the pendulum arm crosses to the other side it indicated the most important factor.

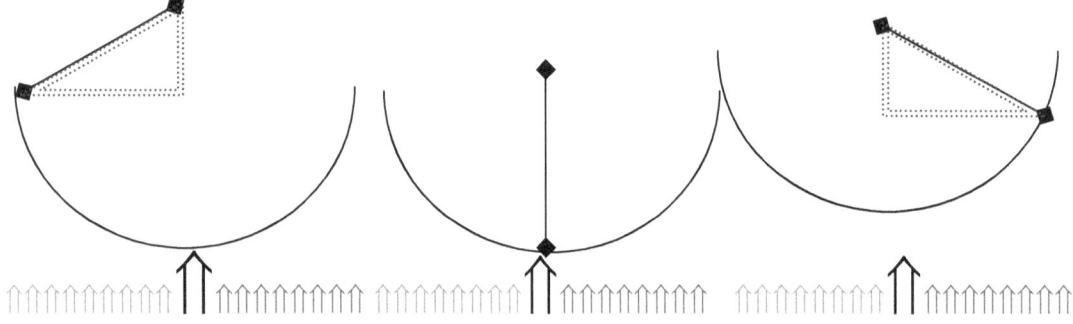

Every swing the pendulum arm does it brakes through the factor holding the universe in place.

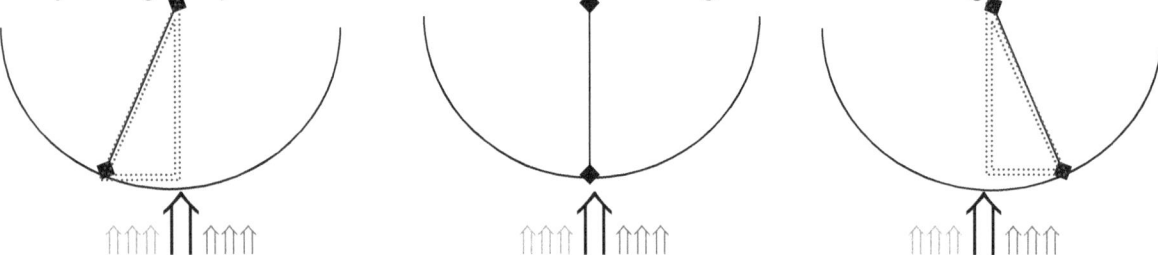

The pendulum not only crosses the singularity the earth dictates at that given time and the pendulum not only points at the factor maintaining space-time on earth.

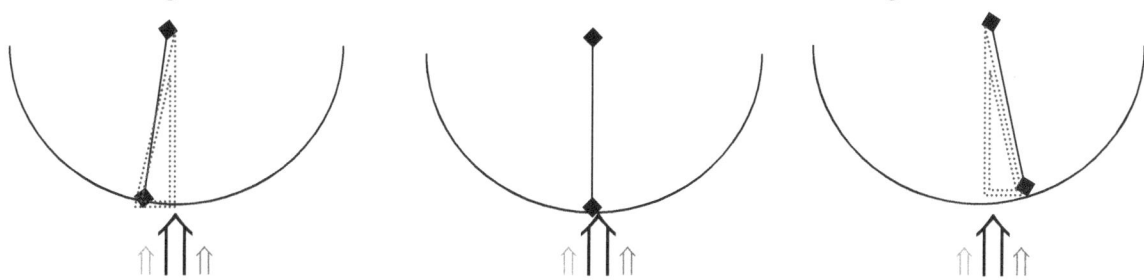

The pendulum not only stops at a precise point that the earth holds as singularity presenting the singularity the earth dictates at that given time. The relation there is in what Kepler discovered and that which Galileo discovered has gone by without many too my knowledge seeing such an extreme direct link. Kepler formulated space-time and Galileo implemented space-time. The space the pendulum swing through is representative of the space captured by the anchor singularity formulated by Kepler as a^3, the pendulum arm becomes the indicator **k** the swing distance of the arm becomes the time T^2. This is the recopy for time keeping since coming into the light from the dark ages. It is a half circle indicated by a straight line forming two sides where each side is holding a triangle in relevancy. This is in sharp contrast to Newtonian claims that the cyclic repeat the Earth has with the sun and all the numeral Equalities derived from that by Using Kepler's formula still after one year comes to nothing.

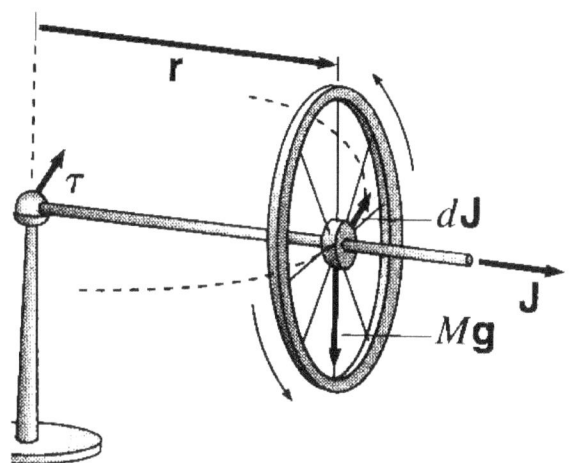

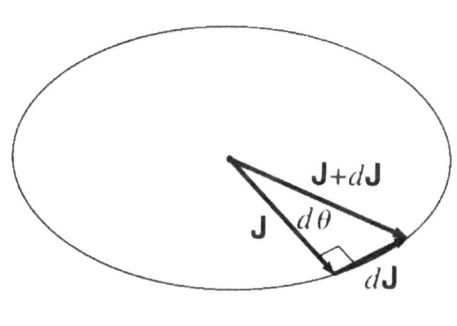

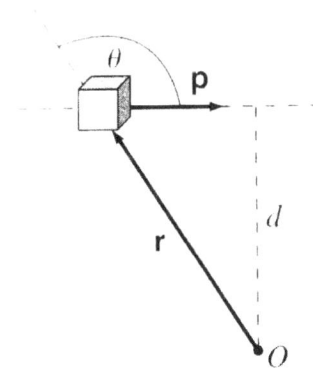

$$\Omega = \frac{d\theta}{dt} \quad \tau = \frac{dJ}{dt} \quad d\theta = \frac{dJ}{J}$$

$$\begin{aligned}\Omega &= \frac{1}{J}\frac{dJ}{dT} \\ &= \frac{\tau}{J} \\ &= \frac{rmg}{J} \\ &= \frac{rmg}{I\omega}\end{aligned}$$

$$\begin{aligned}|\mathbf{J}| &= |\mathbf{r} \times \mathbf{p}| \\ &= rp\sin\theta \\ &= rp\sin(\pi - \theta) \\ &= pd\end{aligned}$$

$$\frac{d}{dt}(\mathbf{r} \times \mathbf{p}) = \frac{d\mathbf{J}}{dt}$$

$$\left(\frac{d\mathbf{r}}{dt} \times \mathbf{p}\right) + \left(\mathbf{r} \times \frac{d\mathbf{p}}{dt}\right) = \frac{d\mathbf{J}}{dt}$$

$$|\mathbf{J}| = |\mathbf{r} \times \mathbf{p}|$$
$$= rp\sin\theta$$
$$= rp\sin(\pi - \theta)$$
$$= pd$$

$$\frac{d\mathbf{J}}{dt} = 0$$

All spinning matter has the point where the spin is still there but the radius is to small to measure by any means. That point is standing still in relation to the rest of the spin. In relation to that logic I do not except Newtonian science holding the radius of s spinning object unaccountable in the spin, whether the spin is applying or not.

Applying Newton's second law F=ma

One arrive at the formula
GMm / r² = m (ω²r)

By replacing (ω²r) with 2Π / T we obtain Kepler's third law
This law predicts that T² = a³

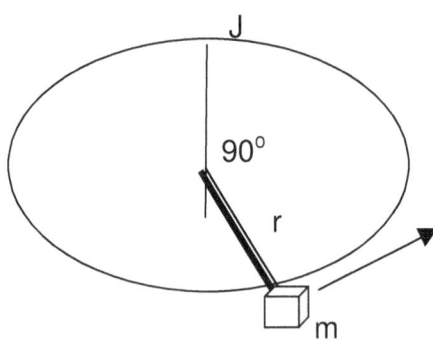

p = m .v

The mass (m) multiplying the speed (v) forms a new value J AND THEREFORE j CONTINUOUS TO IMPLY J = I ω
= r X p where p = (v =r x ω)
J = r.m.v = m.r² .ω = I. ω and becomes interpreted as J = I ω

This establishes that r = dJ / dt

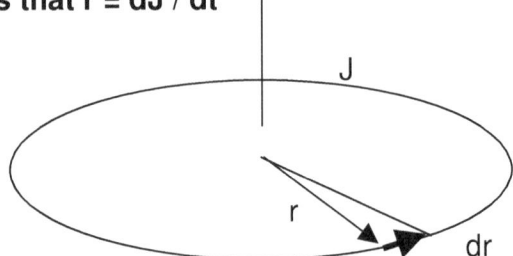

r = dJ / dt In the case of planets in orbit around the sun r forms a value of zero because dJ / dt = 0.

What this statement implies is that r does not exist. When anything has a value of zero it is for all purposes non-existent. Only when an object is following s straight line can the radius be non-existent because the radius alters value through time development.

Taking the argument back to Kepler's law,

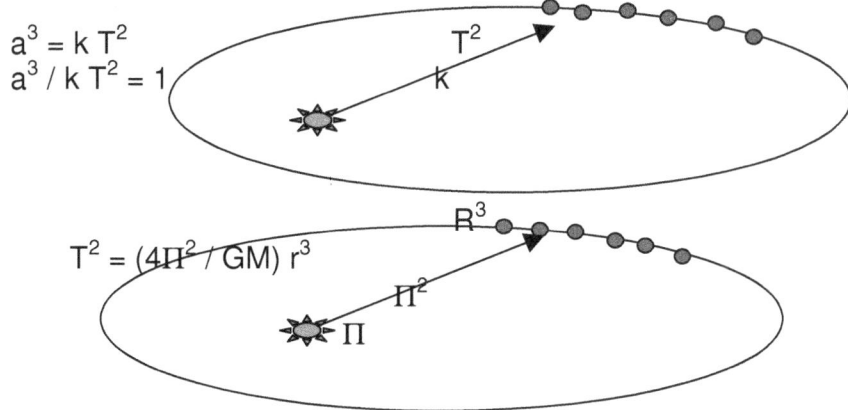

The spinning or not spinning is not part of the issue because at the point of absolute singularity the object never spins. Therefore spinning or not spinning does not apply to the point of singularity because singularity never spins in any event.

<u>Since Newton became an institution forming the King bee of the academic cartel world wide The Brainy Bunch had Newton's vision written in the minds of the future generations almost at gunpoint...well definitely at an academic gunpoint. Any student not readily accepting and understanding Newton's claim packed their bags and went home.</u>

If they did not respond correctly by interpreting the Newtonian claim as put above the student failed the examination and all further studies failed. If that does not come to black mail and threatening by gunpoint, I do not wish to see what then is black mail. In a picture this was what was expected from any and all students to admit should they wish to proceed with further studies in the field of physics. There is a wheel that spins but by spinning achieves nothing.

$r = dJ / 0$ or $r = 0 / dt$ <u>You cannot remove a factor of a relevancy no matter what argument you bring to the table, it then becomes senseless.</u>

$$r = \frac{dJ}{dt}$$

In the case of planets in orbit around the sun r forms a value of zero because $dJ / dt = 0$.

<u>I am not the brightest in the world that I admit, but one thing even I know no one can do, not even if you are the one and only Isaac Newton is that you cannot place any relevancy in a relevancy and then claim it not to be in a relevancy because such a relevancy does not suit your taste.</u> Once in a relevancy the relevancy may become irrelevant by pushing all factors to be one, and that should be the strongest indicator of a non-relevancy, but saying in the event that the relevancy holds a factor of zero goes beyond mathematics.

You cannot put something in relation to another object and then decide there is no relevancy in the relevancy

$r = dJ /$

$$r = \frac{dJ = 0}{dt = 0}$$

$dJ / dt \neq 0$. If $dJ = 0$ larger than then $dt = 0$ That is a mathematical principle, much even Newton

It is as if one then must claim in affect that Kepler held $a^3 = T^2 k = 0$. If the sun and the Earth have a rotating relevancy of zero either the sun has gone away or the Earth stopped existing. One cannot claim there is a wheel and then remove the spokes because according to you taste, too do not like the spokes

In the same manner the ring cannot remove, because the spokes will then still imply where the ring must be. The only way to cheat yourself out of the situation is to remove the wheel and spokes altogether, and you are left with what you say there is: NOTHING. But that does not apply in cosmology. The object rotates the cent restructure and therefore there has to be a radius holding the circling obiter in relation to the cent restructure.

Removing the radius of a circle does not remove the circle, because the circle is there, securing the ring If the universe started from a point of singularity, then there was initial spin at a pace where the spin did not apply and that spin included the entire universe, still in non-existing. That is singularity. That is the only singularity there can be.

The spin was going on for eternity because the spin does not apply, it has a value of infinity and infinity was running along the line of eternity.

Π^0 By receiving the command, singularity received a value outside eternity as Π^0 received edges. Granted the fact that the edges were so small there still was no r to present a circle.

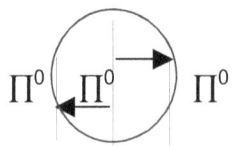

Having edges where Π^0 duplicate to present the edges singularity lost the value of Π^0 to the value of Π^1 with the same value singularity had being Π^1 to the one side and Π^1 to the other side, the cosmos received the eternal value of the first dimension outside eternity. It was the square of Π^1 being Π^{1+1}. That was the first dimension outside singularity Π^0 where singularity has a value of Π^1 in the form of $\Pi^{1+1=2}$. The first claim to space had a value of Π^2. This applied to both sides of the claim to space outside singularity, and the double proton became the dominant factor on matter.

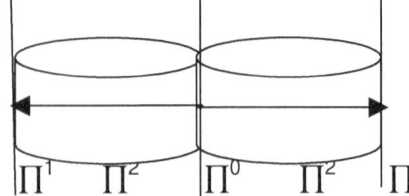

As singularity burst out into matter forming space as much as occupying space inside singularity, the protons started flying around, spinning around singularity, as each individual proton occupies matter in space

It truly makes me feel bitter thinking about the many times I tried to explain the facts to the Brainy Bunch with no luck. You know you are correct, but that person holding the establishment secured for Newton just push your argument aside, because he has the authority to investigate and lacks the interest to initiate change. Newton, and science, made one enormous blunder, from this stance. They took the radius of a wheel not to have any influence on the wheel. In doing that, they removed the very fact that keeps the universal attachment together.

$\frac{dJ}{dt} = 0$ This disputes mathematics. DJ / dt can have any number from eternity to infinity, only excluding one; it cannot be 0. By placing the one in division of the other, you bring in relevance. You cannot then say there is no relevance. By doing such, you proclaim that one of the factors is non-existent.

$$\frac{dJ}{0} = dt \text{ or } \frac{0}{dt} = dJ$$

In both cases, one of the factors then does not exist. Such a claim is incoherent, because you proclaim that a circle has no radius, or a radius has no circle. When calculating a circle, you multiply either the square of the radius by Π, or the quarter of the diameter at a square by Π.

$\Pi \times r^2$ = CIRCLE

If you remove r it then is $\Pi \times r^2 / r^2$ = CIRCLE.

You cannot then say $r^2/r^2 = 0$ and therefore $\Pi \times 0 = 0$. That is nonsense. $\Pi r^2/r^2$ will always be $\Pi \times 1$, and that is the eternal circle.

When looking at any rotating object, there has to be a point in the infinite middle where the one side rotates in one way and the other rotates in the other direction opposing the opposing direction. That point in infinity is the point of no rotation and no rotation means "no rotation", not no existence. No rotation means a factor of 1, not zero.

Quite the very opposite is true and the rotating wheel in fact is the moving wave.

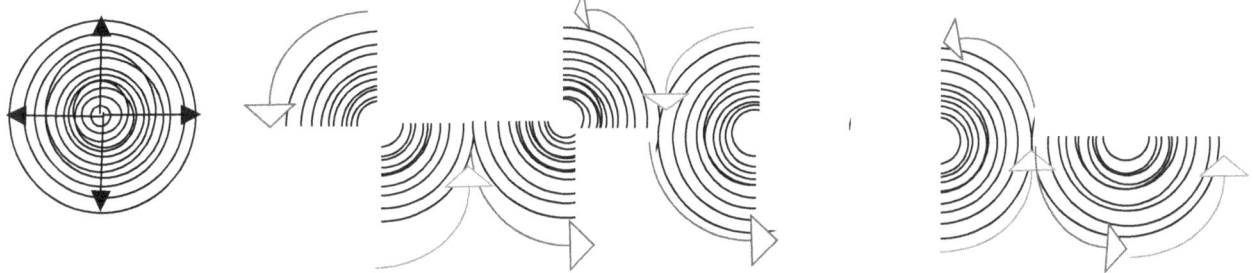

Every quarter of a rotating body is opposing the opposite sector directly and completely

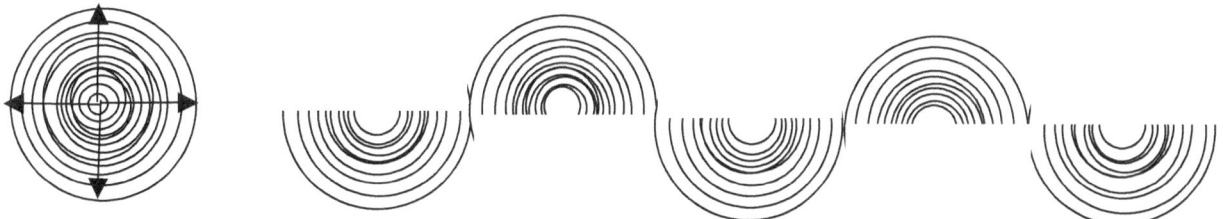

That proves that every rotating object holds the form of a wave and mostly it etiolates Newton's claim totally that no relevancy exist between the rotation and the axle which is pointing toward motion. That proves that while spinning an object holds an absolute relevancy of one in order to maintain equilibrium as to ensure rotating motion.

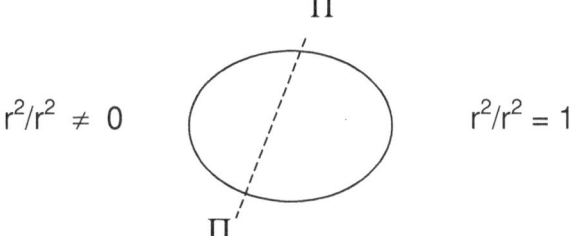

That then is singularity. The eternal Π, the Π that may not have significance but still it is a Π of value. The relativity remains one, eternally one, but it cannot be zero. Therefore, dJ/dt cannot be zero.

> dJ/dt can become eternal or infinitive or at the worst it can become one
> dJ/dt = 1

When explaining this to any child, they can immediately see that. Explain this to any Newtonian High Priest and he may have you removed forcefully from campus. I cannot find one Newtonian, of any significance being large or small to accept that.

By not having a wheel rotate, the wheel becomes the factor of one, and the rotation becomes zero. The wheel does not disappear. In the cosmos, everything is rotating because nothing ever stands still. Therefore the mean equilibrium, the common factor there is to share, has to be one, eternity, the eternal Π, because all rotating objects has Π in singularity, and sharing singularity, gives every object in space a relation with all other objects in space. After trying for many years to bring them the candle, I concluded that Newtonians are incapable of realizing that mathematical principle as reality. The comet rotates the sun, and the sun by itself has a point of singularity where Π remains without r. The comet, holding the orbit, also has a point of singularity, but since there is space separating the two objects, they cannot share a mean point of singularity, the very point of existing. Since singularity means just that, being single, there cannot be two. The comet and the sun have a mean point of singularity but the space they occupy divides their common singularity. That is why they orbit in an oval path, a path where the one structure holds on to more space from its point of singularity towards the space it claims. Since they do not claim equal space, BY THE DENSITY they hold, the space will not be in proportion.

They do share in the common fact of singularity a point away from their individual singularity proclaiming their cosmic individual reason to exist in the cosmos. That point of common singularity holds space between individual singularity and that point of mutual singularity saves and protects the points of individual singularity. Since the start of time at moment-Alfa where both found the space they occupy, in the space they hold, maintaining a time to that space in accordance to the singularity they hold that point will be their individual eccentricity from singularity. The two objects are holding eccentric space around their individual but common singularity. That point of singularity is Π the circle without the radius because the singularity removes all forms or values of r, discarding r to infinity and leaving Π to be singularity.

That is why gravity is a fixation of Newton's mind making Newton bullshit, and his $F = G(M_1 M_2)/r^2$ is utter nonsense. The moment you say Newton or any of Newton's laws, the Newtonian brain stun. Not once did I find one Newtonian surprised at this, I could not once find one single Newtonian to see this. It always leads to an argument and the argument is about Newton being in use for centuries. One Professor even answered me by saying that I should realize Newton formulas brought man to the moon, and if that is not proof of his correctness to me I will never obtain proof. That is beside the point. That is miles from the issue. If you say Newton is wrong, you commit the worst blasphemy possible. One may swear at God and all is understood but mention your not excepting Newton's gravity and they all fall on their knees, cover their eyes in the ground, start stuttering and moaning and you cannot make them see anything but Newton. Dare say there is no such a thing as gravity because Newton is wrong, they run outside and hide the woman and children from your rage of mental instability. Because I have had unmentionable arguments that I in the end lost because the mental Newtonian block all Newtonians hold covering their senses, where I could not reach a single spot of healthy logic within their minds, I wish to run through the facts once more and find what is so incomprehensible about the issue.

This in fact, is the very same findings that brought Johannes Kepler his own everlasting fame when he declared that the planets stand to a value of **$a^3 = T^2 k$** as they orbit the sun. Never once did he mention the presence of a force or gravity. Newton came up with this bogus idea all by himself without the help of other "giants" as he called *Galileo and Kepler*. In a later stage I indicate that I might prove the possibility that Newton did not have enough information to draw conclusions about Kepler's work. Newton saw a circle in Kepler's formula and there is a universe of information hiding in that formula because that formula depicts the key to science namely singularity.

Newton made the formula one big blunder as far as the cosmos is concerned. Newton works perfectly well where there is equilibrium and unchanging in space and time as we find on earth with the earth forming the basis for space-time. Taking Newton to outer space is a blunder and Newton created the blunder by re-adapting his original formula of $F = r^2 / (m_1 \times m_2)$ to fit Kepler's vision of $a^3 = T^2 k$. This very same bogus idea helped Einstein to ignore the space factor of R^3 and place a relative value of one to space. This he stated (without stating it) when Einstein put the universe in a single dimension property at the point where gravity was stretched to the limit. Einstein put the universe to a three dimensional value of matter, space and time and then out of the blue he places space at a factor of one when gravity supposedly destroy time. This notion stand totally unrelated and divorced to reality. I do admit that Einstein is absolutely accurate when saying this, but the space he refers to, as outer space and the space disappearing in time are as far apart as the cosmos is wide. Einstein was the one that said that space and time could never be separated because it was the very same thing, a point I agree with in all my findings. The difference between my point holding the universe and being the universe is within every atom because it is there where singularity is. From singularity through the atom space has the relation between Π^0 as singularity and Π forming space inside singularity holding time $T^2 = \Pi^2$ in relation to the triple value of $\Pi\Pi\Pi$ forming Π^3. I am afraid that Einstein made much more sense when he was still an amateur, working as a clerk in the Swiss patent offices. Then he landed himself under the spell of the Newtonian disciples and all his initial ideas that were factual, became integrated and confused with delusions of the "Xepted scientific Newtonian High Priests" called "acclaimed scientists" and their mesmerizing fantasies about gravity.

There is a way to explain space-time by finding **space-time**. Space-time is not some force well and truly out of our reach, the one we may dream about and wonder why we have to adhere to it with so much respect. Einstein the master of physics was completely lost in his physics. He went looking for a flat universe, he saw singularity in the dark of the night, where he saw gravity lingering around stars with nothing better to do than to wait for passing light and bend seven types of shit out of them. Singularity makes every atom rotate that makes every cosmic object rotate that apply the overall rotation to the universe. **THAT IS TIME**. Each time I try to share the idea of mine with the **"Accomplished Scientists",** I do not get farther than the phrase: *"Newton and Einstein are wrong."* After completing this sentence, I get treated as a raving lunatic with extremely dangerous hallucinations indicating a murderous tendency. Why would not one academic listen to the rest I wish to say before bluntly denouncing me?

Nobody even listens or pretend to listen to the rest of my case. Every time I see the light in their eyes go blank and they sit patiently and wait that the motor mechanic to finish his senseless rambling. It is so obvious they consider me as mindless with arrogance and having a nerve to criticize the two highest-ranking Newtonians of all time! I can assure you I am not mentally disabled! It took me twenty-one years of research and another six years in compiling and writing this book. In any of the pictures on the next page one does not see space, because you see a space filled with particles. It is the atoms holding the space secured that forms the picture. Why on earth would nobody realize Einstein was seeing the universe from a wrong perspective? What Einstein saw was one hundred percent correct but Einstein saw what he saw in the space of the atom and not in space at large.

In this first part of the first letter forming the book, my aim is not to confuse the reader with a lot of complicated detail, however in the rest of the book "MATTER TIME IN SPACE" I DO take on the challenge in addressing the Chandrasekhar -Shonberg limit as well as the Oppenheimer- Volkoff limit. The consequence of the above-mentioned limits effecting, the mass of the star is space-time related and has nothing to do with a factor of mass. It is all about density applied by the value of individual singularity in conjunctions to form mutual and shared singularity. As the time in the massive star increases, the space, which the atoms occupy, reduces to match the atom time value in space relation to the point of singularity in order to maintain a matching value of one.

That is the relation matter has outside singularity. $R^3 / T^2 = 1$.

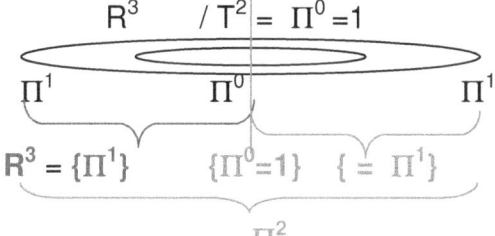

That is space-time holding every aspect the universe hold to a specific relevancy

The space outside singularity holds the time outside singularity because everything in the universe is spinning. Science know there has to be a difference because in space an object might be weightless, although it retains its mass, and no one can say the difference, except to put it down to "gravity". Us, the tax paying public, is letting these Master Minded Academics get off the hook so easily, because every one is to scared to ask "why and how". In the pages above, I pointed to the most basic mistakes about the "gravity" which science ignore, because the answers they do not know. Even Nobel Prize winning work is blatantly misguided. I challenge any person to prove how an atom can collapse on itself, by force, by weight, by pressure or any other means. No atoms will ever touch one another let alone compress to diminishing space, and if they do the result is a nuclear reaction

IF IT DID NOT SPIN, IT WAS NOT ROUND, AND NOT BEING ROUND THERE WILL NOT BE SINGULARITY.

According to Einstein, the speed of light is a constant throughout the universe. The speed of light results from two factors, being distance (kilometres) and time (seconds). This speed is accepted at 3×10^6 kilometres per second. Scientists know that it takes sun light 10^6 years to reach the surface of the sun, and we know the sun is not thousands of billions of kilometres in diameter. The "Xepted scientific Newton Mistaken" explanation about this fact is that the sunlight "bounce against matter" and this retard the sun light all that dramatically. When light hits matter, (except in the case of glass), it joins singularity immediately. Therefore, the sun holding matter on the inside has to be all-glass, or the "Xepted scientific" explanation is not very scientific at all. It all comes down to the density of matter in space valuing the time in that space away from the point maintaining singularity. Why can nobody but me see that?

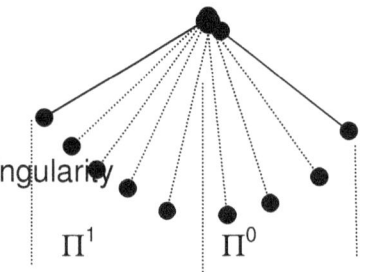

The pendulum arm instant it swings. This is Singularity in position a^3
The space a^3 holds minus the compromise singularity increase in heat.

covers a specific distance per time unit, every because of during time T^2 in instant k precise accordance to the time T^2 that it takes claims from k by reducing space to the

Π^1 Π^0 Π^1

Space-time depends on the relevancy of matter occupying space change position in accordance to all other matter relating or relevant or even only influenced by the space a^3 in the e duration of the time the matter changes position T^2 in the instant of changing. **Space-time is everything including singularity diverting from singularity** and that is what Galileo recognised without realising in his observation of the pendulum. Where Π^0 is singularity and Π^1 is the diversion from singularity forming $\Pi \times \Pi = \Pi^2$ being gravity or time.

FROM THE SIDE-ON POSITION THE CHANGE IN SPACE, DURING TIME SHARING WOULD SEEM THIS WAY:

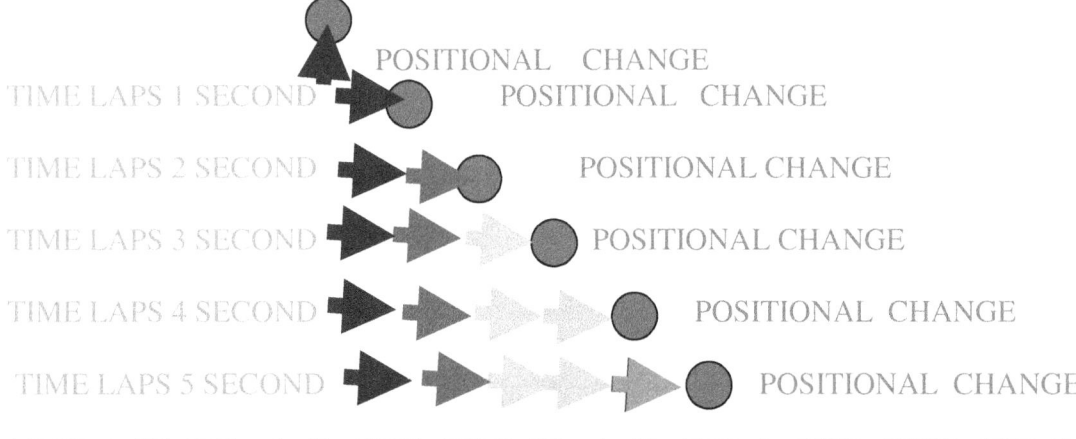

THAT IS AS SIMPLE AS SPACE-TIME IS

That is what Kepler (again I cannot say whether he wittingly or unwittingly) declared by using the formula $a^3 = T^2 k$ he announced space-time in a formula, the formula Newton raped to his advantage because in $\frac{M_s \times M_c}{r^2} G = F$ there can be no pointing to singularity in the cosmic sense.

In his initial formula $F = r^2 / (Mm)$ singularity point at every aspect because as matter fall to earth matter continue down a precise path that singularity provide holding that specific position that leads the way. What it does point at is the motion caused by the earths singularity applying on much lesser objects holding or not holding singularity. I change a to R and T to T holding k to Π^0 which is singularity in the instant

FROM THAT POINT SINGULARITY IS IN EVERY PROTON HOLDING SPACE AS RELATIVE AS TIME. R^3 / T^2 = one

This in fact, is the very same findings that brought Johannes Kepler his own everlasting fame when he declared that the planets stand to a value of $R^3 = T^2$ as they orbit the sun.
Illustrated it would be represented as follows:

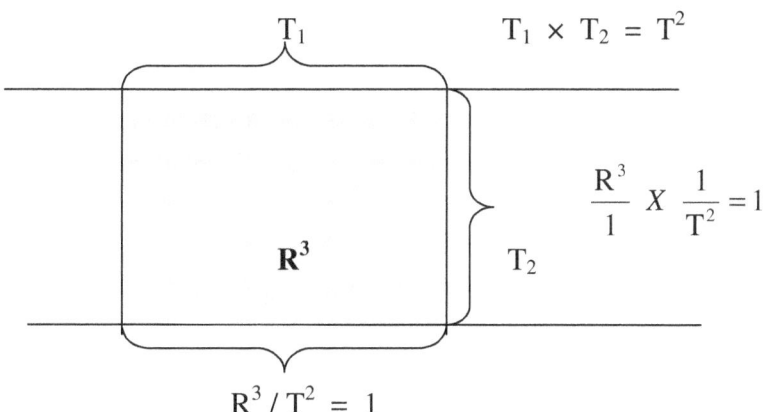

That means when the time that a structure relates to, is effected, the space will be effected pro-rata.

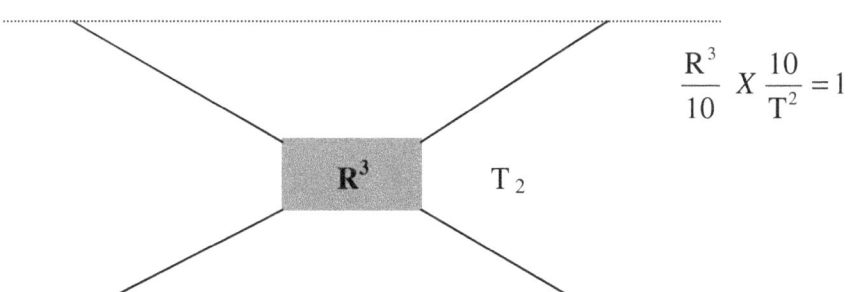

If one can illustrate the universe and its relation with space-time, the following illustration would fit like a glove.

In the search for time in space, the most obvious place to look for the factor, time as such, must be where it is excluded from the space factor and stands alone. Therefore, one should find the place where space is zero leaving time to be eternal. At such a point, it would be impossible to locate and place a value on time. However, in the cosmos at large, there is no such a place, because there is no such a thing as zero time or zero space.

This book "MATTER TIME IN SPACE The Theses" proves the following:

1. The Universe is in a constant balance between positive space-time displacement and negative space-time displacement. The heat distinguishing matter particle's individuality has a definite and crucial position in this motion of gravity. All bodies particles down to atoms apply both positive and negative space-time displacement in relation to each others position in occupying space in the duration of time.

2. There is no force, other than "life". I have manage not only to bring prove of The Big Bang and indicate that the universe at that point went from singularity, but how matter and space came about from the original value of time. In conjunction to two newly defined, but very well known cosmic principles, I manage to take the Creation of space-matter-time back to previously unproven territory. Through that evidence it becomes clear what time is and how time changes in accordance to space-time developing.

3. Gravity, which I prefer to call time for reasons that will become clear, is no force, but a balance $R^3 / T^2 = 1$ consisting of two undividable factors comprising of linear displacement (R/T) and circular displacement (R^2/T). Derived from Kepler's $a^3 = k T^2$. The R has no relation with the symbol used to indicate radius, but R relates to the Afrikaans word **ruimte**, indicating **space**. T stands as **tyd** or in English, **tim**e.
Together these factors form space-time $\$ = (R^3/T^2)$. As there is no official symbol to indicate space-time, I introduce one, which I name estee $.

4. Space has no value, other than that which time lends it. Therefore, time determines the value to space. This also is a deduction, which I prove from the formula $a^3 = k T^2$. Under normal circumstances, the major component having the biggest proton value also carries the circular displacement value or positive space-time displacement value. The bonding value of gravity runs through the heat separating particles, and that mass by number of particles alone does not carry gravity. Gravity comes about from the density that particles have in their proton mass. Hydrogen and helium contributes no gravity, and iron $_{56}$ holds the universe in space-time.

5. The atom comprises of three parts being:
   5.1 Unoccupied space-time
   5.2 Occupied space-time
   5.3 Densified space-time
   5.4 Frozen Time (possibly the GRAVITON?)

6. The space-time, which, is to the "outside" of the electron, is as much part of the atom as the space-time between the electron and the neutron, which, again is as much of the atom as the space-time between the neutron and the proton.

7. Time located in space is moving to time located in matter. Matter is growing while space is diminishing. The growth matter obtains, are depending on the amount of protons forming the concentration of time in that particular location of space.

8. All stars developing in this era has to have an iron inner core which forms the last link a star has to space-time, before becoming a proton star, consisting of the single dimension also known as singularity.

9. Time is located in the proton and not in the universe as a whole. Through this I show how the atom influence the cosmos by means of the dimension changes 3 to Π, Π to Π^2, Π^2 to $\Pi^2 + \Pi^2$. When multiplying all these dimensions the answer is 1836, the mass difference between the electron and the proton.

10. The speed of light is not a constant, but depends on the ratio between space and time, whatever the value to that might be in any location of space allowed by time. The speed of light also has a linear component $3\Pi^2 = 29.6$ (the photon) and a circular component ($3^3 = 27$) the wave.

11. The speed of light consists of a combination of linear displacement, which the photon carries and circular displacement, which the wave carries. The speed of light, holds a photon value of $\$T = 3\Pi^2 \times 10000 \times 24 / 7\Pi$ where as in the wave carries a value of
$\$T = 3^3 \times 10000 \times 24 / 7\Pi$

12. The fastest an object can fall through the sky is
$\$T = 3 (7\Pi^2) \times 24 / 7\Pi$

13. The fastest any object can displace space-time with in the concentration of the earth is $\$T = 3(7\Pi^2)(\Pi^2 / 2) \times 24 / 7\Pi$. This value is also associated with Mach 1 and is incorrectly associated with the speed of sound by means of the Doppler effect. The speed associated with sound, is merely the point where the travelling object's linear displacement surpasses the linear displacement of the earth, reaching the velocity of the earths circular displacement.

14 The maximum velocity any craft can reach travelling in a horizontal position is below three times $(7\Pi^2)(2\Pi^2) / 3$. This limit comes about in accordance with the Roche limit of $(\Pi / 2)^2$.

15. The Doppler effect is a manifestation to the true cosmic law the Titius-Bode principle that I renamed as the Titius-Bode law. The Titius-Bode derives its space-time value from the 7^0 inclination all spheres holds to space-time, forming the linear contingent to space-time and Π^2, which is the circular displacement value to space-time.

16. All aspects of space-time have to consist of both a linear and a circular contingent to space-time, to proclaim existence.

17. Electricity is the revaluing of space-time from a linear stance to the speed of light in a circular stance, forming an alternating of space-time between the linear and circular space-time in and between atoms of the conductor. Therefore, lightning is not electricity, but wind in its most advanced form.

18. Fusion is the product of linear and circular time revaluation pushing time back to a position space holds that predates moment-Alfa. As time prolongs in duration it reduces space by the same margin. When time goes eternal, space is zero.

$S\$ = R^3 / T^2 = 1 = R^3 / T_1 \times T_2$
T_1 = Linear factor; T_2 = Circular factor

$$S\$ = \frac{R^3}{\frac{C}{C-V} \times \frac{C}{C-V}} \quad \text{where } T^2 = C^2 \text{ and } R^3 / \Omega = \alpha$$

19. $S\$ = \dfrac{R^3}{\dfrac{C}{C-V} \times \dfrac{C}{C-V}}$ is the space value of space-time where as

$\$T = (4\Pi^2)(7)(3\Pi^2)(\Pi^2 / 2) / 3600$ is the time aspect of a body seeking individual space-time.

20. The Hubble constant is not a universally applied single confirmative value to times expansion, but each galactica performs to its particular time value to space. The growth rate is in accordance to the Titus-Bode number arrangement of space-time (7/ 10) and in the square of mater it is 7 + 7 / 10 with the square relating tom space as (10 / 7) By adding the relation therefore forms its three dimensional value 3; 6; 12; 24; 48; 96 etc.

21. The universe came about not from one "Big Bang" event, but seven individual cycles of light and darkness, and that the solar system and more in particular the earth went through seven periods or

as I refer to it as solar days (**weather you like it or not**) just as the Bible indicates by merely applying the four cosmic pillars correctly. The seven different periods (that are mathematically proven) from the **inner-Core-value** confirming the time related to each era where the value to Π changes every time.

22. The earth and the solar system also came about from seven individual separated periods of light and darkness, positioning every planet in accordance to the Titus-Bode rule **NOT APPLYING** WHERE AS IN NORMAL CONDITIONS, IT SHOULD APPLY.
I send you the copy of **MATTERS TIME IN SPACE**

Academics, which I contacted in the past, treated my work with great scepticism. That forced me to find more proof and every time I went back to the basics again and brought about further proof IN THE HOPE OF ACCEPTANCE to prove my case. At first, I did not wish to criticize Newton or science, but eventually through the persistence of some academics, I was forced to do just that. It is not that I wish to replace Newton in the least, but when Newton formed his formulas the background to science was very patchy and dark in comparison to what science know at present. I have tried to publish the book through the commercial route by finding a publisher. The response I received was that of publishers not willing to be involved because it falls outside the boundaries of accepted science. The battle continued as I then took a more direct approached by going to the Internet. This was no solution at all because those that seek information on the Internet did not make head from tail to what I was saying, and those persons I wished to draw their attention do not look for information on the Internet. I have spent too much time, money and effort in this work to dismiss it just because people do not understand what I am saying. I am too sure about the correctness of my case too let go.

aanplasing, verplasing, versnelling and inperking

As this book is a translation from Afrikaans originally, some terminology and expressions I had to revise to accommodate my Ideas. Where I could I used modified English words to express a thought or an idea. One such a term is gravity. I had so much criticism about the word, which I feel I do not deserve. There is a certain notion clinging to the idea represented by gravity. Gravity links to a force that is all compelling, but I do not agree with SUCH A COMPELLING FORCE, such as the word gravity implies. Gravity I introduce, works on two principles, but gravity to Newtonian standards is a single force. When I refer to gravity the normal reaction is that I am referring to the force I deny. By declaring that gravity THE FORCE CONTROLING ALL OF THE UNIVERSE AS A STANDARD CONSTANT IS NON EXSITANT, I bring the wroth of the scientific world upon me.

When I make the statement that there is no gravity, every person considers me mentally unstable. Of course there is a movement of energy keeping all objects attached to the earth, but gravity implies work, and with that work principle I disagree, because that is NOT WORK, THAT IS A COSMIC BALLANCE THAT STARTED AT A POINT OF ETERNITY AND WILL END AT A POINT OF ETERNITY. ONLY LIFE, STANDING ALONE AND DETACHED FROM THE COSMOS AS THE ENERGY THAT MANIPULATE SPACE-TIME CAN COMMIT WORK, THE REST IS A BALALANCE RUNNING CONCURRENT FROM TIME'S BEGINNING TO TIME'S END. I HAVE UNBELIEVEBLE DIFFICULTY IN RELAYING THE DIFFERENCE BETWEEN LIFE WITHIN THE COSMOS, AND THE COSMOS WITHOUT LIFE. In the entire universe there is no work, it is a balance running concurrent through time and space. The balance shift in some cases to favour space whereas in the other cases the balance shifts to favour time component more than it does space in that sector. But in all of that shifting, a continuous balance strikes every aspect of space-time.

This applies new ideas never brought to light before and the new concepts clashes with the conventional names that science applies to current ideas. I had to divorce the science ideas from those I introduce and the only way was with new terminology. I have to start implementing the newly created terminology, which will apply to the rest of this book. This stems from my lack in ability to find suitable words in the English language that would define the concepts as they are, in order to establish the difference in meaning from the current words, which convey the existing misinterpretations (or if you wish, to my view incorrect applications).

Firstly, we start with the word ***densified***, which is not a normal English word, but a word I had to produce in order to make a comprehensible statement. The correct word that applies is concentrated, that much I do know about the English language. But concentrated has not the correct meaning or the expression that I would like to bring over. Concentrated can apply to any substance, be it gas, liquid or solids where one of the ingredients become more than the rest of the ingredients. In that way, matter as a solid substance produced from the eternal substance which is heat, cannot be concentrated. Nothing in the entire universe can compare with the density of pure heat that spins at a rate in which that very heat can produce a value and which has a density far beyond anything else. Therefore, I chose to use the concept of concentration in a position where it makes a lot more sense.

A star is concentrated space-time, but there is a huge difference between a star's concentrated space-time and the value of pure matter. When a star does therefore become densified space-time, it can only be at the end of the Big Crunch eternity , witch I prefer to call moment Omega; that is when space becomes eternal and time becomes Zero. In this light I chose to call matter densified space-time. Densified space-time should therefore be in a definition where matter or substance has reached a point in density that will last one eternity, but has no limit. Concentrated space-time, on the other hand does have a limit, which is at the point where it becomes densified space-time.

The second word I created is ***Aanplasing,*** which is the ongoing redirection of heat as in matter to heat as in time and that connects to a circular deepening of the separation that matter undergo transforming to time as it discards heat for the cold of fusion. Later (I hope) it will be clear enough for every reader to comprehend and to distinguish between the various factors that bring about **aanplasing** as should the reasons be clear why I prefer to have created this new word.

In this case however, there was no English word that could merely be altered and then be re-applied. With a choice to my disposal I chose to alter an English word as was possible with densified. A more suitable word that relates to a better meaning in the case where I brought in **"densified"** would have been the Afrikaans word **"verdigting"**, where "verdigting" stands in relation to "konsentrasie" (concentrated). The fact of the matter is that I am not wilfully forcing Afrikaans down the throat of the Anglo-American and in the case of density I was able to adapt and modify a known English word that could adopt a new concept. Unfortunately in the case of aanplasing using an English word would mean that there is no liberation from the "misleading" focus that depends on gravity, nor can it liberate the feature of this "misconception".

As for the Afrikaans words: **aanplasing, verplasing, versnelling** and **inperking**: there are no such words or concepts in existence that the precise meaning can derive from the English written or spoken language. Should any such words exist, the misconceptions that remains connected to the original English words, would not bring justice to the concept which I wish to apply to convey the meaning that lies behind the correct value of the thought. If I stuck to the word "gravity", the concept I wanted to introduce would forever remain confused with Newtonian application. To that end the new realization would then never come across in the way I intend it to be.

The R that I use in the formula has nothing to do with Radius as a term, except when used, to calculate the value of a circle or a sphere. The R is derived from the Afrikaans word **Ruimte,** which means space. In the Afrikaans word: **"Ruimte" the "u" and the" i "** is used in conjunction, which is pronounced the same way as " ai" is used in English words such as in **pain, drain, train, vain, rain, etc**. So spelled incorrectly it should be pronounced as **"Raimte" and the "te" is pronounced "huh". The T stands for the word tyd, which incidentally is time.**
This is what I named negative space-time displacement, which results in **verplasing. The word verplasing is pronounce FHERPLHASHING (FHER – PL –HA – S – H-ING),** which means to "relocate" without destroying or changing the composition in any way, as the object is moving away from a certain position.

The word **aanplasing is pronounce AHNPLHASHING (AHN –PL – HAS – HING**) and literally means to relocate without damaging or destroying the composure or structure of the objects, as the object is moving toward a certain position. This very same value was previously mistakenly confused as being gravity. It is the effect on matter where space-time is in motion and matter is motionless or "stand still".

Both **aanplasing** as well as **verplasing** cause time differentiation and matters structural re-valuation. That means the duration of time is re-valued and the space compromised. The excelling of the time factor is; versnelling and the reduction of space are: inperking.

The **word versnelling is pronouncing FHERSNELHING (FHER –SNEL – H-ING**) and means to speed up. If one is placed in a star it would seem as if time inside the star is accelerated while time on the outside of the star would come to a standstill. This concept is explained in far more detail, in a later stage in the book. The **word inperking means reduction or containment of the structure or scaling it down to a different size without penalizing or altering the shape in any way.**

Both inperking and versnelling is how matter relates to change **in space (inperking)** and **time (versnelling)**. The generation of the heat is within the structure and relate to time in space.

Inperking: This stands apart from the idea of curtailment because in curtailing. The *In* part is pronounced as in English where the *per* one pronounced in the same fashion as the sound an English sheep makes BHE placing a *H* sound before the *E* with *king* already explained. Sorry but that is as far as the Afrikaans lesson goes for the day.

Inperking: This stands apart from the idea of curtailment because in curtailing something or someone, means that object's or person's movement or moveable motion is deprived. This then brings over the misconception in the accepted notion of an expanding universe. In due course I shall explain the concept, but inperking involves the same value that was there at first and will be there in the end, only the location in the balance shifts to favour one or the other part of the same coin. Because of the fact that none such a thing applies when space-time "accelerates" (versnel), and where this brings about inperking, it does not apply. Instead, all functions and factors still apply when inperking becomes valid, therefore the meaning of inperking becomes more applicable and this word describes the process much better. Inperking relates to time, where the duration of time extends, but not the value of time as such, as time applies in the cosmic sense.

One should realize that the entire atom, as well as its surroundings including all other surrounding atoms are reduced in space-time volume, so the atom is not actually curtailed, nor is its surrounding which means the word curtailment does not really apply. All aspects of occupied and unoccupied space-time are in reality, re-focused down in the true sense and above all, remains to the precise relative relation value it had for one entire eternity where the relation between such times, only refocus. However, scaled down would neither apply, because that would not refer to the time involvement, which lies at the hart of this revaluation. Where less time applies, inperking would be more severe and where more time applies, less inperking will apply.

In this reference to time, one second would remain one second to matter inside the star but the duration of that second, compared to geodesic time validation, would appear to stretch enormously. All words in the English language by implying its dictionary meaning, will inevitably lead to more language confusion, seeing that the explanatory meaning does not cover time enhancement and space reduction, heating and slowing of time lapse. By introducing a new word to the reader, I hope to screen out any misconceptions. Hand in hand with inperking, goes versnelling. When the reader encounters the concept of inperking, it should accompany the idea of versnelling.

Versnelling: It carries exactly the same meaning as acceleration, but the meaning or concept connected to acceleration implies to matter as the matter increases its own positional change in space and time. That is not the impression I wish to relay when referring to versnelling, because it is exactly the opposite of that meaning. In this, the actual meaning is more applicable to the true connection. These I must explain carefully, not to convey confusion. When a person stands outside an explosion of some sort, the time laps seems instantaneous, quicker than the senses can relate to. However, inside the explosion, time is almost standing still.

Any person, who is inside such an explosion, would relate to time on the outside as being instantaneous. Whether this statement is accepted or not, the truth is that a person in an explosion cannot die, although his body is shattered in a million pieces. Such a person is sealed in a period separated from the period he and we lives in. This I explain at a later stage. The time duration slows down immensely, but from the outside, it accelerates immensely. Therefore, time versnel to the outside of where ever one relates to.

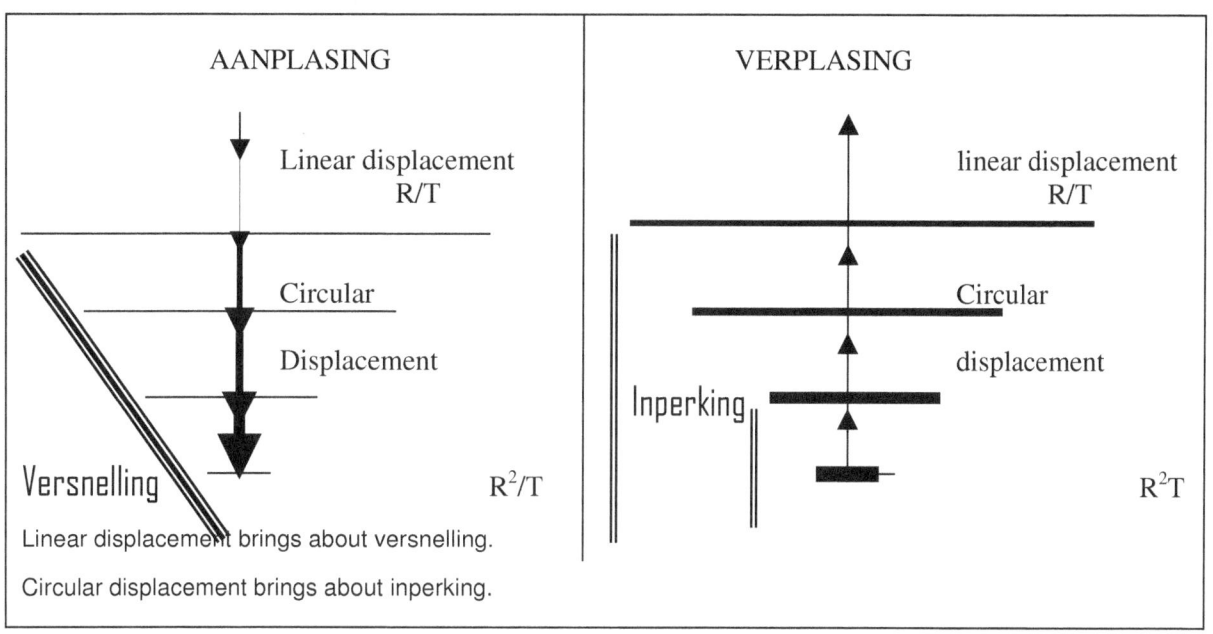

I wish to bring over the fact, as just been said, that the concept we have, is quite the opposite. Versnelling implies that the motional increase lies with the transfer of space-time, regardless whether matter occupies it or not. As aanplasing (not gravity) and versnelling bring about inperking (not curtailment as the body remains free to do as it wishes) the space-time that the body occupies and the surrounding sphere are in constant state of versnelling. The increase in motion has an effect on the matter, but the matter stands weightless as its specific density applies the time in that particular space.

Verplasing: This word is preferred to that of displacement, because although the matter in motion is displaced, time and space are implicated in the process. Verplasing is in fact the transferring of newly created magnetic space-time by matter, as a body composed of atoms has to replenish the space-time it occupies in order to maintain its position, place and structure in space-time in time in space, according to its geodesic positional allocation within the star's space and in time. Verplasing comes in effect as matter progresses in position, but the time-affect of verplasing that it has on matter, comes into real effect when an object reaches Mach $_3$ depending on its shape and altitude. In short: **Aanplasing** is relatively where matter is in a geodesic motionless position as space-time carries the motion component of the two values. This means that aanplasing is relative to positive space-time displacement.

Verplasing on the other hand has to do with the motion being with the newly created space-time in relation with the matter and the geodesic space-time remains relatively motionless. In both cases inperking and versnelling is a consequential result of the process. The difference is in the application of the time component itself.

A practical example of the difference between aanplasing and versnelling is as such: a body in **aanplasing** is in example a skydiver is falling towards the earth and **verplasing** is where a body, such as a rocket is on a trajectory path as it fires into space. Both bodies will comply with the linear and circular displacement, but the circular displacement will relate oppositely in each event.

This aspect, Newtonian science disregard in, as much as they disregard that there is any connection between the atom and the cosmos

I deny the fact that a star can have winds, although winds are as close to that concept which the earth can provide. As you will later see, winds are the transferring of heat, but so is electricity and lightning, and one cannot call lightning wind. Neither can one call lightning electricity. It is altogether different product of the same transformation of heat, but the applied principle separating the products of heat transmitting stand in total different areas.

As far as ordinary physics go, nothing changes at all. Every aspect of physics remains the same, except the way science view the cosmos. The formulas I show, has NO CALCULATION ABILITY. The only value in the exercise is proving what no person ever proved before, AND THAT IS THE INFLUENCE THE ATOM HOLDS ON THE UNIVERSE, AS THE ATOM INFLUENCE EXTENDS BEYOND ALL COSMIC BOUNDARIES.

To me everything makes perfect sense and while saying this I do admit full heartedly that I am not a Master such as yourself with the knowledge you possess. In that light, should you feel there are aspects I do not explain to a sufficient standard, I am willing to work on it.

What is the universe? This is such s simple question every one gets wrong because of the relevancy we humans place on the universe and the relevancy what the universe truly is. Official Policy Protectors really get tide in knots with making all about nothing so complicated it absorbs everything holding back nothing.

The universe is a compliment of matter, ranging from as little as a photon and radiation to as large as a Black Hole and Galactica. still it is a compliment of matter, holding space. There is no open universe and there is no singularity running free in the wider universe looking for some space to be within. There is only one universe holding many, many, many compliments of parts in the form of atoms to one space that does not even exist. This means every particle holding singularity is the universe and the complimenting total of that is the universe. Light is only part of the universe and not time as such. Light in as much as the photon is space in the form of heat that we can see and space in the form of heat is light we cannot see because it diverted from our direct line to singularity. Light is the smallest particle that maintains singularity while running away from a mutual singularity amongst photons to join a singularity within larger matter. The darkness of space is light we cannot see. Light is the darkness of space we can see. There are three components in the universe, one is time, the other is space and the third is matter. Looking through the looking glass there is only matter that matters because we can feel it and matter that so far did not matter because we cannot see it. Then Einstein brought in light as pure energy and with that he was correct up to a point. Einstein led us to believe that energy was some fluid magic where "gravity" on the other hand was some solid magic. If that is the case, then what is gas magic? If there is two of the three forms identified the third form also must be somewhere. By introducing singularity as the third magic is surely not on. He introduced singularity but never identified singularity. That brought about that singularity was also some witches brew of magic and ghosts and all the rest. There is not even a universe at large because take away matter occupied or unoccupied and you are left with nothing. That is precisely what space is. Space is nothing. I have a huge problem with the fact that Einstein saw singularity, but never could place singularity, while the position of singularity is so very obvious. I AM VERY AWRE OF THE FACT THAT ALL NEWTONIANS BRAVE ENOUGH TO ENDURE MY VIEWS UP TO HERE ARE STEAMING WITH ANGER AND I AM GREAT FILL NOT TO BE PRESENT IN THEIR MIDST. BY NOW EVERY NEWTONIAN PURE AT HEART WILL INSIST ON MY EXECUSION. At this point I have Newtonians stretched to all limits with my going on with the second largest Newtonian that ever walked the earth. Newton was off course the biggest Newtonian of all, and the rest of science filling all the other positions from top to bottom according to personal importance and devotion. I truly consider the world of science a conspiring collaborating Brother hood. Blame me if you wish, but the Newtonian religion and devotion to Newton has no bounds. Every one is supporting the other while covering for the group as a whole. If they wish to castrate me for saying that, then answer some simple questions No Newtonian knows time or space by definition or place, yet they want to bend, stretch and tie it in knots. I should think they must first locate the objects before they intend to torture it with such tenacity. What is the universe and where does it end? That is questions that one should address before going on to the more significant major issues like time travel and galactica infesting. The universe begins inside every atom and ends where the atom's influence on space ends. The universe runs from atom to atom and the lot holds the universe in line with individual singularity The universe is the total of all heat holding space occupied or unoccupied in whatever form heat may be. The universe ends where one will find the last of matter occupied or unoccupied wherever that may be and that will be nowhere because of the vastness of the universe. The centre of the universe is in the centre of every proton, where singularity is. There is no collective universal centre because the universe is holding together atoms, which is holding together space in different dimensions.

Xepted science has that also the complete opposite to what light is. No object can at the present geodesic space-time be in motion at the speed of light or to any related value, therefore no object can move towards the earth or away from the earth at even a fraction of the speed of light. What this indicator reveals is the geodesic space-time value during that specific era when the light was permitted to brake free from the time concentration during that point. The higher the geodesic time-space value was, the more time was to space and the less time was committed to matter.

IF THE MATTER WAS NOT SO SERIOUS, A PICTURE ABOUT The beginning WOULD HAVE A REMARKEBLE SENSE OF HUMOR TO IT.

In the beginning, there was time Zero to moment Alpha. There has never been a Big Bang, as such and there were too many Bangs and too numerous to count. Everything is a variation of time duration in space. During the period of time Zero to moment Alpha the value of 1 second was equal in duration to about 1 000 billion, billion, billion, billion years (I am only stopping with the billion part in order not to bore the readers), measured in geodesic space-time values that currently applies. It could be even billions times this duration because the value of time then, was measured far beyond the speed of light, since light did not yet exist. We have no way to calculate the duration of time that applied during this geodesic space-time era, except to put it down to eternal. When a bowl of soup is boiling, have you seen the bubbles of air rising from the soup? Has any Newtonian ever taken the time to explain that process in detail? I think not, because such explanations would be far too "everyday-like" to bother their mighty brains. To find the mass of Black Holes and finding the mass of the neutron, that is what such mighty brainpower can cope with but it cannot bother with small events.

Well, that boiling soup tells the complete story about the creation. Poets and painters and writers always wishes to say how "they created their creation". That is rubbish; they created nothing. They brought nothing new to the cosmos, they only rearranged what was a small part of the cosmos into a new order, that one can detect a distinction from. Creating is producing what never was before. When looking at the boiling soup, one sees bubbles rising from the soup at the top. In the soup's brew, there are only liquids and solids before the heat came. No one placed air in before the event or during the event at any time. Yet from the brew of liquid and solid rises gas, or if you wish space. That space was not there previously. That SPACE WAS CREATED.

That space is energy en energy is the interaction between heat and space. As space becomes a part of the soup, a part not there before, with no room to be, it moves out. We refer to that process as boiling. That space creation is applying heat to time, and time in singularity will respond as space in singularity. The space created will vanish just as it came, back to singularity. By applying heat to time, brings forth space, and from the three components, only the heat factor is not in singularity. It removes space in singularity from time in singularity to establish room (space) for heat (time).

That is how creation started. Time in singularity overheated and the product of that was space. That is the 180 ° of the straight line as much as it is the 180° of the half circle.

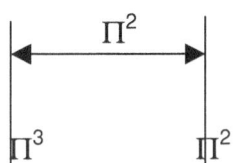

The Π^3 is time in singularity
The Π^2 is matter or heat.
The Π is space in singularity.

Because time and space both, is part, of the same thing time became space and space became time. I prove that the repeating of this process happened about seven times, in seven different ways and explaining the other five ways is rather complicated and tedious. Therefore, I shall only give two explanations in this book being part one. One is as I explained Π^3 (singularity time) parting with Π (singularity space) leaving "gravity" Π^2.

What happens to space happens to time. Space holds three parts with six sides, of which only three sides directing towards any object at any time. Therefore the sides hold 6/2 (half of the six sides are valid at very instant, only 3 has an effect.). Time in singularity holds a line directing to time. A straight line is 180°. Matter Π^2 holds Π^3 from Π, being valid in forming $\Pi \times \Pi = \Pi^2$.

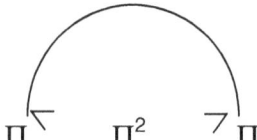

The half circle is 180° placing matter in a circle but because space only applies, to one half, 6/2 and matter holds space to value 6/2=3 only half the circle comes into effect. Half a circle is 180°. Because space has three parts in effect, it also becomes a triangle.

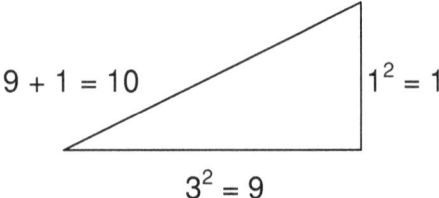

That means where space holds three and time is one, the heat within that space becomes another dimension, the fourth dimension holding space-time $(3^2+1^2) = 10$. That changes the matter inside space in singularity at ten and "gravity" at Π^2. This is why "gravity" Π^2 is space (10) losing one dimension (Π^2). "Gravity" is all about space (occupying matter and heat) losing one dimension back on a long journey to singularity.

As time is in singularity, and space is in singularity and both are the same thing ($\Pi^3 \to \Pi^2 \to \Pi$) the 10 of matter (heat) that affects space (10 Π) will also affect time (Π^3) and therefore time carrying heat will become 10 (Π^3) with space 10 Π. Anyone with a simple calculator can divide 10 Π^3 by 10 Π and see where Π^2 fits in.

The reason why time in singularity is Π^3, lies in the two components of space-time occupation or "gravity" manifested in the Roche limit. All object spin and spinning is a circle Π^2 while all objects are moving in a direction $\Pi^2/2$. Again only, half of Π has any dimensional validity at any given time, therefore the dimension surrounding an object is Π. Because the "gravity" of the surface of the cosmic object extends from Π^2 to Π but only half of the circle of Π (180°) can apply to time (Π^2) being in a straight line $\Pi^2 \to \Pi$, "gravity" will form at that point of $(\Pi/2)^2$ giving the Π in space the "gravity" to hold.

$\Pi^6 (\Pi^2 + \Pi^2) / (6 \times 10)$

That places Π in a total of Π^6 with 6 sides in space (10) affecting the proton $(\Pi^2 + \Pi^2)$

That is why space will forever comply with $7 / 10 \Pi^6) / 60 = 112$, (the Π^6 is $(\Pi^2 +\Pi^2 + \Pi^2)$) and time forming the line (180°) between the half circle (Π to $\Pi = \Pi^2$) at a 180° will form the triangle of space in half (180°). The matter component of the Titius Bode law effectively apply to the value of space, therefore 7/10 comes into the calculation. That places any atom with an existence in space at a premium of 7/10 (Π^6) (6/10). The reason why plutonium at $5(\Pi^2+\Pi^2) (\Pi/2)^2(3/5)=244$ is at the element limit is obvious; when dissecting the relevancy in detail. The complete element holds the very edge of what an element in space and time can endure in this era, but two or three era ago it had the function cobalt has at present

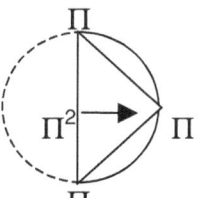

That will produce time in singularity a value of Π^3.

Explaining the other five stages of gravity (Π^2) development is extremely complicated and for that there is no room in a book meant to introduce new ideas such as this. My motto in this book (part one) is "Keep it simple"

With time in singularity, time was eternal.

Π°

Time is the spin rate of heat in space. That means the way the movement changes where matter and heat relate to other matter and heat in space All the movement is relating to a circle (Π^2) going somewhere (Π) in space 3. The Π will form the radius to the circle (Π^2). Any novice can see that the longer Π becomes, the wider Π will be and therefore the longer change in the repositioning of matter will be.

If any person wishes to cling to Einstein's view that the speed of light is the fastest that matter can apply velocity, explain how the Black Hole works. Inside the Black Hole must be matter, because there is no space, yet time does apply because it takes the particles spiralling inwards to the centre time to move from point to point. Matter in motion is time. However, no light can return to the surface, therefore the light is slower than the moving particles within the star. The only thing about the star is that it maintains a higher relevancy than the relevancy the speed of light can apply. By accepting the existence of a Black Hole, the Einstein claim about the speed of light being the fastest that time can travel and becomes zero.

Another place where the speed of light becomes obsolete is within the centre of galactica, where the accumulative movement of matter exceeds the speed of light. That is where doctor Hawking saw a Black Hole that is not a Black Hole, but the precise opposite. Light matter and heat, moves inward in an effort to maintain cooling as the group of proto, proto stars two era to the future) still claims their share of heat maintenance. Those particles in such close proximity, establishes a time well above that of the speed of light. Everything in the cosmos is all about relevancies. Time started at such a high velocity, it had to be eternal. Nothing diverting from eternal can become more than eternal so it has to be less than eternal. It is fragmenting eternity into parts making eternity smaller. The smaller eternity becomes, the lesser eternity will be. That means that time started at eternity and became shorter with the introduction of infinities that broke the monotony of eternity. The more infinities there are affecting eternities, the shorter will eternities be.

I prove in a later part in "*Matter's Time in Space* – The Thesis" where Einstein went wrong in his theory about "The curvature of space-time". There is the space-time complying with singularity and filling the space-time in singularity is heat and matter valuing space-time. Space-time (Π^3 to Π) cannot bend, cannot curve, forms a straight line, but what fill space-forming time is matter (Π^2) and heat (3). That part changes. The atom cannot be gas, or liquid, or a solid, because the atom is densified in occupation of space-time. It is the heat in unoccupied space-time that produces the gas, liquid, or solid that all substances can form. It is THE HEAT in SPACE that produces TIME, that can and does curve, bend or whatever. That HEAT in SPACE forming TIME that forms the relevancy of space-time and that does bend. If, by applying the forming of gas, or solid or liquid to the element instead of the space between the elements, of course you will get the incorrect vision Π, where the space-time (matter holding singularity forming singularity) is doing all the bending that apply to the curvature of space (validating time) and time in singularity (a straight line) will be solid. Einstein placed the relevancy incorrectly on singularity, instead of heat.

I do admit, IT IS A LOT MORE COMPLICATED THAN WHAT I ALLOW IT TO BE AT THIS POINT, but the motto is, Keep it simple. If you wish to keep time in space constant, everything in the universe will be oblong. That is why the Newtonians have an absurd view of the cosmos, and they present facts in the cosmos in a way nobody (least of all the Newtonians) can understand. The first time I can detect space / time and matter is when the proton came to a relevancy of $\$T = 7(\Pi^2 + \Pi^2) = 138$

That produces space at the square in relation to matter at 7.

In that this relativity indicated that at such a point the proton was already demanding space ($\Pi^2 + \Pi^2$) in time set to the protons conditions, and controlling Space-time not yet established, but creating the correct environment for controlled space-time to be. Time was slow, time became faster and faster because by extending the position of Π, Π^2 will produce speed.

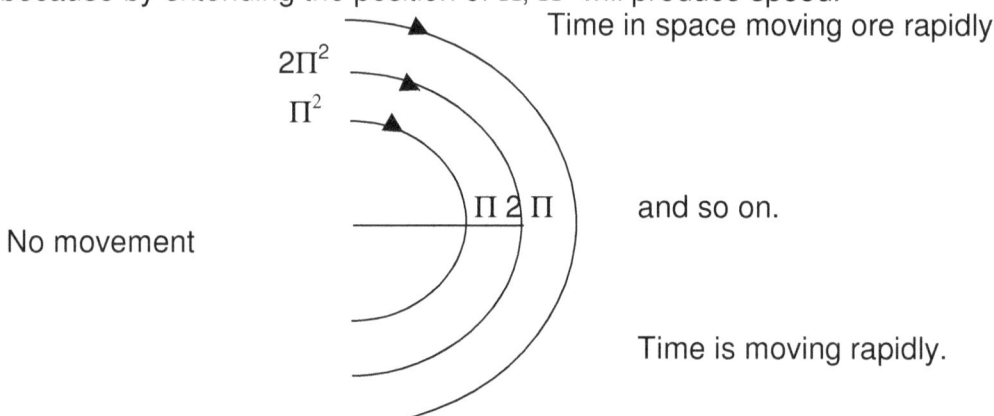

Time started at what ever point you wish for it to start, and we can take time back indefinitely. That will be pointless.

Time started by placing the double proton in space and in matter, where space and matter will always be in a sphere. The sphere always forms 7° from one point to another running outwards.

$\$T = 7(\Pi^2 + \Pi^2) = 138$

Explaining is as follows:

$\$T$ is space-time ($\$$) in the time sector (T)

7 refer to the 7° of any sphere.

($\Pi^2 + \Pi^2$) refer to the double proton.

At the same time space formed as a consequence to the Roche limit and so did matter. Forming the Aanplasings-Atomic-Epitome (the point where matter breaks in singularity)

$\$T = 10/7 \, (\Pi/2)^2 \, (\Pi^2 + \Pi^2) = 139$

Explaining as follows:

10/7 Forming the space factor and

7/10 Forming the matter factor in the application of the Titius Bode principle.

$(\Pi/2)^2$ the interaction that matter forms in sharing space and time by applying the Roche principle relating to matter-to-matter.

($\Pi^2 + \Pi^2$) the double proton

This is what the universe consisted of, everything that is today, was then, in a dimension that only holds "gravity" or the "Graviton".

The only way new space can form is by unleashing the forming of heat. Some particles still overheated which evidently forced more space. From the overheating and the consequential space that came about, particles broke down allowing matter to dissolve to matter (neutrons) and space (electrons). It must be very clearly stated that neutrons and electrons did not yet form but only a suitable set of rules formed that later would apply, where matter could supply matter to initiate a sufficient heat supply as heat, locked in time, converted to space. The process that happened at this point, lies far beyond that of a nuclear reaction, it lies far beyond that of a "Black Hole" or as I prefer to call is a proton star. All of this still fall under the same aspects and parameters we find energy or the application of energy to be at present.

Heat converts time in forming space, that is energy back then, at present and to the end of the future. In order to keep matter short, I shall proceed to the "Big-Bang" explanation, the period of nuclear heat forming space. Space formed for the very first time an identity separated from matter, apart from time, becoming the fourth dimension. In other words, the atom as we see it came into practice. The last element formed and had one proton, but still the temperature was out of control. Then the final element came into place, the cosmos.

$$\$T = 7/10\ (\Pi^6) / (6 \times 10) = 112$$

7/10   is the matter has the dominant value.
Π^6 matter has the six sides it holds in the fourth dimension.
6   is the six sides to space occupying matter.
10   is the value or dimension in which space holds a ratio to time.

The cosmos began, not to a specific space, because all the space that was there, initially is still there at present. Any atom above 112 cannot apply to the fourth dimension not then and not now.

A proton with a "mass" of say 12g/mol on earth will have a "mass" in accordance to earth standards of 25g/mol on Jupiter and it will hold a comparable "mass" of 100g/mol within the sun. This "growth" in mass of any molecule within the structure's potential occupation of space-time increases. With this in mind, one cannot merely bring in such a relation to the "mass" of the proton in the beginning or within a star, or as it is on earth. As the atom's spin increases or decreases in the relation to $\Pi^2 + \Pi^2 \rightarrow \Pi^2\ \Pi \rightarrow 3$, and with it, the "mass" will subsequently alter.

What that says is that the square of the proton diverted from singularity by the square as the proton claimed the square from singularity and hold an overall control of space in a sphere (7) and that was the form matter from that point on would always be. Space is the condition set by the neutron, and with out a neutron there is no space. To understand this one must firstly understand the principle behind the theory I introduce. At the most inner point one find time or if we can supply it with a completely fictional name: "The graviton". The graviton carries the value of Π^3. This value determines time in eternity, a position matter has no space, but is occupied in singularity. Taking the neutron position to that of the proton we find the value created when the three dimensions (six sides) came about $\Pi^2 + \Pi^2 + \Pi^2$, which carried to the fourth dimension in cosmic or geodesic space-time becomes Π^6. When relating Π^6 to singularity it becomes Π^1 (space) x Π^1 (time) x Π^1 (matter).

I do realize this explanation does not suit normal mathematical principles but we are working in dimensions and Π^1 (time) in a straight line is 180° and Π^1 (matter), which is half a circle is 180° and Π^1 (space) which is a triangle is 180° as each Π^1 represents one dimension establishing another dimension and providing that dimension's existence.

In relation to the "graviton" space through a straight line will be Π and through the half circle matter with heat positioning space ($\Pi = 1$) to 3 sides 3. Relating the graviton Π^3 three will always be a Π^2 and a Π combining 3. The half circle will be Π^2 and the straight line Π.

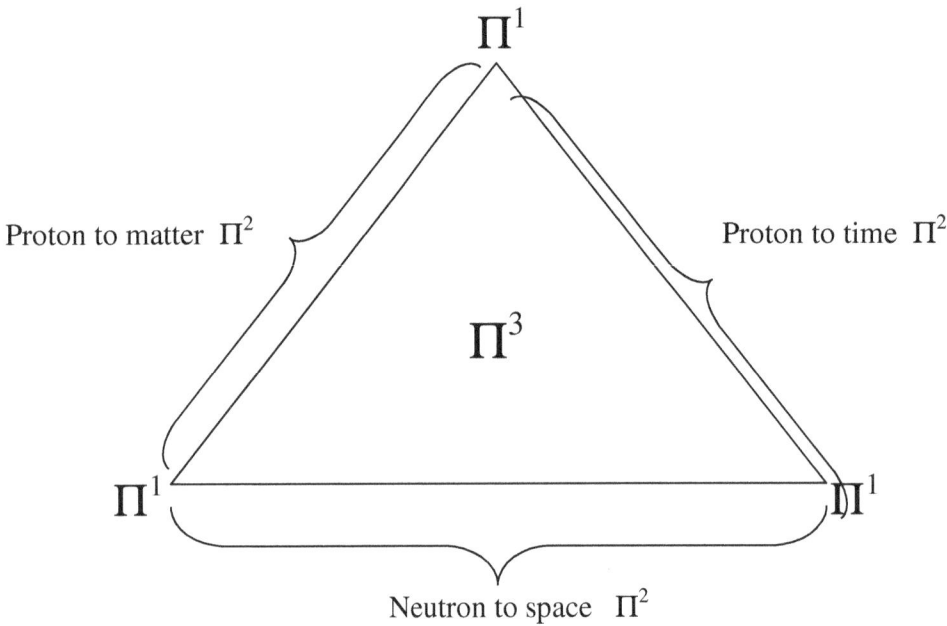

FROM THIS SPACE HEAT AND MATTER DEVELOPED

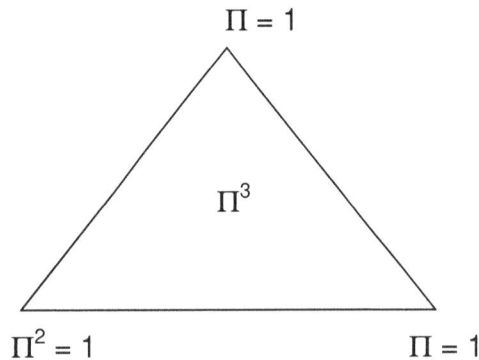

Behind all of this explanation is one obvious rule, the one subatomic particle positions in such a way as to displace space-time in the form of heat breaking down the value of space in order to meet the requirement of time. It is again all a relation between space and time and the less space has heat, the less the value of space becomes, because space has no value. Without heat in space and without matter there is no such a thing as space, therefore space does not exist but for matter and heat valuing space to form time.

THERE WAS SINGULARITY

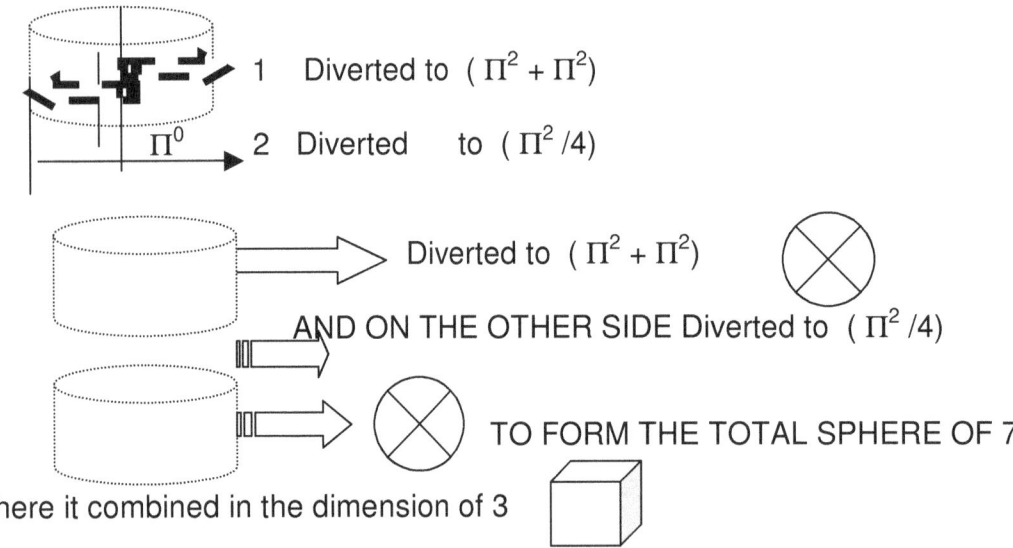

Where it combined in the dimension of 3

Π^0 singularity diverted to $(\Pi^2 + \Pi^2)$ diverted further by $(\Pi/2)^2$ in forming neutron space

When saying this, I wish to include the following explanation for those that may have an interest in the technical aspect.

Energy is a term for a power or a force, an effort to get work done. Energy relies on movement. The influence one object holds relating its position in accordance to another position.

When an object remains in one position relating to the rest of the surrounding objects, the time it remains in that position is unchangeable. Therefore, for that duration it will remain eternal. To reflect this relation to a formula will be $R^3 = 0$ and $T^2 = \Omega$. Once the movement starts the position changes, therefore the space relation changes. The movement relates to time because any movement takes time, even with an uncontrolled explosion. Changing space always takes time.

The misconception claims its incorrectness in the way Einstein placed the speed of light. The speed of light is not time, but merely particles in time occupying space. The Universe is not in singularity: the universe is the way matter divert from singularity, claiming space and time within the constraints the universe in singularity will allow. Singularity is time, holding time at an eternal value and allowing space to bring about the infinity that interrupts eternity. The time aspect is in the matter that keeps space in singularity apart from time in singularity.

TIME IN ITS ORIGINAL VALUE DID NOT DISAPPEAR. TO THIS DAY AND UNTIL THE COMPLETION OF THE COSMIC ETERNITIES, TIME WILL RELAY ITS GROWING VALUE TO TIME DUPLICATED.

Time started ROLLING AND ROLLING IT REMAINS. **TIME CANNOT STOP; THIS IS CRITICAL IN UNDERSTANDING THE COSMOS. THERE CAN BE NO MEASURING OF T, ONLY OF T^2 AS T^2 INVERSELY RELATES TO R^3.** Therefore the formula to determine **time $t = \sqrt{(1 - (C^2 - V^2))}$ is nonsense**

The first question that one can ask is why would there be the value of $(\Pi/2)^2$ between orbiting structures positioning themselves in a time relation to space.

Every person knows about the entry restriction-orbiting spacecraft find that forces the craft to comply with the entry maximum of 21,991 and the minimum entry of 7. This is without doubt, the number of Π (21,99/7). The earth holds its value to $4\Pi^2$ and when an object is not part of the surface of the earth, even say a mountain; it becomes a holding value of 7. Later in this part I explain the sound barrier in more detail and the 7 will then become better understood.

At this point one must see the earth $(\Pi^2+\Pi^2)$ all particles will in relation be either Π^2 or Π. When water is in a vapour form, it will have a value of 3Π, having heat separating the water to the factor of 3. By dislodging the thunderbolt, the 3 receive a square value and displaces to the earth in the linear light to time stance of 3^2. With heat (3) grouping by initial spin value, it will remove from space leaving the water to the value (Π to Π) and this will then give the water a relevancy of Π^2. The factor of Π^2 places the water no longer amongst heat as gas, but heat as a liquid (rain) or solid (hail).

The relevancy of the water will change from 3Π to Π^2 placing the water's position from space (3) to liquid or solid (Π^2). Where does the Π that one find in the Roche limit $(\Pi/2)^2$ and the vapour(3Π) find its relevancy to gravity. Every particle that enjoys space-time outside the earth's structure $(\Pi^2 + \Pi^2)$ will hold a neutron position of $(\Pi^2\Pi)$. The Π^2 end will be at the point where heat passes through the object directly to the earth and this position of space-time relates to the neutron time link of Π^2. The space link of the neutron will then form the Π link. The value of the Π link we find to be $(\Pi/2)^2$, but the explaining to why it is $(\Pi/2)^2$ is rather more complicated.

In nature there is the Roche limit placing a limit to the reduction of space and the inflow of heat to sustain proton cooling. At a point of $(\Pi/2)^2$ the reduction of space disallows any object the cosmic object cannot reduce, an entry to its area of reducing space.

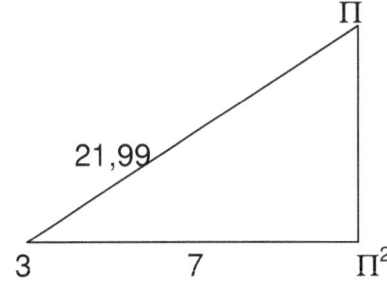

At the end of the space relevancy 3 where matter occupy space (21,9 / 7) is a border Π. That border is the exact point where space reforms to a square of time placing all matter (occupied heat) and heat (unoccupied matter) to a value of the square of time. That specific point is in relation to the square of the diminishing shield around the earth. However it takes matter (R^3) from the 3 dimensional position to the square ($Π^2$) in relevancy to time in singularity. With time holding space in singularity the 4 sides of Π truly relates to half of the total square value. Let us take the "Sound Barrier" from a point we see phenomenon apply laws that matter complies with.

The "gravity" factor of $Π^2$ becomes one and only holds the square to the Π position as it holds space to singularity at a square (time dissolving space at a square) and the time value (Π) remains dimensionless in singularity at (1). It then is $(1)^2$ where the one becomes the space position (Π/2) representing time (1) at that point. This makes the position time (normally $Π^2$) but now directly links to $Π^3$) relating to the singularity position of space diminished from the three dimensional to times single dimension in the square. That makes the Roche limit hold the position of $(Π/2)^2$ when the neutron position of time ($Π^2$) links directly to singularity (1). This may only represent figures, something to accept through intellect but lying far outside the reality surrounding our everyday understanding. I may try to the best I can, bring across how I can see it.

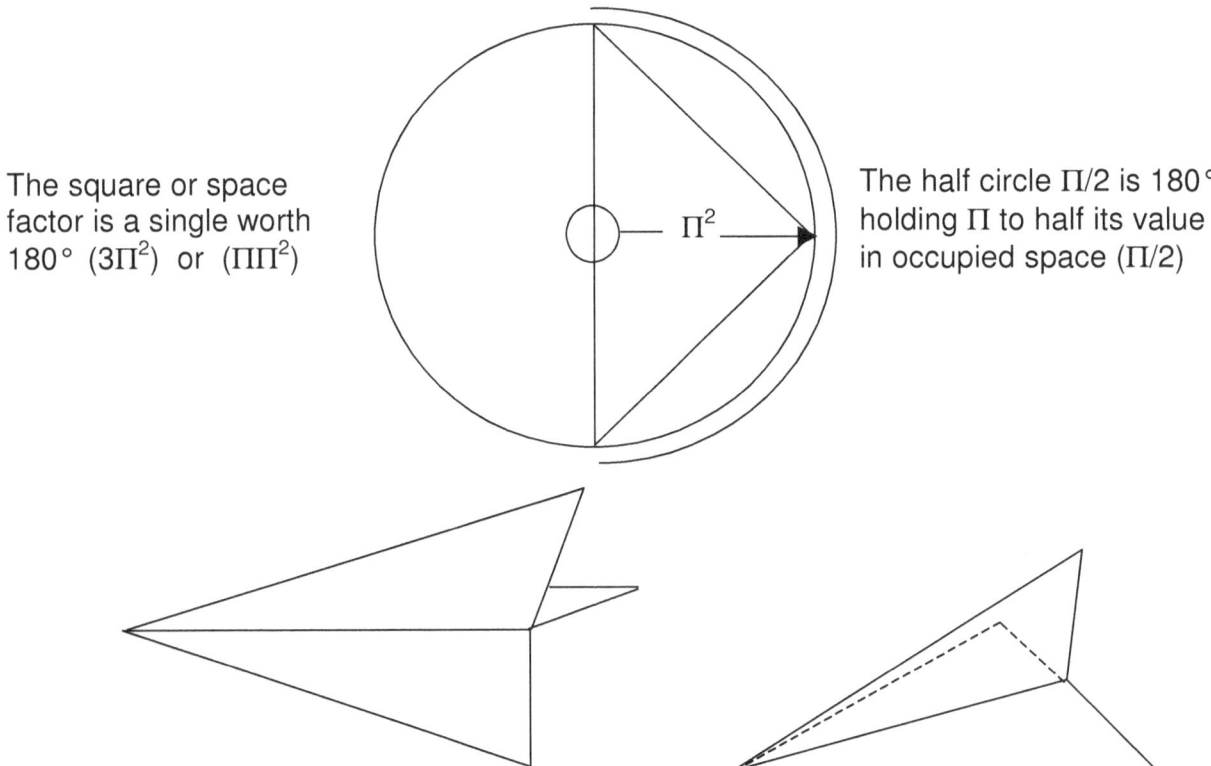

The square or space factor is a single worth 180° ($3Π^2$) or ($ΠΠ^2$)

The half circle Π/2 is 180° holding Π to half its value in occupied space (Π/2)

That is the shape needed to fly in the surrounding of the earth that we presented with a name THE ATMOSPHERE. In outer space a flying object can hold any shape because any shape fits the three-dimensional forms. It could be round, oval, flat or any combination of all shapes and forms. The dimension is 6 with a linear value of 3 applying in one direction. This implicates the fact that wherever an object directs a direction in time revaluating positional change in space, it will forever be heading in three directions in the same time. The only control to direction in is the concentration of heat at a point opposing the direction of travel. By applying the release of heat, one is applying "anti-gravity". One is producing space that never existed before, because the release of heat will produce space. To the object in travel another dimension brings about influence to his three directional space application. This application of the heat brings about a fourth dimension that applies directly to time. I shall come back to this point duly.

Without the application of specific heat, the object remains in the three directional moving of six possible directions. The value of space unoccupied therefore remains $\Pi \Pi^2$, as it was before the "Big Bang" event, whichever "Big Bang" you wish to refer to, because there were many. But space unoccupied holds time to the value of 10 to 1, and as

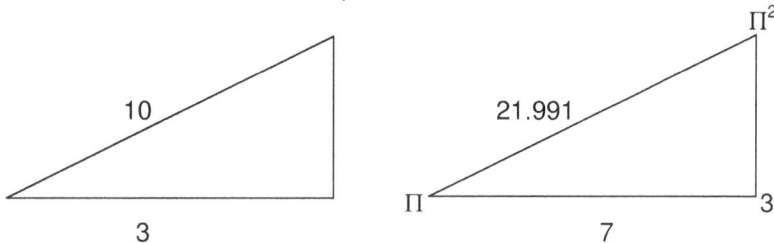

also indicated, holds space to Π. Therefore unoccupied heat holds the relation to space in applying 3 directions of influence $(10)^3 = T^2$ $(R^3 = T^2)$. Always part of this equation is the dual function of space in $R^2 = T$ while at that very instant one have $R=T$. Therefore in space in time you have $10(10^2) = T^2$. Applying the fourth dimension does not bring about another part in the six dimensions, but actually cancel the influence of the one dimension by favouring a specific dimension. Bringing about the fourth dimension will lead to halving one dimension because of favouring the direct opposing side. Then the equation of influence become $(10/2) (10^2)$ and this means the implication of any heat, be it heat, be it light, will apply under the factor of $5(10^2) \Pi$. The implication of this may not dawn on one the very instant of realizing, but to scientists, there is no greater shock than just that. To any application of movement, the factor will be

$3\Pi^2 / 5 (10^2)$
Pythagoras be comes a factor

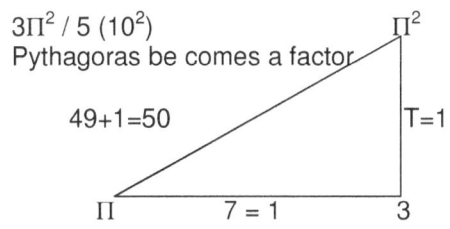

$50 = 5 (10^2)$ where the complete is $2(5)(10^2) = 100$

$\sqrt{100} = 10$ the value of space.

The fact of this comes as 49 plus one becomes 50 and that is in the three dimensions of space R^2/T where T holds the relation to one and R/T again where T relates to one. At this point it is most important to remember that Pythagoras works on the application of the sum of the square of the two sides. When seven has a direction in the fourth dimension applied to it, the opposing dimension will be one and this applies in time relevancy, therefore the interchanging in time between infinity will place matter at $7^2 \times 1$ relating to circular and $7^2/1$ with $7^2 \times 1$.

The square in matter holding 49 plus one (singularity) always being a factor of one. Space in time however, never can be a cube, it will always be a square with one side pointing the direction of time from time to the past (1) to time to the present (1) to time to the future (1). This means with space in singularity in relation to time in singularity there will always be $R^3=1^3 / T^2 = 1^2$.

It is the value of matter, be it surrounded by heat solid, liquid or gas, that differentiate between time in space and space-time. Without applying heat release the space in time would be $10(10^2)$, which is the circle we think we see space to hold. This is where astronomers make their biggest judgment error, they look at space thinking they are viewing the cube of space, but they are in fact viewing the half of the base times the square of the cube $(10/2) (10^2)$. What they think they see is 1 000 instead they are seeing 500, but as I will later explain, they are not viewing $5 (10^2)$ they are not viewing light setting the time factor $3\Pi^2$ but they are in view of light in space (Π) being in time ($\Pi^2=1$) the connection to space is Π/500. That means whatever the radius between the objects the light reaching it will have a comparable velocity of the distance of travel divided by 500. This will be discussed later in more detail as well. From this there is a measured point where the dimension of space in heat forming gas ends and space in "liquid" forming the neutron begins. The dimensional time application in space diminishing starts at that point. We think of it as the earth's atmosphere. This no person, no force, no money can recreate. I have read articles about some Newtonian Wizard that plans on building some ship where this ship will "create artificial gravity by centrifugal spin."

Again and again Newtonians show just how little they truly know science. The depletion of space through spin increases the earth brings on acting as the nucleus of an overgrown atom. The point Π indicates where 3 lose one side of its dimension becoming Π. In that space all objects are at the mercy, not of a FORCE called GRAVITY but a change in dimension At that point a dimensional change take place where the universe sacrifices one dimension from 3 to Π (in reality it is 6 toΠ.). In that space objects which wishes to fly, need a replacement for the dimension sacrificed. The aircraft needs wings to fly. Hot air balloons and the Zeppelin Hydrogen flying cylinder use the natural tendency that hydrogen provide by enlisting concentrating heat to maintain at the value of Π and not reducing to $Π^2$. Aircraft of any kind, that wants to maintain a vertical flight would have to apply three directional wings for stabilizing because of the loss of one dimension. There may be the possibility that with enough application of heat at various points, the introduction of the fourth dimension (heat application allowing space creation) may even bring about that stabilizing may occur in such a manner. Although the prospect seems light, however with no research to widen the view on the possibilities available, it is hard to tell.

What is overall important though is that the point the Roche limit indicates, being $(Π/2)^2$, the dimensional change takes place. Any object holding space-time in own value, will either reduce space occupation, or enlist time duration enhancing, or most probably a little of both as it will cover itself with heat. This is the result form the atoms on the outer edge of the atom, repositioning space-time occupation where the normal application will be in the dimension of the Titius Bode law developed to a certain degree. This change will bring about that the atom will revalue its position to incorporate the Roche limit in protecting its own space it holds in the time of that specific space. The value of the space-time of the atom of the aircraft therefore will reduce its electron position, as it locks itself in a position under the cover of the heat shield. The point of the Roche limit is as much part of the Macro Cosmos as it is of the Micro Cosmos.

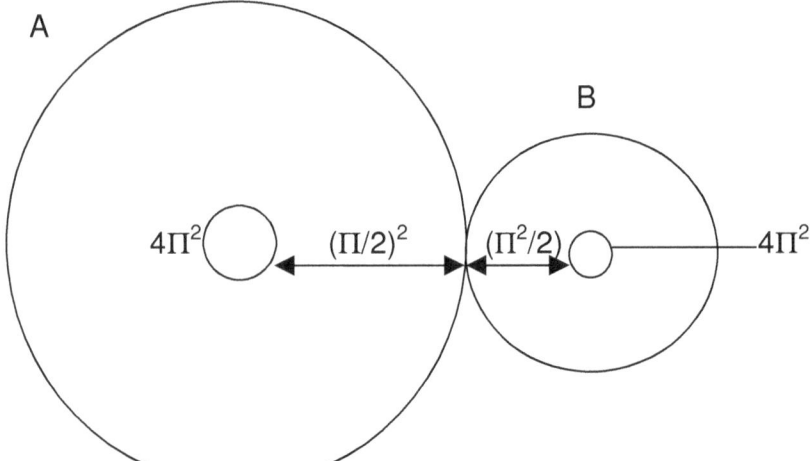

Both structures hold eternity to$Π^0$ in singularity as time $Π.^2$ and space Π diverts from singularity at Π. The deviation of $Π^2$ and Π. Will be in accordance with the measure of deviation that heat occupied and heat unoccupied holds in relation to both objects claiming space from singularity. .
At this point of $(Π/2)^2$ the reduction of space through positive space-time displacement will start. Any other cosmic solid that will not comply with the flow of heat to the object will remain outside that area. This means in more suitable language, that the second object will block the flow of heat to the first object (A). To object A everything outside its proton sphere of $(Π^2+Π^2)$ must either be $Π^2$ or Π.

If the object is in a neutron time position it will hold the value of Π^2 and if it is in a space position it will hold a value of Π. Should the object B maintain a density where it will become a singular value to the proton value of object (A) in as much as maintaining its own proton value of $(\Pi^2+\Pi^2)$, it will cause overheating in the proton heat flow of object A.

The demand in heat supply of object A will either remove the heat object B needs for cooling, thus increasing object B's chances of overheating, or it will start overheating itself. By overheating the space object A occupies, will increase because the overheating will demand more space occupied. By increasing object A's space-time occupation, the demand on heat will increase as the point of space reduction shifts further away. The increase to the value of the $4\Pi^2$ factor, will increase the $(\Pi/2)^2$ linear factor. As the demand for heat rises, the competition for heat sustaining will also rise in the occupied space of both structures' time zones. That is the Roche limit and that is the purpose of the Roche limit. At a point where the space reduction and demand on heat supply starts in all earnest $(\Pi/2)^2$ seen from the view of the object in question $(4\Pi^2)$ it will either reduce the other object to a suitable value of (Π^2) or Π, but it will not allow any object to compete with it's demand $(\Pi^2+\Pi^2)$ on the supply of heat. It will reduce the competing object's value of $(\Pi^2+\Pi^2)$ as it relates to its own value down to a mere manageable (Π^2) or (Π). The reducing method object A will provide, is removing as much heat flow from object B in order to get object B to overheat. Thus object B will increase its space-time value from a solid $(\Pi^2+\Pi^2)$ to that of a jelly (Π^2) or even better still to a liquid (Π) by removing all gas (3) from object B in its demand for heat. If object B can comply with its own demand for heat, it will remain in space of it's own, providing it's own time within the cosmos $(\Pi^2+\Pi^2)$.

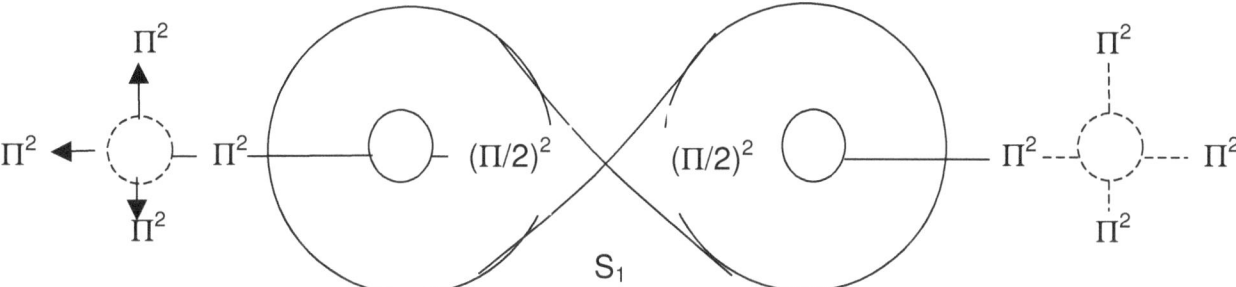

The fact that it takes the space-time back to what the universe was when it was in total development $(\Pi/2)^2 (\Pi^2+\Pi^2) = 48,7$. At present that value supersedes the space value of 10Π completely, so it holds a relative position far above the geodesic value (10Π). That is not all, because the one object acts as a neutron to the other object, both claim a proton flow of heat from space, holding a time of $(48,7 + 48,7) = 97,4$. That places the active space-time development with the combined structure to an atomic era that was relevant just after the "Big Bang". Heat became a gas at the atom value of $7/10(\Pi^6) / (6 \times 10)$ and darkness parted from light at the atomic value of $3^2 (\Pi^2+\Pi^2) = 98$. As you can see, the two objects pushes each other back in space to a time that applied three eras ago.

Therefore one can observe that the Lagrangian point of S_1 is in fact the electron position because the two objects hold the status of a compound, sharing space-time and heat flow. In other words, the two objects become two super sized atoms forming a compound in space. By the same measure a Lagrangian system becomes an enormous five-electron atom with all the cosmic objects working in conjunction as a unit with one mean goal and that is self-preservation through group preservation. Since this falls outside the spectrum I allowed this book to have, I mention that only to fill in a picture that may otherwise seem empty.

If the superior object does not take the minor object as a threat blocking its supply of heat flow, it will consider the minor object as a neutron that can help it grow. Even if the two objects maintain a relation of $(\Pi^2+\Pi^2)$ it will hold the other object to a value of $(\Pi^2\Pi)$ and outside the combined effort they will jointly relate to the rest of the universe as 3. Therefore seen from the two separate views there is the following atomic order:

$(\Pi^2+\Pi^2)$ $(\Pi^2\Pi)$ S_1 $(\Pi^2\Pi^2)$ $(\Pi^2+\Pi^2)$
Proton   Neutron   ↓   Neutron   Proton
                Electron

But placing the two in a battle for survival the relation becomes super strained where the one will not give any way to the other object's demand for singularity dominance and then the relation becomes that of

$(\Pi^2+\Pi^2)$ $(\Pi^2/2)^2$ $(\Pi/2)^2$ $(\Pi^2+\Pi^2)$
Proton ↓
       Space-time establishing

This will bring about space-time development that applied when the universe was fresh and new, hot and lively and full of spinning power. As the objects cannot destroy one another, there is the Hubble Constant affect that came into place. Each structure has to build a Π value, to maintain and secure its relevant position until the end of the final eternity arrives. Said with using other words, each object will secure its place in future as a Black Hole or Proton Star. By producing a value of Π separating the Π^2 from three and thus ensuring itself an own atmosphere it will have to revert to space-time development that even preceded the "Big Bang". I will produce once again the value of Π from its position of Π^2 (time) and matter. Matter is seven and time is Π^2. Both hold claim to the same space diverting from individual singularity, therefore the development of matter and time will produce Π and through that the Titius Bode Principle.

They join space-time therefore the matter factor is the same. This is where one can visually see the one object, filling the space of the other object's atmosphere.

$$7 \times 7 = 7^2 = 49$$

That is matter Π^2 (time) times matter (49) = 483,61.

As this is all under the law of Pythagoras the law will evidently place a square root to that value of 483,61 and therefore $\sqrt{483,61} = 21,991$. This leaves the space value of the Roche-limit, as it develops into the Titius Bode law giving them a shared value of 7 (matter) and 21,91 (space) the value of $21,991 / 7 = \Pi$.

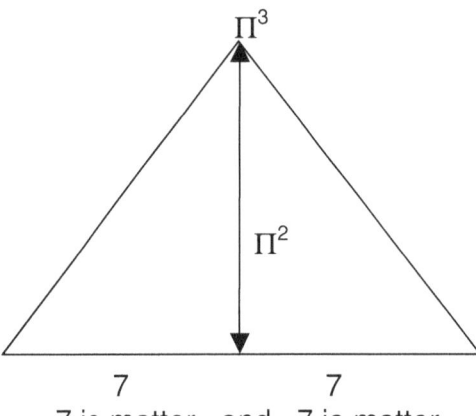

7 is matter   and   7 is matter.

Then the relation becomes

$(\Pi^2+\Pi^2)$ $(\Pi^2\Pi)$ $(\Pi\Pi^2)$ $(\Pi^2\Pi^3)$ holding space (3) still outside. They therefore will share space and that sharing will continue till times end. We know by now that matter is 7, and space is 3. holding time to a relevancy in singularity of 1. Sharing the space means that 21,9 will become (10) space to the one side

1   to the instant position of time (k^0)
,99   lost to space depletion $\Pi^2/10$
7   the relation to matter.

Through that the Titius Bode law comes into affect of 10/7 or 7/10, depending on whether space or matter holds a superior position to time. From that stance, all objects will relate to one another by the value of $\Pi^2\Pi$ and seen in a whole sale total 7/10 or 10/7.

We on earth take our measurements per second, minute, hour and so on. This time relation we derive from the ever-changing position the earth holds to the sun first, and thereafter to the universe. That is time in space and more precise space-time.

Time is the spin of matter in space. **It is the changing relevancy that particles have in the eternity Π^0 To matter relating to the outside but also relating to matter holding the earth Π^1.**

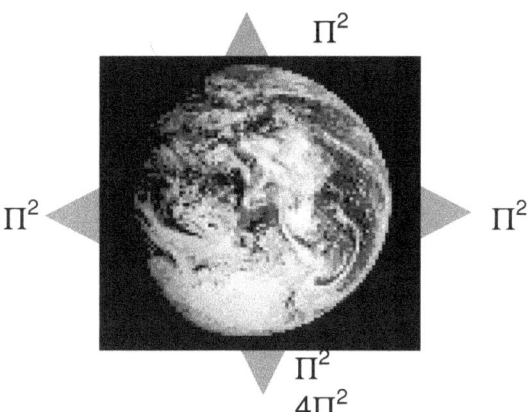

Every position matter holds, irrespective of the different place it is on the planet, relate to an angle of 7^0 to the next position, in relation to the centre of the earth. When the aircraft breaks the $7°$ relation, it has to come into the calculation as it affects the space position surrounding the aircraft.

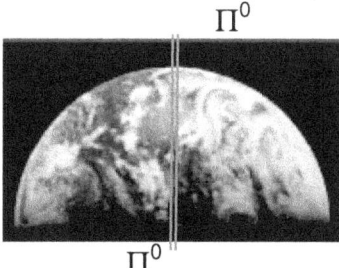

Then after all this lecturing, you may correctly ask how would one define time. We have been staring time in the eyes as long as man could calculate.

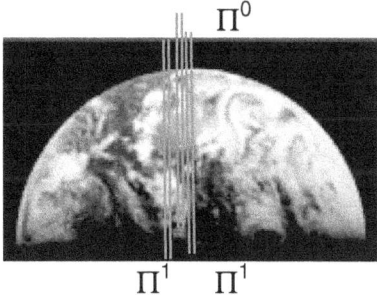

All objects are either outside the dome or inside the dome.

When a spaceship enters the earth's atmosphere by more than 21° it will burn out. If it enters the earth's atmosphere by less than 7°, it would bounce off into space. That value of 21 to 7 is also Π. That puts objects outside the dome running from the point of singularity beyond the 7 and 21 degrees of entry that relates to Π the dome or atmosphere. That dome is the point where "gravity or dimensional change starts at $\Pi^2 / 10$. The factor of 7^0 forms one of the three space-time values determining the Titus-Bode principle.

The 7° comes about by the concentration of the atmosphere and this bounces the spaceship outwards. However, when a spaceship enters the earth's atmosphere at an angle higher than 21° it will in fact meet less air than in the case of the spacecraft entering at an angle of more than 21°.
At any one point the relativity of the atmosphere extends to $\Pi\Pi^2$

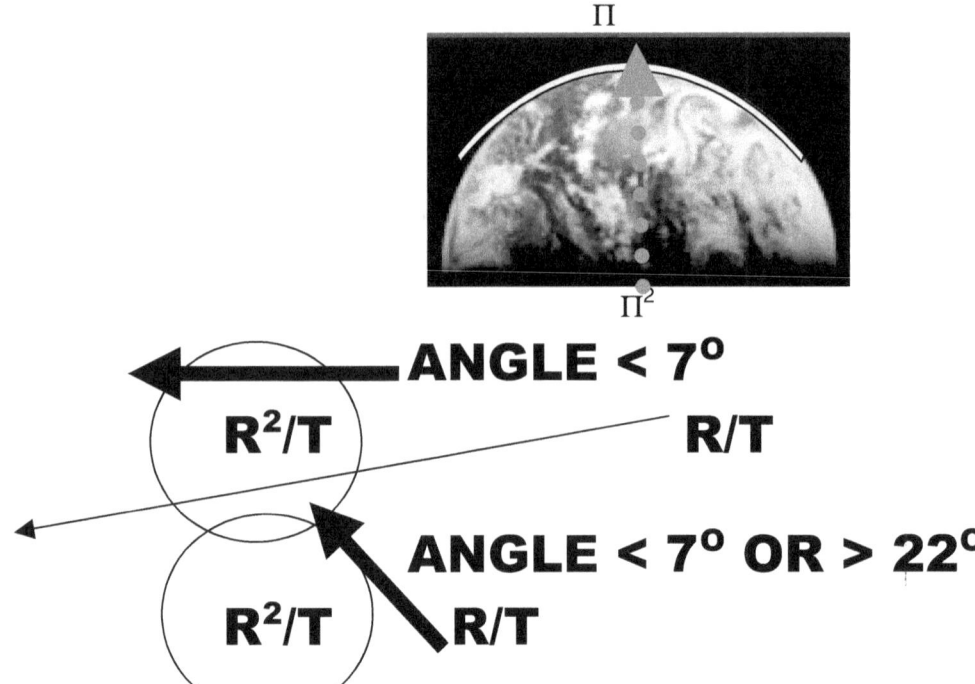

Any person can see that by going straight down, the spacecraft will have much less contact (less friction) with the air than in the case where the angle is only 21° degrees (much more friction). So obvious as it is to any person, this cannot be the reason. I shall explain the actual reason why a spacecraft has to enter at an angle of not more than 21°. What becomes clear from these two illustrations is the following: Where there is a direct entry by the spacecraft, the time factor is not sufficient in duration. This will not allow the time-duration needed for this structure of the aircraft to revaluate its space occupation an therefore the space factor of the spacecraft cannot adopt to its new position in space-time occupation as it has to relate to a new value in accordance with the value that is determined by the earth's concentration of space-time. The 7° angle is known for a very, very long time. Ptolemy concluded this observation hundreds of years before the DARK MIDDLE AGES, yet no one ever tried to connect these findings to the true laws of nature.

The dome or atmosphere is under the conditions set by singularity running from any given point inside the centre of the earth Π^0 to Π^2 and on to Π.

Working from the point of singularity the value of matter-to-matter is $\Pi^2 + \Pi^1 + \Pi^0$, which is 9.8696 + 3.1416 + 1 = 14 with only half the universe acting(14/2) = 7

Every person knows how Ptolemy came to the knowledge about the earth being inclining at 7° about Ptolemy and the way he came to realize the seven degrees inclination the earth has. Yet, once more, this value has been lost to man for thousands of years.

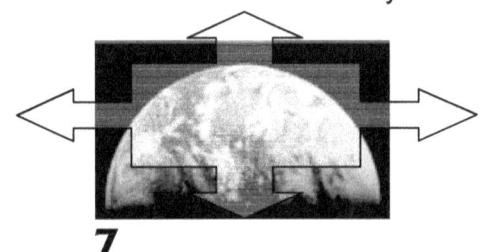

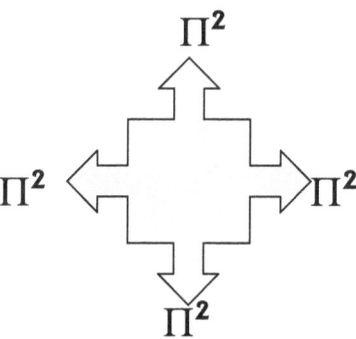

What ever is in the earth's atmosphere holds a relevancy of $7(3\Pi^2)$

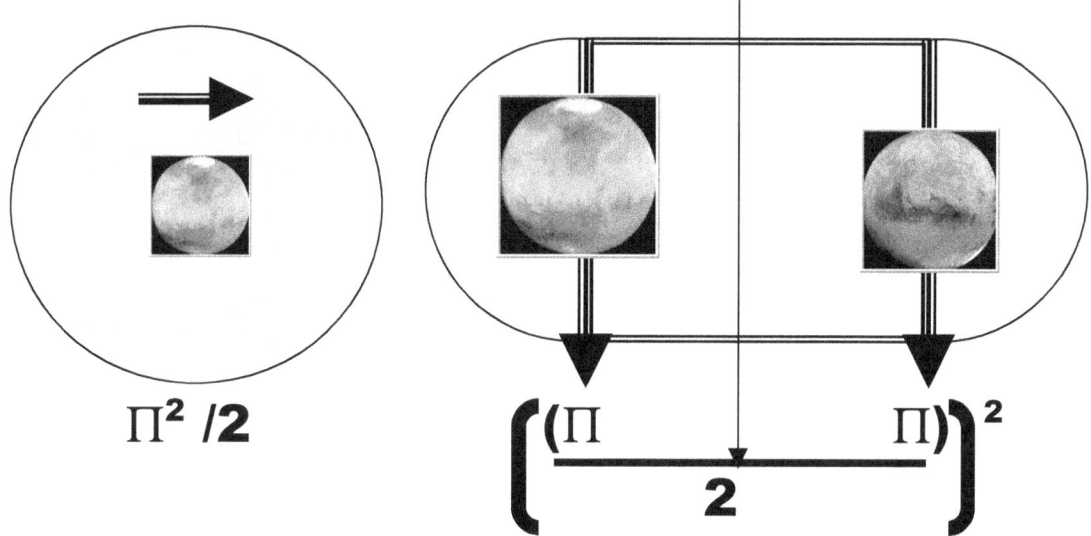

THE ROCHE LIMIT APPLYING:

WITH IN THE EARTH'S ATMOSPHERE	SEPARATING TWO COSMIC STRUCTURES OF EQUAL TIME VALUE

It would be far too complicated to explain why space-time and water share so many characteristics but they do and as the book progresses the explaining will surface one by one. I have, to some extent, tried to explain it but I am aware that the explanation falls short of satisfying until round about part 5. I will repeat it once more, well aware that it cannot bring acceptance. It is heat that produce gas or liquid or solid and the period before light, everything about the cosmos was a soup cocktail, heat was liquid and space in singularity was liquid flowing heat that appeared like water

It seems very ironic that science with all its bravado, money, wisdom and splendour, can only begin at the point where light came to the universe, while the Bible explains the creation in detail, long before the "Big Bang". The reason why there was no light before the "Big Bang" was that the spinning matter exceeded the speed of light, being $3\Pi^2$. The Authentic Author of Genesis refers to this as a mighty wind and this leaves a question. What better name can one give to this occurrence?!

The Bible says, when God made the earth. Concerning the words use, one must accept that in the time the original author of the Bible wrote these words (May it be Moses or someone thousands of years before Moses) there was no word for matter, particles, elements, atoms and what not. I faced the very same problem when writing The Thesis. We have only one concept for eternity, but I came across many forms of eternity. At this point in time there is only one name to use, eternity, bringing over the consensus of time without end. Mathematically I can indicate it to e 10^{500} or even an enormous figure of $10^{50\,000}$ still will have an end. What do I name a period that HAS NO END, as time was in singularity and the very first instant, where time was the longest period one may imagine minus the very shortest period that one may imagine. Mathematically one can state it with ease, R^{Ω}_{α} but bringing that notion across linguistically the concept becomes hazy.

Well, after all is said and done accepting the age of advanced science we find ourselves in at present, spare a thought for the Authentic Author of the Bible, in using words that did not exist maybe five or even ten thousand years ago.

Where modern science still cannot accept what took place before the "Big Bang" as it happened during creation. In that regard, spare a thought for the poor sole that faced the daunting task explaining his vision without understanding any of the explanation and worst of all, confronted with a linguistic problem lacking any form of vocabulary to extend or support his effort in explanation.

By only accepting this logic human flaw and replacing earth with matter, then what he said was that (Earth was without form) matter was in a state of singularity. By mentioning that space was void, again confirms the state of singularity. Even Einstein could not describe singularity as well as the Authentic Biblical Author. Matter is the only factor that changes space from singularity to a dimensional factor. The relation matter as occupied space-time holds to unoccupied space-time ($\Pi^2 \Pi \rightarrow 3$) because half of the double proton ($\Pi^2 + \Pi^2$) is always in singularity. Because the neutron ($\Pi^2\Pi$) and the electron (3) did not yet exist, all matter (Π^3) was in a void, a massive void, a universal void we humans are incapable of anticipating.

All atoms hold an electron to the very outside of the space which that atom occupies. All electrons are negatively charged. No two sub atomic particles that have equal charge can touch; because of the way, they will repent one another. Should there be a way to force the particles to touch, say with "matter" versus "anti-matter" a nuclear discharge will result from such a manoeuvre. Therefore, short of causing a nuclear discharge, no matter can come into direct contact with matter because of the negative relation the electron holds to other electrons. NO ATOM CAN EVER COME INTO DIRECT CONTACT (TOUCHING) WITH ANOTHER ATOM.

This is what the universe consisted of, everything that is today, was then, in a dimension that only holds "gravity" or the "Graviton".

This piece of logic shows that matter places matter in relation to the heat separating matter. There is never any empty space keeping matter apart. I hope that this bit of minor logic will end all theories about the fact that "Gravity" has the ability to contract or draw matter closer. If you wish to stick to the terminology using "gravity" as a term, then the best way in explaining the way "gravity" works is the ability of the element to reduce the heat between the particles and as such reduce the space. Explaining the factual working of "Gravity" would be rather complex to use in this book and the intentions in publishing this book does not cover the most intimate technical details. Time and space were still parting in relation to the Titius Bode law growing from the Roche limit.

$10/7\ (\Pi/2)^2\ (\Pi^2 + \Pi^2)$ and
$7/10\ (\Pi/2)^2\ (\Pi^2 + \Pi^2)$

Explaining as follows:

$10/7$ Forming the space factor and
$7/10$ Forming the matter factor.
$(\Pi/2)^2$ the interaction that matter forms in sharing space and time.
$(\Pi^2 + \Pi^2)$ the double proton

The only way new space can form is by unleashing the forming of heat. Some particles still overheated which evidently forced more space. From the overheating and the consequential space that came about, particles broke down allowing matter to dissolve to matter (neutrons) and space (electrons). It must be very clearly stated that neutrons and electrons did not yet form but only a suitable set of rules formed that later would apply, where matter could supply matter to initiate a sufficient heat supply as heat, locked in time, converted to space. The process that happened at this point, lies far beyond that of a nuclear reaction, it lies far beyond that of a "Black Hole" or as I prefer to call is a proton star. All of this still fall under the same aspects and parameters we find energy or the application of energy to be at present.

At first, the universe was small dominated by the density of unoccupied space-time; the value to unoccupied space-time was little less than the value to densified space-time, leaving no chance for occupied space-time. Ever so slowly time in space evolved from a continuous eternity, flowing in undistinguishable periods of eternity in duration forever weakening the strangle hold of time on matter. Then the age of the atom arrived, and light, which brought fore ward the atom encircled by electrons that is at this stage quite unknown to man. By the time, the atom was a recognizable unit, more eternities passed than we have sells in our brain.

The cosmic dynamics are all part of proportions and relevancies changing as relations change the applying dynamics. One must see the cosmos starting from the size where what ever was in the cosmos at the time would currently fit on top of a needlepoint. However that does not make back then small and at present large, because what ever were in the cosmos, is still in the cosmos, and grew with the cosmos so what ever was small had grown with the cosmos and is in practise as large as it was. The dinosaurs are a vivid reminder of that and in Seven Days Of Creation I explain this extensively.

These first atoms were small producing very insignificant positive space-time displacement in the sea of negative displacement. The atoms were small, but to our great fortune growing as they extended their size and influence. By the time, time was a distinguishable factor the nucleus was presumably less than the size of the electron. The electron's orbit was right next to the small nucleus. Because of the size of this tiny atom, it could participate in the geodesic space-time value back then, which was a little less than the photon's velocity value back then. I have to emphasize that by explaining in this manner, I put the relevancy on the human aspect in view that it concerns the Bible, and therefore mainly the perception that life holds on events.

Heat converts time in forming space, that is energy back then, at present and to the end of the future.

In order to keep matter short, I shall proceed to the "Big-Bang" explanation, the period of nuclear heat forming space.

Space formed for the very first time an identity separated from matter, apart from time, becoming the fourth dimension. In other words, the atom as we see it came into practice. The last element formed and had one proton, but still the temperature was out of control. Then the final element came into place, the cosmos.

$\$T = 7/10 \, (\Pi^6) / (6 \times 10) = 112$

7/10   is the matter has the dominant value.
Π^6 matter has the six sides it holds in the fourth dimension.
6   is the six sides to space occupying matter.
10   is the value or dimension in which space holds a ratio to time.

The cosmos began, not to a specific space, because all the space that was there, initially is still there at present. Any atom above 112 cannot apply to the fourth dimension not then and not now.

A proton with a "mass" of say 12g/mol on earth will have a "mass" in accordance to earth standards of 25g/mol on Jupiter and it will hold a comparable "mass" of 100g/mol within the sun. This "growth" in mass of any molecule within the structure's potential occupation of space-time increases. With this in mind, one cannot merely bring in such a relation to the "mass" of the proton in the beginning or within a star, or as it is on earth. As the atom's spin increases or decreases in the relation to $\Pi^2 + \Pi^2 \rightarrow \Pi^2 \, \Pi \rightarrow 3$, and with it, the "mass" will subsequently alter. In explaining all of this, it is quite impossible for me to give it a value in mass, or time as both these factors alters in space-time occupied.

There were two options, the universe had. The matter which is densified space-time could join the geodesic space-time and spin itself into another eternal oblivious, or its positive space-time displacement could withstand the geodesic space-time displacement could withstand the geodesic space-time negative space-time displacement.

Obviously, our luck was in and the positive space-time displacement within the atom stood its ground against enormous odds. With this effort, the expanding of the universe began, as space-time transferred from the universal value to the densified value. The more time transferred this way, the less time had a value in the geodesic sense and the more time had a value in the densified sense. In this, the space value grew in a geodesic sense as the time value in a geodesic sense declined. This brings about the eternal expansion of the universe that will only stop once the geodesic time value becomes zero and the space value becomes eternal. This then will be the second eternity, to which the Bible also referrers.

This very process was the inspiration, which lead the universe to be what it now is. In the chapter, designated to moment-Alfa, an elaborate explanation shows how the value of Π has changed fore times already, and will change another three times. The development through all the different era brought about that the universe had to endure because of this very principle. I do not wish to go back to the old cliché about gravity and all the theories that were thought up to explain the means which will enable the universe to retract again. The retraction of the universe lies in the growth of time as it is transferred to matter. This process consists of the transformation of unoccupied space-time to occupied space-time to densified space-time.

In this lies the value that all stars have in their role they have to fulfil to bring about a successful conclusion to the life story of the universe. Objects are growing more massive as time transfers from one value (Verplasing) to the other value (aanplasing). As R^3 is showing an increase in both the occupied as well as the unoccupied space, T^2 reduces from the unoccupied value to the densified value. In the star R^3 reduces as T^2 transfers to the densified space-time. This is what stars, are made of, when they grow up to be stars.

With time in eternity, space in zero and matter being time and space, what would ever bring about that this situation changed? No Official Policy Protectors ever came forward to explain this. Why did the "Big Bang" start and what brought the "Big Bang" about? Only the Bible produces any logic to this question. Time, matter and space froze in one; there was no reason in nature for things to change, since this situation lasted eternally. Nature with all nature's laws did not apply; therefore one cannot say that nature started it. Nature was frozen. Nature was not even solid; it was in a state beyond being solid. Nature was nowhere!!

If the spoken language can refer to the earth as Mother Earth then surely there is all the right in the world to address the sun as Father Sun, for that is precisely what the sun is to us earthlings. If we have the sun as a father, then it is a very young father. Us, as humans, have an infant star that is, as far as stars go, just become a toddler. A fully-grown star that has not yet reached its middle age development, will have discarded its hydrogen and most probably its helium layers. As it ripens in age, it will discard its carbon and all helium related layers to have only silicon, an iron and an inner core that has outgrown the element value altogether. This process leads the star to a route that takes time back to its origin of time = eternity; space = zero.

The spin in the universe slowed down, up to a point where the spin was equal to that of the speed of light. At the point where the universe spin equalled that of the speed of light, the universe was still in total darkness. The light (photon) was there, but did not yet produce light. Light only came about as the spin reduced to below the speed of light, and only then light became obvious. The universe grew away from darkness as this event lasted many eternities, during the period where the light separated from darkness.

How did this "Big Bang" take place? The best way to examine the reason is to see why anything in the universe expands. To get anything to expand one has to heat it. All matter expands when overheating. Science may come up with whatever brilliant theory, the fact of the matter is that when matter overheats it expands. The bigger the overheating, the bigger will the expansion be, it is as simple as that.

This means whatever leads to the forming of the "Big Bang", whatever preceded it, it had to come about from matter that overheated. With the event of the nuclear age, the proof came about that matter is heat in some frozen form. Unleashing heat from its frozen form brought about a jolt of heat, never yet experienced by man. By breaking matter from the frozen state, of which it is in, within the atom, heat produces light and heat. Where this process clearly shows how new space-time forms is where the releasing of heat caused winds that stun man's logic.

The nuclear explosion shows quite clearly what the "Big Bang" was, with the nuclear explosion being a very minute form. Yes, we have all heard the rubbish about matter and anti-matter. What can anti-matter is, since matter is heat, defined to a certain space occupied for that time. With matter being frozen heat, what would form anti-matter. Anti-matter means the opposite to matter, and if matter is frozen heat, anti-matter must then be overheating heat. This in itself is quite ridiculous. Anti-matter can only be matter with an opposite spin to that we think of as matter.

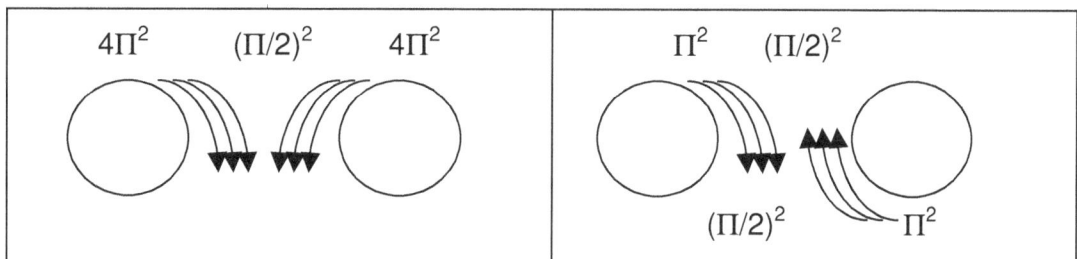

In the second sketch $\Pi^2 (\Pi/2) / (\Pi/2)^2 \Pi^2$ the Roche limit cancel each other and with that the space of the neutron effectively disappear. The two protons touch destroying each other and the neutron as well as the proton demolish and became heat 3^3.

The process where matter then touches matter, it will bring about a reduction in the feeding process of heat,. where all matter in that space will overheat and expand, producing unfrozen heat. This means there was heat occupying space, and matter with both in relation to time.

The proton with a positive space-time displacement less than 136 placed its displacing properties in negative space-time displacement. In short, to substitute for mass shortcoming of less than 136 grams/molecule and still finding sufficient cooling properties for the proton to survive the overheating deficiency, it has to spin more rapidly, therefore by spinning it makes contact with more heat than it would otherwise do. One may consider this as "breathing". If the proton does not find adequate supply of heat by being motionless, the atom has to substitute the movement through motion. Forming an object that has an increase in heat supply through work commonly uses this fact. In nature, just the opposite is true because of the motion by movement, as one find in the case of wind.

There is no "force" in the cosmic because everything is a "force" in one way or another. The proton takes heat from space in an effort to maintain temperature and stability. This flow of heat brings about the reduction of space by increasing the heat in that space. The flow of the heat, through the electron by means of the neutron to the proton is time. The amount of heat taken by a proton is a constant throughout the universe but relative to the space reducing effort of all the protons influencing that specific space. That is "gravity" (a term I denounce and reject). "Gravity" is nothing else but additional application of time. The higher the gravity is, the slower the time will become by prolonging the duration of time, NOT TIME ITSELF. TIME REMAINS A CONSTANT, BUT THE DURATION DEPENDS ON THE SPACE THE HEAT IS CONFOUND TO.

By taking this statement and introducing that to a Galactica, the shining luminous middle part, holds time to eternity through movement. The atoms in the middle admits light because it holds time duration to the speed of light, the longest duration that time can be and still remain in the fourth dimension.

The centre of all cosmic structures determines the time that applies to the structure itself. As the cluster of protons supply the density that influence the space of occupation, the density is a collective reducing of space with the increase of heat. This can apply through an object relating to space through movement by the object and by the object reduction of space through the density of the accumulative effort of the cluster of protons we named elements.

As we earthlings had a double advantage to the process of binary stars, I would like to ponder on the binary stars, as they are the key to time in the cosmos. The first influence of a binary system we benefit from is the sun and some other minor star, which by braking into fragments formed the non-gaseous planets. The second encounter was the binary to which the earth and moon developed, and from which LIFE then was able to take full advantage.

The two stars develop in the galactica in close proximity as they help each other in transforming negative space-time displacement to positive space-time displacement. With the combined effort, they can grow extensively in supporting each other.

In spite of all our Official Policy Protectors teachings, stars do not form as result of cosmic dust storms. It is just not possible to form that way, because even in the present era we find ourselves the iron era, only elements with a space-time occupation value exceeding 17,5 can reduce space sufficiently to apply "gravity".

Planets never form naturally because a planet has no role to play in the universe. Cosmic objects do not form in any way or means through the application of "gravity". Matter occupies space presenting that specific space a time value in accordance with the time in space the matter holds. I realize this is rather a mouth full to understand. To us humans, planets are a necessity, which one cannot do without. We humans have a measurably small place in the universe, so small the universe do not recognize life as a substance.

We see around us a planet overflowing with life. Life is so abundant, we can hardly distinguish between nature and life because to us nature is life. We have to think of nature without life because cosmology is without life. An easy way to distinguish is thinking of a mountain burning. The mountain cannot burn, it is plants that burn and the plants are not nature. The plants are life, therefore the mountain cannot burn. Less can the mountain move. It is only life than can manipulate space-time. Other than that all other objects may extend the space-time they occupy by natural converting heat from a gas to a solid through proton growth.

Whenever you wish to distinguish between life and the cosmos, the rule of thumb is: Can it move on its own accord? Envisage a mountain that decides to go on a vacation to a nearby object to lift from the ground while still in a solid state, can only happen when life takes a major role, and life has to do all the work. No mountain can fly to the next continent for whatever reason. That action belongs to life, and is a piece of metal must accompany life, it will do so as an extension life bring about. That is where Newton's physics apply and apply they do. The comet obeys rules applied by the cosmos, in terms the cosmos provide. Newton's physics do not apply to the cosmos however which way Newtonians try to explain it by applying Newton's laws. A comet cannot leave the earth and journey around Pluto, and come back with a detailed report. This action belong to an object that is acting on behalf of life as an extension of life. I am very adamant about this, if science wish to understand the cosmos, do not make the cosmos one big boundary less earth. Up to now, such an application only brought miserable results that apply to nothing and helps science establish the joke it became. Life is the extension of God, the Universe is the Creation of God, do not confuse the two.

The sun is not an earth on steroids. Life cannot be on the sun because the sun holds completely different relevancies. the relevancy has a comparing formula but the end result will be totally different in every aspect one wishes to apply.

By referring to "gravity" only one aspect of the space-time relation of any elements apply. There is no mention of the second and crucial part of "Gravity" where the motion of the object brings about the space-time relation, or if you wish, providing the cooling aspect. At first, the proton cluster's total positive space-time displacement has an insufficient "gravity" to secure a stable cooling effort. This spinning of the element clusters is inherent from an event, even predating the "Big Bang". To find proof of this statement I just made, look at the photograph of any galactica. In the centre part heat is a liquid flowing like a river and taking matter for the ride. This is one part of the relevancy. The other part is that matter is spinning at such a tempo, it is capturing space in the form of pure liquid heat. The proof of this is obvious in every galactica.

Galactica is living proof of cosmic development and what is much more is the fact that it proves cosmic development in stages. In the centre of a galactica one can see that the Big Bang is presently arriving for what ever matter holds space-time within. Then as the circle expands, it is clear that the widening brings about less heat occupying more space producing darker regions holding bigger bodies of matter. Every galactica proves the Big Bang theory in precise detail.

The spinning motion of the element clusters (or proto stars or if one wishes to use the name of future stars, it will be just as applicable) hold their relation to heat secures by maintaining motion. As the time value in the clusters space occupation (mass) secures an era related value, the structures that were spinning, reposition in such a fashion as to 4apply a new linear displacing value. At first the motion is such that the linear position is negligible, but as the mass grows, the linear distance grows accordingly, placing the revolving structures further apart and at the same time, "pushing" the rotation of the objects in a wider revolving orbit. By widening the rotation circle, the objects rotate at a "lesser" pace and this pace coincide with the space-time occupation ("mass") of the totality in the effort of all the protons put together. In this one will not find a "force" but it will be a complete balance between matter, space and time. By securing an ever-increasing space-time occupation (mass) the future star will reduce its negative space-time displacement (motion) and increase its positive space-time displacement (gravity). The higher the positive space-time displacement (Gravity) becomes, the lesser the negative space-time displacement (motion) will be. At present only stars holding an iron$_{56}$ inner core can maintain a star status, and any object with a lesser element in the inner core will not bring about fusion, or in fact, any form of luminosity.

For instance by the time the "Big Bang" arrived, only elements with a "mass" of 112 had the ability to release from the Galactica luminous core and during the Era of the Quarks, the releasing mass of the time determining elements carried a combined proton-cluster "mass" of 88. As the single proton's time holding value increased (molecular mass) the time grew less and the universe "grew bigger".

This was
AN OPEN LETTER TO SELECTED ACADEMICS
ISBN 0-9584410-9-X
as part one of
MATTER'S TIME IN SPACE: THE THESIS   ISBN 0-9584410-8- 1
© KOSMOLOGIESE EN ASTRONOMIESE TEGNIKA

FOR ANY OTHER RELEVANT INFORMATION THERE IS THE FOLLOWING ADRESSES TO CONTACT
PLEASE SEND ANY QUESTIONS ABOUT MY WORK TO:
MY PERSONAL ADDRESS IS gravity@btrwtr.co.za

This formed a letter I sent to Academics in charge of Physics in the most influential positions without receiving one response. I include another part to fill a void and bring some other concept about the force life is as a non cosmic energy and in this the piece highlights the differences there are between the Cosmic Life and The non-cosmic life.

PART 2 INSTATED TO FORM THE CONCLUSION TO
TO SELECTED ACADEMICS
ISBN   0-9584410-8-1

All rights are reserved.
No part, parts or the entirety of this book may be reproduced by publishing, electronically copied, duplicated by whatever means that form reproduction or duplication, without the prior written consent of the copy rite owner.

BY
PEET SCHUTTE
FROM THE ORIGINAL AFRIKAANS: "MATERIE SE TYD IN RUIMTE" I. S. B. N.   0 – 6 2 0 – 2 7 0 4 1 - 1

An open letter

Man-In-Motion, Man-In-Mind, Man-In- Motive, Man Is Blind.

I wish to state here and now un-emphatically and categorically without any reservation of any kind there may or there may not be that it is totally against my personal religion as it is against the religion my Congregation upholds that I belong to in converting whom ever for whatever reason and never, never has this article or any other reference I may make any purpose in converting any body to my way of view about the spiritual in any way. It is not what you believe or not believe but how you live by your convictions and what ever your convictions may hold making you man or beast. My remarks about atheism are to show those practising atheism the foolishness of exclusion and to have an open mind because teachings of what ever nature has a positive and a negative connotation and the individual sets the standards. I am not intending or have any intentions in converting or changing any person's outlook on the spiritual side even in the slightest way imaginable. Me, living by my conviction to its full believe no bigger sin can there be than converting a person for that more than any other fact holds the highest epitome of sublimation there are. Secondly and in line with the first is my belief that the Bible has a base derived from ancient Egyptian teachings and I refer to that as I go along exchanging thought in this article. To this day we with all knowledge of splendour can still not understand how the Egyptians erected the colossal structures and the manner in which they did it. It is not realistic to consider a civilisation being that advanced in one area exclusively and with no other wisdom in other advances.

The hour in thinking has dawned where we humans must come to terms with the cosmos, with creation and with life. Mixing and matching was fine up to a point in the nineteen sixties where thinking about the cosmos and creation was a smart way to show superior intellect but being rite or being wrong was only a case of honour and pride that can hurt. Since the sixties life loss results from incorrect principles and maters got far more serious than it had since man had a first time ever look at the night sky with a conversation beginning from that. In another book as part of THE THESES I show briefly why I am of the opinion that man became human when he saw the funny silver dots in the night sky with a degree of admiration and recognition to the splendour of the unknown.

Now there is no longer only splendour in the unknown but a quest to find the unknown and gallant as they ever may be, it is fool heartiness to send brave men and women on search and not know what are there waiting on them. What dangers will establish the outcome of their fate? It is not philosophising for the pride that we should find evidence in distinction where distinction should be but crucial to man and to machine, machine because man's life interlinks with machine as it did at no other time in development history of man. A great philosopher I may not be and I will never be but thinking does not hurt and some advances may come from the weakest of thoughts when the thought try to unravel a thread running in thoughts. With this I too wish to connect in sharing thoughts of woven patterns as I see them and for what they ever may be worth.

I WISH TO TAKE NOTHING OUT OF THE UNIVERSE AND SHOW IT AS THE UNIVRSE IS: AN OVERLOWING CONTAINER FILLED TO THE BRIM WITHOUT THE SMALLEST FRACTION OF EMPTYNESS ANYWHERE.

As I stand on earth holding my first dimensional space-time displacement of our planet I can observe by using the second dimensional light source of the sun where my surroundings are made of three dimension atoms holding space-time in the forth dimension in time and space. I wish to move from point A to B and think in consideration about my planned action as my brain sends electric impulses to my muscles and that brings my muscles in mechanical motion.

Arriving at point B I think to stop (not necessarily by thought or mental planning) and my brain stops sending electrical impulses to the muscle fibre concerned with the action in applying my body motion. At that point I come to a stop and my thoughts go to a rugby match played at Loftus a South African provincial rugby team's head quarters where a game is in progress at that specific moment. As the proverb goes: I am there in spirit and my spirit being at Loftus are then some 400 km. south of my body where my body is on my farm. The duration in time it took my thoughts to travel is beyond human measure

The very next instant my mind goes much farther back in time and space as I travel by mental motion to a game played in Christ Church New Zealand a week prior to the day in question. My thoughts took me not only half way around the world but out of the present time dimension. As I am standing in thought I see my next-door Neighbour (to us in South Africa your next door neighbour lives normally 30 km. from you) coming towards me and my thoughts return not only from New Zealand but also from a weak past to the present in the very current time span my body was occupying all the time while I find my body moving towards my neighbour without my actual realising of this moving motion. I use the second dimensional system in the wave to transmit sound by means of repositioning the three dimensional atoms between Old Neighbour and me in applying motion to the atoms between us by the fourth dimension of space in time to convey thought harboured in the fifth dimension to him being in the fourth dimension of space and time. He then uses the same system to convey a thought by massage to me. Please note that it is a thought from the fifth dimension that I convey with the applying of organs in the forth dimension through ordering electrons in the third dimension to control matter in my body placed in the fourth dimension of space in time. While my words carry towards him, he drops down like a log as a result of not fighting the first dimension called gravity. A thought from the fifth dimension prompts me to respond in the forth dimension by creating electrons in the third dimension while my body stands supported but restrained at the same time by gravity in the first dimension. The thought from the fifth dimension orders a response in all the other dimensions ordering my atoms in the third dimension to use the forth dimension of space in time to act by fighting gravity in the first dimension on what I see in light holding a place in the second dimension which is restraining my motion in the forth dimension.

My response comes from some emotion and as it is not part of my mental reasoning or thought pattern it is directly conveyed form the fifth dimension to respond. I feel his pulse and find no beat. His breathing stopped and my next action is to look into his eyes. There is a dullness in his eyes that was not present moments ago. Something went that was. My observation consists of thoughts relaying massages that is transmitted by my physical body in relation to my senses receiving and responding to electronic massage translations about Old Neighbour in the fact that he somehow relinquished all earthly responsibility and problems of an earthly nature to his next of kin that is now saddled and burdened with his last remains.

The heart shows has no beat indicated by the absence of a pulse. The longs lost all ability to provide oxygen for transmitting and burning food. His eyes became stony marbles. I communicated with him moments ago, but his ability to respond by hearing and speaking has gone absent. My thoughts and breathing, heartbeat and hearing are still there. I can speak to him but it is his ears that have gone deaf. I can squash air down his longs but he is unable to use it through his voice box. I can hit his chest with fury and support a heartbeat, but the blood will carry the oxygen but can no longer create heat to live. The air I force down his throat still has the ability to produce sound because my shouting to him creates sound, but his ability to establish a method whereby he can create sound from the air I push down into his longs has gone away. His ears are still connected to his head and all the required tools equipping all previous aid that use to enable him are still there, all intact, but also gone forever. All the biological organs needed for hearing and making sound is still unscathed in the right places not damaged in the least, but the use has gone. The electrons needed to translate whatever requirements enabling body function must still be in there somewhere, because I saw no discharge of any sorts flashing from his body.

Even by giving him electrons through an externally generated flow of electricity will not create any of the required but lost electron flow to generate life back in place. I may shock him till he hops around all over the place, but motion is denied for brain activity to function once more. His brain has gone empty, although it is full to the scull. His thoughts are no longer with us or with his body. It is no longer Old Neighbour lying there, but it is his remains. Even an atheist will tell you there is a difference in what is there on the ground and what were there moments before he dropped to the ground. No heat or electricity can revive what he lost. It is a body without life.

Minutes ago there was life to talk think and reason, discuss and argue, be angry or glad, but that, which now is lying on the earth has no more such ability. The source of energy giving life to the cadaver is no longer present. Science proved that energy cannot go lost but has to go from one form to another form. Energy can never destroy or vanish but has to replace form or attachment. The body is there, and it is holding all the organs and the organs has still got the required heat to perform because Old Neighbour has not gone but a few minutes ago and in the South African sun bodies do not go cold through lack of heat because we are use to temperatures of forty degrees Celsius and more. The cadaver has all the essence to sustain life and if life was electricity, then I should be able to recharge him by connecting leads somewhere and call an ambulance. But supplying any form of current at any voltage rate will not bring back life once it has gone. You can heat him with a blowtorch while shocking him with a cow prodder (and does those things unleash electricity!) it will revive him as much harm him or do him bad or good. He has become apathy in every sense.

His lifeless body will never carry his mind anywhere again because although the brains are still there holding all the mass it had when Old Neighbour was still with us, the brain is thoughtless and that has taken Old Neighbour away from us. Our dearly has departed although his physical remains stayed with us to rot if we do not take care of the cadaver and the sooner the better for everybody involved. We that are part of the living now have to move Old Neighbour because he no longer has such abilities. Minutes ago he still had the abilities but from him went energy. It must be energy that he lost because all other necessities in for filling such duties he still has (that is if you consider his body as Him) But his body cannot be him because his body is there part of the fourth dimension in space and time securing all his abilities to function as a human but that abilities has gone vacant. All the effort he may muster will not allow a wink.

The only visible something he lost that makes him less of a human being than he was this morning when he woke from a nights sleep is the energy of motion. His body with all the parts still hold dimensions in the first the second the third and the fourth dimension, but clearly it is the fifth dimension that has gone absent. The cadaver is still part of every dimension excluding the dimension of life and life then has to be a dimension above and beyond that of the fourth dimension in space and time. The cadaver is at present what we refer to as being lifeless and dead. The generator or power source or dynamo or what ever you may consider it to be but that dynamics providing energy in sustaining motion has gone away never to come back.

Whatever any person may try to do the machine that gave drive to motion is no longer able to provide motion. All the wonders that the human body possess in motorised function is no longer in motorised function although it should be if it was only a matter of replacing the lost energy by providing an electrical shock or some fuel of some sorts. Nevertheless no fuel can get that motor running again therefore the energy lost is not a replaceable kind as in the case of ordinary heat from fossil fuels, food or electricity. The machine of human motion has gone for good. Surprisingly the problem of energy and life becomes far apart when logic replaces Newtonian atheism and illogic.

Shove a ton of coal down his thought and it will do him or you no good at all. Roast him with electricity and see how far that will convince him to return to life. Push a gallon of pure glucose into his veins and see what the reaction will be. Energy is not merely energy and once again Newton got every thing very wrong. With Newton's incorrectness all the sheepish atheists go about an echo one can hear for miles around, but all the echo is only echo after all with no substantiating individual thought about and amongst the lot of them atheists. When energy is not used it becomes latent or so does science proclaim in any case. One cannot ever consider a rock rolled up a hill having the same latent energy because the rock needed the same life that has gone absent from Old Neighbour to role uphill in the first case. When inspected closer life is the energy keeping the human body running as a motor and by distinguishing life the motor stops. Something went latent and not vanished. Life was part of the fourth dimension up to the moment it went latent. It shared time in the body and space with the body thus it was part of the matter of the body it no longer uses. Without doubt is the fact that life was the indisputable source of energy driving the body through the fourth dimension? Where the space-time sharing then ends, life cannot end because it is the functions of the body ending and not the energy driving the body while inside the body when it gave the body a function of movement the body had an ability that no other cluster of atoms enjoys in one construction in the universe. It gave the body the means to displace space-time not only by gravity and motion as all other structures have but it gave the body a means of changing the space the body occupies in time that the body occupies. No mountain can move a little in the morning to avoid the blistering sun and shift to another place at night to escape a blistering cold wind. The human body including all life on earth can shift position as to suit the needs and requirements of space enjoyment in time duration. This means is very exceptional as nothing ells known to man in the cosmos can achieve such motion by pure will power. A plant may not be able to run to a better position but when in competition for sunlight it can try to outgrow its neighbouring plants and claim a larger share of the available heat the sun has to offer. That effort is completely out of the domain of any rock. A plant can grow its seed in such a way as to ensure distribution and gain advantage over the spread of its space it holds on earth as territory. There is no chance that a puddle of mud can run after water to keep wet. Life can manipulate the space-time it holds to its advantage in the sense of bettering its chances on survival as well as its species chances on relocation.

That is the overall advantage life holds and is not merely an energy that does some work in relation to the growth in the universe. Try and measure the time a mountain holds space and compare that to any one form that life holds measuring from birth to death being on this planet. Then after getting an unbelievable answer a person can appreciate that life is the energy and without life the structure becomes the equal to what a mountain is from the onset of the lava flow. Life is the manipulation of space-time and the higher the degree of advance is, the more life can manipulate space-time. An aircraft flying may be as dead as the next mountain is, but through the aircraft, man as the ultimate form of life can manipulate space-time far outside the reach of lesser species. I do not wish to start comparing life as being advance or more advance so I leave my argument at that as far as life development goes for now. In the very beginning I stated that through the way the mind travels, it has to hold a higher position than what the body holds because I showed how easily I could travel around and even half way around the globe in no time at all. Sure I was not there in person, but my thoughts conveyed some understanding of what was happening on other places outside my range of vision.

The mind sets a norm that the body can follow or not follow but the body never sets a norm that the mind cannot follow. All sells even those holding life has an electron a neutron and a proton and very deep within the very deep within next to the truly unknown is a structure that holds position in relation to singularity. When a sell holds life it is different from a sell not holding life although when the sell not holding life still constitutes of the same composition it had when holding life, something changed, something is different.

A life-carrying sell not carrying life has gangrene a most deadly disease that kills as none other. A sell absorbing heat normally is showing growth whereas a sell in abnormal heat intake is cancerous and again is deadly. I can go on and on about this but it is apparent that as soon as life looses control over heat the stabilising factor or thing go abnormally wrong and such conditions can, may and will lead to the vacancy of life occupying the body more permanently. To understand the way I wish to direct the argument please allow me to indicate how I see the normal as we will find in the cosmos in life carrying and non-life carrying matter.

The line =180^0 The half circle =180^0 The triangle =180^0
Since almost before serious recorded history dating became scientific principle mathematicians knew that the straight line holds 180^0 degrees matching the half circle as well as the triangle. But never have I read any definition about this phenomenon and how it comes about or what may cause

such odd connection. Heat occupying space has the cube that can apply r, as a straight line bringing about the cube with all its other names that may find attachment to specific form but nevertheless still remains only a six-sided cube with angles changing in some cases.

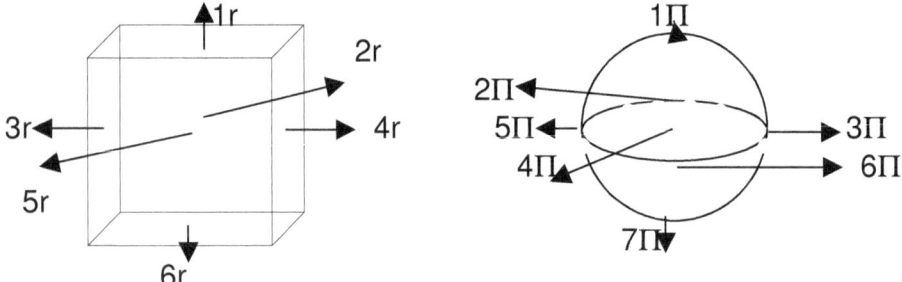

In the sphere there are no radius but only the extending of Π from the centre Π in six opposing directions relating to one another by the square but remaining Π because of the unity the matter holds in relating to space. It is not possible to draw a precise line that would form a precise ring and not cut some atoms in parts. Because there will always be an atom disallowing the precise positioning of the circle the circle continues on a solid basis holding Π as a positional reference and not r. In every sphere there then are the seven Π relating in precise dimensional and positional equality forming equilibrium to the centre Π as well as to one another by 90^0 and 180^0 implicating the dimensional positioning. Therefore the sphere holds 7 ^{Π} and the cube holds 6 r^2

Where space comes into contact with the sphere the cube loses one of the six dimensions it has to the more dominating seven dimension of the sphere whereby the seven dimension in equilibrium will dominate the six dimension loosely connected bringing about that the cube then has 5 sides to the seven of the cube. This means that in the cube the "bottom falls out" and without a "bottom" to support objects they fall to earth. Remember that a body "floats" in space, but at one specific point it starts to "fall" to the earth. That is gravity and it is a dimension change much more than any force.

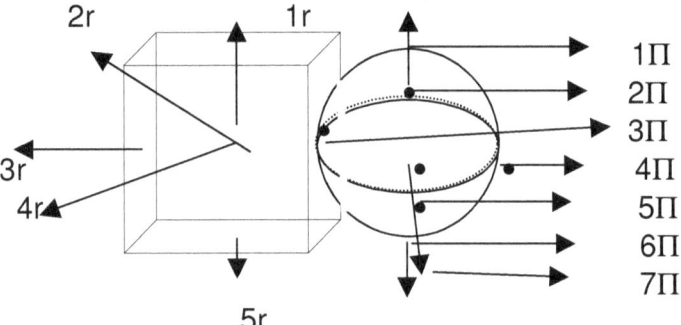

From such a point every other point will be opposing any other point not pointing in the direction to which the first point is pointing, whereby it extends the direction it holds. No matter what the point is or where the point leads, such a point holding a specific direction will be unique in the direction it is rotating because at that or any other specific point wherever, it will be directing not in the direction it spins but in the direction flowing from the centre point outwards.

All atoms are a minute form of a coming black Hole and viewed in the structure composition it is clear why I say this. On the outside there is heat trying to get inside the atom where the heat is needed. On the inside of the atom there is a need for heat and the inside is in constant regulation of the heat flow as to keep stability. In understanding the dynamics of physics we must understand the cosmos where the process begins and where the process ends.

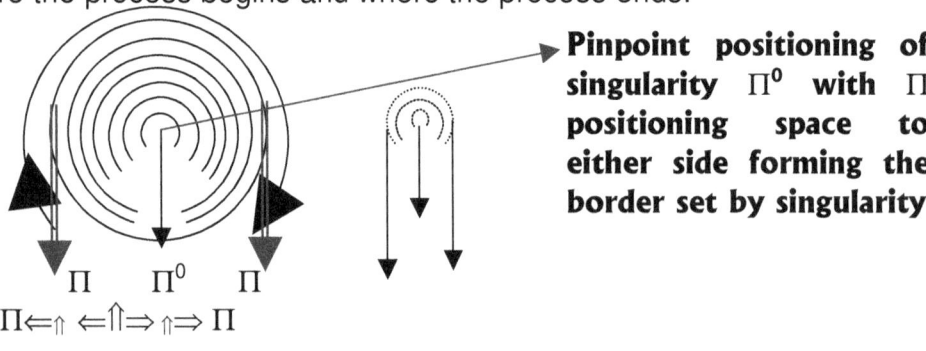

Pinpoint positioning of singularity $Π^0$ with Π positioning space to either side forming the border set by singularity

The atom holds a very unique position in that it links three dimensions to a forth dimension and this part is where I came to understand Einstein's thinking but a with Newton I could not accept Einstein's thinking. Only bringing in religion could I get further about the formulation of singularity because in that I found what connected the universe whereas Einstein left space as space and tried to link time to space as an additional factor. That would be the same as not linking life to the body or exclude life from the body while trying to argue about life being part of the cosmos as Newtonians seem to do.

This occurs in all atoms through out the cosmos with no exception on the rule. But life-carrying atoms in carbon $_6$ commits life as an additional supplement to the atom as life can become absent from the atom leaving still in the normal range of a cosmic structure. In the past number of pages I brought reason to those of reason that there are more to the body supplemented by the presence of life than merely carbon fibre. It runs much deeper than physics can intrepid. As far as pure physics go, nothing changes when life goes absent and yet everything alters when life abandons the atom.

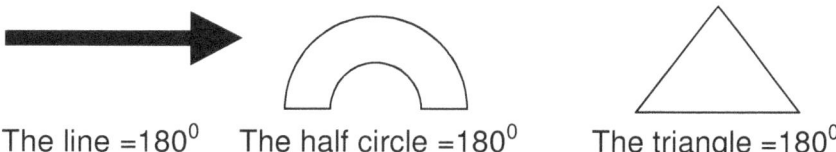

The line =180⁰   The half circle =180⁰   The triangle =180⁰

I saw a very neatly outlined connection that the atom has in its position in the universe as it was the evidence of the smallest all connecting matter tying what is matter into a small container.

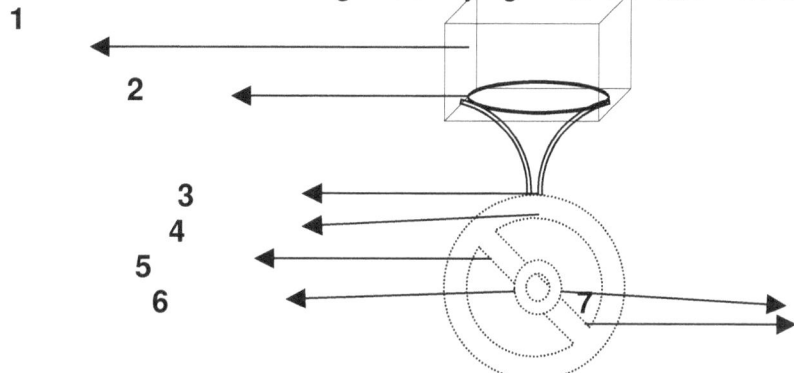

1 The square value of space-time in the fourth dimension holds a positional relevance to singularity by 10 points or places.

2 The space – time in the square of space loses the value of 10 by entering the atomic relevancy formula and become 3 sides to the cube.

3 The space – time holding space as the square loses the value of 3 by replacing 3 with Π thus going on down the line of the atomic relevancy formula and become Π sides to the circle as $\Pi^2\Pi$.

4 The atomic relevancy of space-time displacement changes once more to Π^2 where space goes flat in time.

5 The atomic relevancy of space-time displacement changes once more from the neutron square of Π^2 to the proton's double square $\Pi^2+\Pi^2$ as space unoccupied disappear and time forms the square by the square. The relevancy flowing from this figuration is so very important when archaeology presents facts and with this the archaeologists (may dare to say) became the blabbering fools they should not be as they are presenting serious science as a qualifying joke with the funnies they come up with in setting every one with thinking minds laughing.

I shall return with this argument in a later time when more facts relating to the argument are exchanged between us. For now we still have a cadaver on our hands to dispose of and quickly we must because of dire consequences that may follow if not done in urgency. Rotting corpses bring untold diseases as the great influenza epidemic of 1919 can support in evidence. Then why the danger about the corps all of a sudden if life within the corps have such minor importance as the Newtonian atheist wishes us to believe.

The molecular structure is still the same and so are the chemical composition within in the body and the mind. If it is chemicals making up man then, the man should be there because none of the chemicals went AWOL. If life is about the electricity that runs down the spine, our distinguished atheists should replenish life with some minor application of current as a means of stimulation. All the ingredients are present yet the manipulative nature of life making life so very exclusive to all other cosmic ingredients does no longer function.

Everything the fourth dimension can provide is still present within the body, yet that substance beyond the forth dimension, that ability we cannot detect but with a lot of intellectual thought, that ingredient Newtonians deny them of recognising as very special exclusive to life is not there any more. The ability manifesting as the energy or energy supplying has relinquished its role within body and mind. If it is only a matter of electrons generating pulse consistent to flowing from sector to sector in the providing of artificial current from an external source should supplement life's conducting of the work of the body. But shock the body as much as you may, the body functions disappeared with the disappearance of life from the body. With the exit of life the vital electron distribution seized and the cadaver became just another particle containing what all substance other than life contains. The cadaver became just one more structure in the cosmos with the same ability as a rock or a mountain. The electrons conveying the massage may be replaceable but the sender and receiver and the decoding of the massage has no longer any function within the body. We may send some electrons into the body, charge the body with oxygen via a machine doing the pumping of the air to precise rhythm as did life, we may stimulate the heart by nerve pulses artificially supplied. We may contract muscle by supply of electrons. But replacing life can never be one of our accomplishments once life has left.

All muscles including the heart and longs can be stimulated artificially and in doing it may prolong the writing of the death certificate, but having the person stand up straight once more or pronounce some wish to be for filled or just a simple effort as winking eyebrows when asked to do so is far beyond the abilities of the lifeless structure occupying space-time in the fourth dimension with the aid of all other dimensions excluding the fifth dimension. Previously the body seemed to manipulate space-time with the utmost ease, walking wherever the mind chose to walk through unoccupied space-time, only adhering to the restraint compiled by the other dimensions and inflicted on life to startle the manipulating abilities. Now such abilities disappeared, vanished with life to some place we see as death. But life as energy has permanency that can never become denied and disappears. It cannot become washed away, wished away found to have disappeared for it is energy, a something of eternal power in being.

As energy has linking to eternity it has to come from somewhere as much as go somewhere it came from after it left. If it went latent even then it has to be stored in some place of gathering energy of life's nature. That storage facility is not part of the fourth dimension as the rest of the body is, so it will no longer be in range of the detectable, yet it must be somewhere with the linking energy holds to eternity. We also can tell that lesser forms of life will destroy the composition of the cadaver feeding and preying on it till what is left has no longer use in sustaining other forms of life as food. We learn from the esteemed and well respected Brainy Bunch the food we eat provide the energy we use. Sure that is very to the point and easy to swallow. But what uses the food in maintaining the energy. This is my problem with the SUPER-EDUCATED-WISE-OF-THE-WISE as they will forever give information that has no substantiation but only scratches the surface and leads nowhere. How does food give the energy and what uses the food for energy. If it is the protein filled body of the human flesh, then what makes the intake of food become meaningless after life departs. Should the food be energy then just feed the man and revive him. Food is all he needs to regain life and if such an effort fail to revive the dead then one should seek to find deeper meaning to what is obviously not obvious.

If life only connects to the fourth dimension of space in time through energy supply such as food may supply then the body is there to be nourished back to life. Also in the opening of this argument I showed how life travels time with no limits to boundaries in time. I did admit that such travelling only applied to the mind and not the body but since it is the mind that has gone absent and not the brain, then the travelling time by the mind must still be in affect as it was the mind that did the travelling and not the body. The body did the travelling of the time constraint but it is the same body that cannot even permit travelling to its grave because of the absent of the mind.

Time did not restrain the mind and if it did not restrain the mind time must have little control over the mind. Since the mind is no longer present in the body and time is still doing all the restraining on the body we may conclude with some extelegence that wherever the mind goes in storage time will hold no constraint over it. All the atoms that were in use by compounding a human body would one day again form some flesh of another being. It will ultimately not be human and it will obviously never form a body as one group of solid fleshy matter and it is logic that the compliment will become divided when forming a future body amongst millions of bodies all containing substances previously located in millions of different bodies to form millions of different life forms but it will become use full once more in the future.

With such a remark I do not say that the atoms scattered after destruction of the body of say a dog will form a dog in future. Such a presumption is madness. There will be dogs in future claiming some of the matter and there will be other life forms finding use for the ingredients that once constituted Old Neighbour but the principle is that the atoms that were in use will again find use because of the atoms eternal connection to time.

The carbon fibre is on earth placed for the use to carry and support life and as it did in the past, so it will do in the future. It is part of the eternal qualities of the atom to maintain space-time for the foreseeable future and the foreseeable future I suppose is the duration of time that the earth has to sustain life. To us humans such a concept of time on earth holds all the factors we connect to eternity because to us the earth and eternity is almost alike, but in thinking that never should one forget that in the realms of the cosmos the earth is but a flash in the pan, a wink of the eye and it is gone. But not straying that far into the future we are still measuring the chances of Old Neighbour becoming Old Neighbour once more for he was quite a likable chap and some people will miss him (I suppose).

What will the chances be of resurrecting Old Neighbour to his former self? Well left to the simplicity of the arguments held by the astonishingly brilliant Newtonian atheists we must consider it better than one hundred percent and according to their superb argumentative powers it is as good as done with the aid of a pump, an electric generator and a shovel to push food down his throat. But beware because when gauging by their record of previous successes notwithstanding the simplistic manner they go about denouncing the complexity of life, my prediction is also my advice: if you are a betting man do not bet on a positive outcome because you are about to lose money in such a bet! Your chances in winning will be as good as that of our atheists' wonderful arguments being correct.

Well now Old Neighbour is going to push daisies, or is he? Who is who and what does the daisy pushing? We know very well it is Old Neighbour that is going to push daisies because he is not with us and the "not being with us" part means gone away. Should one force he argument that his body, the compliment and assortment of DNA sells arranged to a specific order matching a pattern profile that belonged to Old Neighbour exclusively is Old Neighbour, or at least that is what our Newtonian-Bright-Boys insist on being the case. You, well any person, makes up a compliment of DNA sells and according to your sell arrangement whereby you become you and by pre-selecting sells and arranging them to a specific order where they form one totality and arrangement by assortment giving any person the prospect of life. Our distinguished atheist loses all other related arguments past this point in order to conclude what they believe to be correct. Considered in the utmost simplicity, yes, that is correct and as that alone it leaves no doubt.

Through this a rat cannot be a horse and a dog cannot be a lion

It is so very simple to understand when explained with such excruciating simplicity that even us living on the other side of the universe where the Lame-Brains belong are can accept without arguments because we are so scared of putting the least of effort into the simplest form of thinking giving the Brainy Bunch the scope of miles around to come up with the most idiotic answers they can dream up and we the Lame-Brains are too willing to accept as long as we are excluded from any form of thinking. Therefore we allow them so gracefully and with all dignity applied to both sides of the intellectual divide, to bullshit us to a stand still and make us feel great full that we were so privileged in accepting they're demising and diminishing mentality bestowed onto us. I say this from a stance where I am part of the idiots ranks and stand amongst my fellow mindless admiring those of the fortunate and privileged with they're wealth of thinking power because they achieved so many a splendid degree and are therefore the rich in thinking making me just one other poor beggar in thinking-power. The human being as with all beings having life connected to the body structure they occupy which are the compliment of arranged sells and such an arrangement exclude my being a horse and it excludes the horses chances of being an ant.

With things that simple and sells going nowhere as they did not go anywhere in the dying of Old Neighbour why are they not functioning? What made them go on a permanent strike? Why can our Brainy atheists not once more persuade or force those sells on strike into accepting responsibility for their work responsibility because all of the world needs Old Neighbour around and the medical profession did not yet receive they're rightful chance to drain his money like a broken dam wall under the banner of keeping him alive for his family. Well at least until his medical aid runs out and his bank account has gone bust. With that simplicity being the case of life the atheist can at least replenish the life to the sells until everybody in line from the chemical manufacturers down to the cleaners washing the hospital floor had they're chance of becoming Old Neighbour's inheritors and not his wife and children. With Old Neighbour circumventing the money draining system it becomes totally unfair and what is more is why did the system spend so many billions in creating a net where they made Old Neighbour so scared of death and disease he will gladly part with all his money as long as the system gets the chance to help him cheat death (should you not believe me look at the cancer and other advertisements and think for yourself who is paying for the brain washing). Why not only tell those with cancer to do the fighting? Why charge everybody up to come out with they're six shooters a blazing in spraying lead. Who is paying for such advertisements and who receives the benefits of such advertisements all done under the banner of securing a longer life for every body.

I am a diabetic and a smoker that does no exercise of any kind but to get out of bed in the morning. I was medically ordered on so many occasions to quit my smoking, and I not sooner did that then they started feeding me anti depressing pills and anti anxiety pills and sleeping pills and stimulants to fight the sleeping-during-the-day-attacks and the.... The list goes on almost indefinite. Once I pick up my smoking habit again I suddenly do not need one of their pills to keep me "normal". While my smoke may kill me the exhaust fumes of the cars in use which pours the most deadly of gasses into the atmosphere being carbon monoxide is not maybe but definitely not only killing me but also nature in every aspect. Carbon dioxide is a natural element on earth while carbon monoxide is a chemical acid eating or more accurately said devouring even the likes of statues chiselled from granite rock as well as things manufacture in iron to rust. That aspect no one ever comes to mention BUT SMOKING is the killer destroying life by the billions! The doctors are reluctant to allow the tobacco industry to kill you because that will deny them the chance of killing you chemically and making the profit themselves either through driving their luxurious cars or stuff they prescribe and you can only purchase through chemists. So the doctors scare the daylights out of you about death (which you will never escape in any case) to feed you pills (so chemically poisonous they can only sell on prescription as they are sure killers and most dangerous) and the system is creating another slave by making another fool so brainwashed he truly believes he will eventually cheat death! And Old Neighbour had the audacity to escape the loose of the system and die still with money in his bank account! Such a dead is outrageous and cannot be tolerated. Believe me if the medical profession got to Old Neighbour before I did, in his dying effort they would have kept him alive for another few hundred thousand reasons, reasons you keep in a bank vault and pester his wife and children with guilt so that they part with the money so willingly they will even pay anybody to advise them to part with the money. (If that is not why you pay the doctors treating a man that is ninety nine percent dead already then why are you paying him in any case). You the reader may not see it but this is all resulting from atheism and a system promoting atheism and is an all out war world wide making every breathing person on earth a slave to milk until death does its part. Convincing people about the simplicity of life will encourage them to fork out money to be kept alive so that the slave will gladly allow more milking.

Slavery so I am told and so I do believe from the bottom of my heart is wrong. But the slaves did not have it so bad in the days of the Greek and Roman Empires. They were much better off than us the slaves of the current World Order. Slaves under the Roman law were fed clothed and accommodated on the Master's account. The law was that the owner of a slave had to feed him and provide accommodation for his slave. Then the slave had the right to ten percent of the income the owner generated from the services of such a slave while the slave had the chance (if he could) to buy is freedom Slaves in the current World Empire of the Hoggenheimers an Mammonites enjoy the pleasantness of a just system where the system does away with the need to bay slaves, the slaves join the system or die. Furthermore they make the slaves pay from their wages for food logging transport and clothes while the Hoggenheimers do not even pay them ten percent of what the Mammonites earn from their services. Under modern law, modern slaves are worst off than slaves two thousand years ago! And to top this Old Neighbour had the audacity to escape the slavery without even paying his last bid for his freedom. How criminal can a man become in such a manner of escaping what was rightfully his dues to pay. With all the simplicity about life and the promoting of escaping death why can the atheist not bring Old Neighbour back to do his last part and fill the already overflowing money caskets of the Hoggenheimers and Mammonites.

There is this wife of one certain pop star a member of a very well known group in the sixties and one of the four members in this very well known group. This wife of the famous pop star made millions on promoting the abandoning of the use of animal meat as food. She told about her and her husband having lamb chops one afternoon while some other lambs were grazing nearby. As she saw the lambs with her mouth stuffed with their friends she then and there got thinking about cruelty and the humane aspect about eating lamb in the presence of lamb nearby. She was devouring the flesh of sheep that was killed for the purpose of feeding the human population and that gave her the idea to make millions on that thought and selling humanity in the process.

For some sake of sanity let us scrutinise the situation and for once go just a shade deeper than just being prognostic in our conclusions. The lamb has carbon$_{12}$ as a mixture of forming the composition that we named protein. What will be that different from eating grain and eating flesh? Both holds life and both holds death after life. The grain is an infant that did not yet start life whereas the sheep is an adult whose life was cut short during life. Both faced death before they received the honour of completing their sole purpose on earth and that will be to feed man. She went on a campaign promoting vegetarian dishes that did not even contain fat as protein but included the biggest variety of plants imaginable. While on the tour of promoting the eating of plants (and selling her book to millions of other fools that run on emotions they do not understand, cannot control and where such emotions totally outsmart their thinking capacity) she stopped far short of explaining why she would consider plants lesser life than what sheep are. Can the reason be that the sheep think nothing of devouring the grain and she allows the sheep to do the thinking on her behalf? Is it because grain does not run around when "chased to become grained" for food. Or could it be that the price of the book and her selling power of the content of her book allowed her to sucker some idiots (and I believe the number of idiots caught in the scam runs into millions) tinted her perceptivity so very slightly in favour of the consuming of plants that cannot make any sound or request any human emotion by running and shouting in protest trying to escape the butchers knife or in the case of plants the sickle.

In the case where we consume fruit as food the fruit we eat is food still alive in the same manner as does lions starting to eat a buffalo that is still standing on all four legs. If someone somewhere came about the promoting of eating animals while they are still alive I would surely go on the same protesting crusade as she did in her bit to fight the food supply in the form of meat. We now are faced with the same cynical questions our friend Old Neighbour left us with. Is it his corpse lying there or is it he lying there. If it is he then I have to admit that we are eating lamb. But if it is his corpse then we are not eating lamb but merely the remains of what was lamb once. I am not wasting any space on arguments about killing to eat because kill to eat we do because we have to do it. There are no other options open to us but to kill or to become killed through starvation.

The bottom line underwriting everything said about what form of food we should or should not eat is the human capability of becoming completely self-absorbed in sublimation. We think we know exactly how God created all around because we know exactly God did not create that which is all around. Therefore it is our claim to right that we may take the place of God and decide what should count where and what is food and what not food, but for god sake keep it simple otherwise we will not understand why we may think ourselves as gods. As long as science portrait matters simple excluding the not very popular complications of thinking every thing thought through decisively we may find that being god can be a very pleasant way of living and un-complicated. If we do not complicate everything we may even think of ourselves as very clever gods without the excruciating effort of being clever gods. Just go about and visualise our brilliance in reason and tell ourselves how kind-hearted and humane we are without any deep philosophising about truth and matters of complexity.

If you are in support of the humane aspect then consider that the deed of eating fruit will be far worse when eating the unborn and defenceless or robbing the unborn defenceless seedling of nourishment so dearly and lovingly accumulated through severe hardship and unquestionable devotion in loving labour by a caring mother than a developed specimen of any specie. Remember that when eating the unborn fruit or the food meant to feed the unborn seed will be denying life the chance to be and that is very unfair! At least the meat eaters gave the sheep the feeling of being sheepish before removing the feeling permanently but in the case of fruit eaters the fruit never had a chance of feeling fruity. I should add that to my mind humanists are the worst practising sublimation because atheists deny the fact of God but humanists are in criticising of Gods way in creating the balance we know as the echo chain. Humanists are constantly trying to show all that are willing to listen to their senseless rambling how much better a job they have in mind for all life on earth than that which God established up to now through giving man reason to think with a mind and not an emotion and forgetting that the methods applied got civilization in such a tested and tried state as those methods did but still they whish to change it because they think they know so much better.

If our pop-star-wife did not have the pop star fame and all the pop press in support and with the wealth of food supply around how far would she come with the cheap mentality and the thoughtless advocating of the shameless theatrics to support her promotion of self enriching by selling books. When any nation is in total starvation as the Germans were just months after W.W 2 I wonder how many hungry men and women with children crying starved to almost death would applaud her madness as greatness. She got through because there were abundant and not because she had sensibility in her quest to make money.

She could manipulate others while the others were swamped in good times and rolling in the fat of fortune fed to burst while gloating about how their humane hearts bleed for the helpless sheep all over the world knowing very well none of them ever had to skip one meal because of want. They never had to live through one night of agony where their children were crying because the children were too hungry to fall asleep. When thinking about such conditions their gloating in self-praise is quite sickening. From me and mine to you and yours I am telling you this shocker: the total destruction of mankind may only be as far away as the swing of a telescope, and the announcement of a funny little dot that seems to grow as it is heading our way but more about this later on. She is merely one of millions making senselessly money without thought of dangers larking

This I say because nature tells the truth about man and the way mankind evolved. All predators on the hunt have eyes pointing foreword to find the maximum advantage in three-dimensional sight. By focussing in hundred present accuracy the predator can pin point the kill and act swiftly and abruptly minimising the chances of the hunted from escaping such an attack. On the other hand when looking at animals that is mainly vegetarian we find their eyes on the side of the head to secure maximum vigilance and response to such an attack. When looking at the human face we find the eyes even more in the centre of the head than in the case of an eagle, famous for his hunting skills and such a small but obvious clue demolishes the entire bleeding hearts cry for passion.

All animals dependent on meat for food sustaining have eyes pointing to the front the very place humans find their eyes to be. The road our humane idiots genes followed took them through a ancestral path with a long range of meat eaters that brought the gene carrier to what he or she is in the modern age, but being smart they make themselves the fools they are. If we humans were fruit eaters only and had no natural inclination for meat then our eyes would be next to our ears instead of being rite above our noses in the centre of our faces. Those placing meat eating in so many disputes should then also change their eyes position to the side of the head and denounce their ancestral trace of meat eating.

Man has a vision allowing 180^0 sight where as animals born to be the prey has a sight range of 360^0 and none of the humane intellectuals ever came that far in reasoning. With such direct and undeniable evidence about our eating of meat, how on earth can those shouting no in support of meat eating show their faces around as intellectual beings. This also goes to some religions denouncing the eating of meat but as long as they keep their religion to themselves without trying to convert me to such rubbish they can believe what they want and exclude me. I say this because on occasions I got into debates with such people that wanted to push their religious ideas down my thought about some Indian god living in India and you send him money with a prayer where he then fixes your problems rite across the ocean providing you do not eat meat because of his say so.

All species on earth are what their history made them. They are moreover the road they followed down to where the specie currently is than what they are at present because when circumstances change the genes with idle qualities will arouse the complexity of the specie and old habits that saved the specie from extinction in the past can come to the front and again save the members from extinction. The Sudanese can survive by eating leaves from trees until the rains arrive to bring about new harvests (although the rains never return permanently). On the other hand the impala cannot start eating lions to keep alive until new vegetation grows again. But even the harmless impala is not that harmless to grass, as grass has to grow meters every year in order to sustain the impala's nourishing needs and at the same time secure the survival of the grass as a specimen of life on earth.

Man too, if need be, can survive on grass and that puts man on top of the evolution ladder and not their misguided impulses in correcting the ways we developed. It is great to play god when God gives in abundance. It is great to play god when God brought your specie this far. But try and play god when God closed the clouds bringing rain and hunger with facing starvation. Then the mind fills only with thoughts silencing the hunger pains and the obsession comes as the hungry wish to fill the stomach with food without filling the mind with cheap sentimentality. How brave will the Super humane then be I wonder. Being humane is closing life to a very single minded approach and in this the massage of the atheists simplistic views about life ring out loud.

All this may be fair but there is another side as all things in creation stand in relevancy In my quest to find answers one question I could never find an answer for is why do the world not import the Sudanese to Britain America and Australia instead of exporting the food they donate to Sudan. Sudan has become a country that will never again support such massive numbers of people and the food will forever be needed. The growing desert claimed the country and it cannot sustain human populations. Declare Sudan uninhabitable and take the people to the countries donating the food. It will be much cheaper to feed them in the countries I have mentioned and at the same time it will please the bleeding hearts, give the Mammonites more slaves and the Mammonists more slaves to drive while not hurting the unemployed one bit for jobless they are because jobless they wish to be. Change the relevancy in the equation and take the people for once to the food and not the food on a yearly basis to the people.

Before every Anglo American starts demanding my immediate and successful castration without precondition let me add why I say what I say. By feeding the population the bleeding hearts are getting their wish but in it they are sadistic and devilish cruel. Before any aid can be requested a disaster must be in progress. Being a disaster in progress means millions are suffering. There has to be an enormous lack of food supply to wake the caller. Babies go hungry mothers weep fathers run off because they wish to find food and disappear in the process. Suffering runs deep as it runs wide and no aid can prevent that as no precautionary measures will ever be good enough. By helping once you are spreading the suffering to last longer and with more pain next time around and we all know there is a next time around because of climate changes going on. Feeling good about your self because of proving once again your good nature, your blessing heart and empathy by the giving aid helps no one because of the coming of the next time. The simple truth is that those in power and those with influence give nothing as much as care for the helped victims. The philanthropist collect money on behalf of the Hoggenheimers from the bleeding hearts while the philanthropist encourage the bleeding hearts to donate in giving for the simple reason the philanthropist share in the spoils of the unselfish act. The Mammonites bay the food as cheap they can in names of companies they own with as little money possible from stocks the donating parties would trash in any case because of poor quality, then bay the food from their private companies with huge profits going to the private company because the selling party is also baying on behalf of the relief organisations with the money the bleeding hearts donated not because out of true sympathy, but the bleeding heart wish to kill the guilt they feel as they know they have it splendid and therefore they need to prove to all but mostly to themselves they're godlike generosity by donation.

The Hoggenheimers take their cut with excessive profits by distributing the bleeding hearts', which the philanthropist collected so unselfishly as proving it by taking their fare share of the profits going around giving the money to the Mammonites, which are baying on behalf of the bleeding hearts from their firms as they sell the stocks they previously bought for next to nothing with excessive profits. At this point the Hoggenheimers bring in the Mammonists to do some slave driving as the spoils has to be sent across the world. In this heart braking act of generosity some more unbelievable profits go the way of the Hoggenheimers and the Mammonites because the firms involved just so happens to belong to a shared venture between the Hoggenheimers and the Mammonites and by some more overwhelming generosity they share crumbs with the Mammonists doing the slave driving.

Now you tell me who is unhappy while all this good heartedness goes around and is there any blame to be where the rich becomes richer as that is no one's fault. If the bleeding hearts were serious about their conviction in generosity they would not bay some guilt relief. If the other parties were serious about their convictions they too would try to find a permanent solution but then there will be less profits to gain. The bleeding hearts are quite satisfied that big planes are used to transport and distribute the food but they know very well that that is the most expensive means of transport and someone somewhere is changing very unselfishly a dime spent to a dollar wasted. In this way the relevancy is getting the rich richer, by giving the guilty guilt relieve and helping the luckless to another round of heart ship in hunger.

Change the relevancy around if the act is in pure kindness and brother love. Take the luckless out of the equation of desperation in cycles by removing them from the problem. In that there are some more relevancies involved. If the bleeding heart were serious they would never mind bringing the luckless to share in their abundance. The other part of the option is to let nature take its toll rectify in natures way and be done with it but then the profit issue stands to lose millions of reasons why neither option is an option. The relevancy will lead to a cheaper solution although more expensive the first time around. Everything is about relevancies. On the one side of the relevancy is the earth became unsustainable to carry a human burden in that part of the world and on the other side of the relevancy is, the western countries have food to donate in tons through baying and selling agents, (and I shall gladly eat my farm if the politicians were not sharing in the bounty of tax money donated in generosity).

The one side of the relevancy is the Sudanese will never be self supporting because on the other side of the relevancy is in the long run a desert means drought and water will never again be abundant. The only solution to the equation in solving the problem is by changing all aspect around in the relevancy and through that finds a permanent solution to an unsolvable problem that will forever remain unresolved until the relevancy changes to finding an answer instead of avoiding a solution. If the cosmos can tell us one thing it is that changing the relevancy brings about solutions. By creating the Big Bang it solved a problem of overcrowding as we have in Sudan and by creating space as we should in Sudan the cosmos separated matter from space as it is still doing with the Hubble constant proving that space is on the increase. But if space is on the increase and all is about relevancies something else must be in decline on the other side of the relevancy to find equilibrium between the problem and the solution.

The cosmos brought in space on the one side and matter on the other side and between matter and the factor of space growing must be some sort of problem solving. If we wish to find the answers to the cosmic mysteries it should be the most obvious starting point because there is one side of the relevancy known to man and then looking on the opposite side of the relevancy must be the solution. Where one thing is growing something else must then be in declining and in that comes the answer of the relevancy that I share with the introduction of my theory on matter holding space in time.

Most prominent in all relevancies there are must be the atom, the one little container giving matter character and different uses in the universe and by adding or removing one small part it changes in character as Doctor Jackal and Mister Hyde never could. Every one knows what is in the container but what is the container in? If someone ever gave that thought the light of day I have missed it.

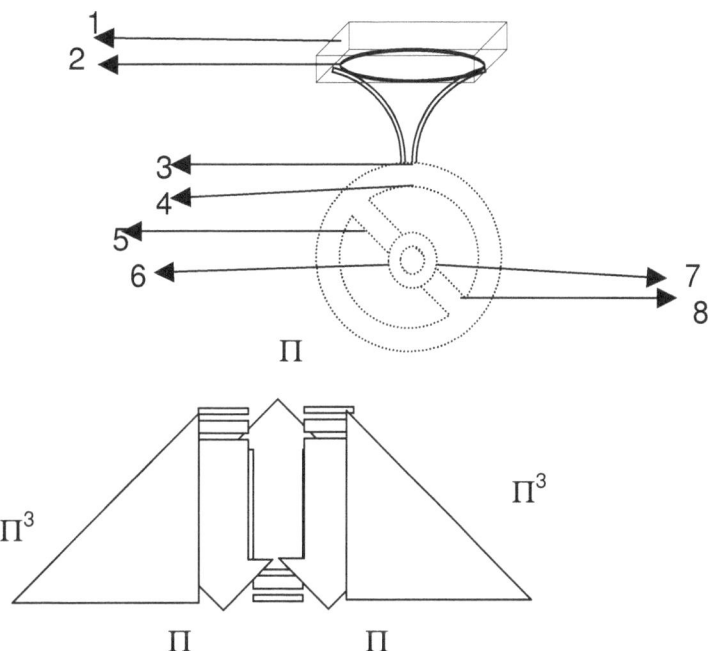

Diverting from singularity as extending Π time claims space from singularity comes about the double proton in space $\Pi^2 + \Pi^2$

Referring to relevancies means one may exclude only nothing from attachment and as such that put whatever there are to consider in a relevancy to whatever there are to consider. The universe is one giant spinning machine holding everything in tune and aligned with all ells. Everything become relevant as all in the cosmos divert from singularity by the line of singularly applying sides to singularity forming the triangle in singularity bordering the half circle to form a position where Π will become r^2 and lead on as a value of C. In singularity the value of space held by lines diverting from singularity forms the space value of $\Pi\Pi\Pi$ which is so close to eternity the **time value applying exceed the dome compliment of Π^2 matched only by the half of the square being the triangle at Π^3**

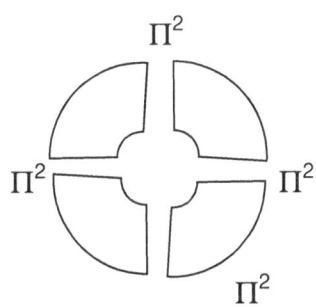 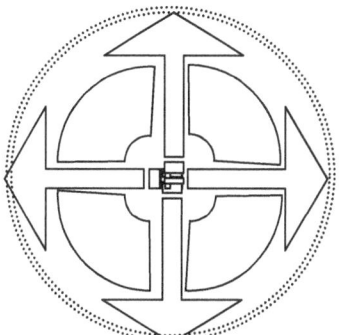

From the square of Π^2 the dome as a half circle by four places the line implicating the line by four to place the triangle relation by double half a square

In this it is clear why the Titius Bode ([10 + 10 + 1 + .991] / 7) and the Lagrangian 5 \\ 7 systems part their ways when applying the different processes they hold. With all the differentiating, the observer must also consider the dual massage that light uses in travelling through the vastness of universal space. The thought of nothing is just what it is, a thought of nothing and although it is in the human mind common nature to present nothing as a value in the recalling of something, nothing is a presentation of the figment in the human mind. There can be no number such as nothing and that was (possibly) Newton's biggest error. Nothing represent non-existing and that is just what nothing is, it is non-existing.

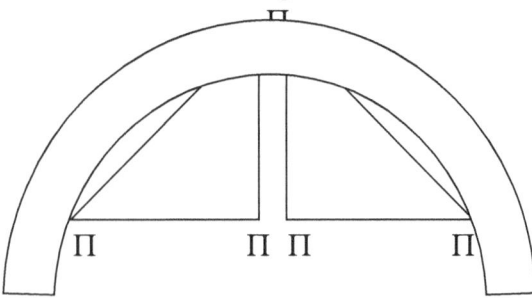

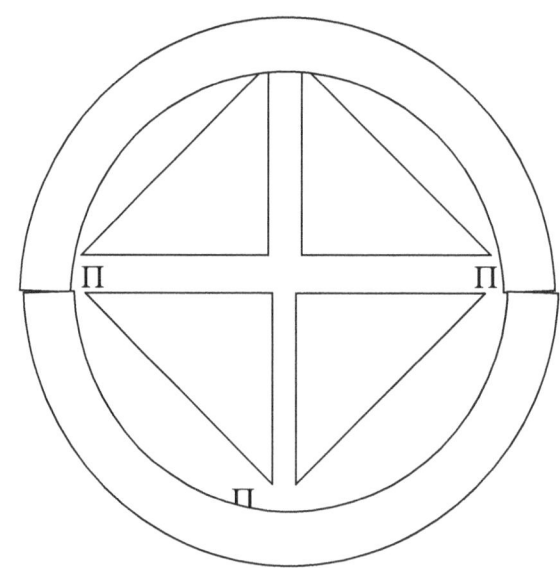

The total value of the four in the circle bringing about the space of the triangle reflected onto the time the circle holds committing the straight line to a value becomes the fourth dimension of space-time.

In order to prove my point I wish to ask the reader to define the shortest line there can theoretically be. If he should answer anything but that the shortest line will be at a point where the beginning and is the very same spot he will be wrong. The shortest line that can ever be anywhere must have a start and finish holding the exact same spot. The line will be humanly impossible to create but we humans are capable of very little.

Stars can and stars do overheat, sometimes and the polar regions where the Titius Bode matter to matter applies holding the square of space (10) in a double relation to the square of time (7 + 7)

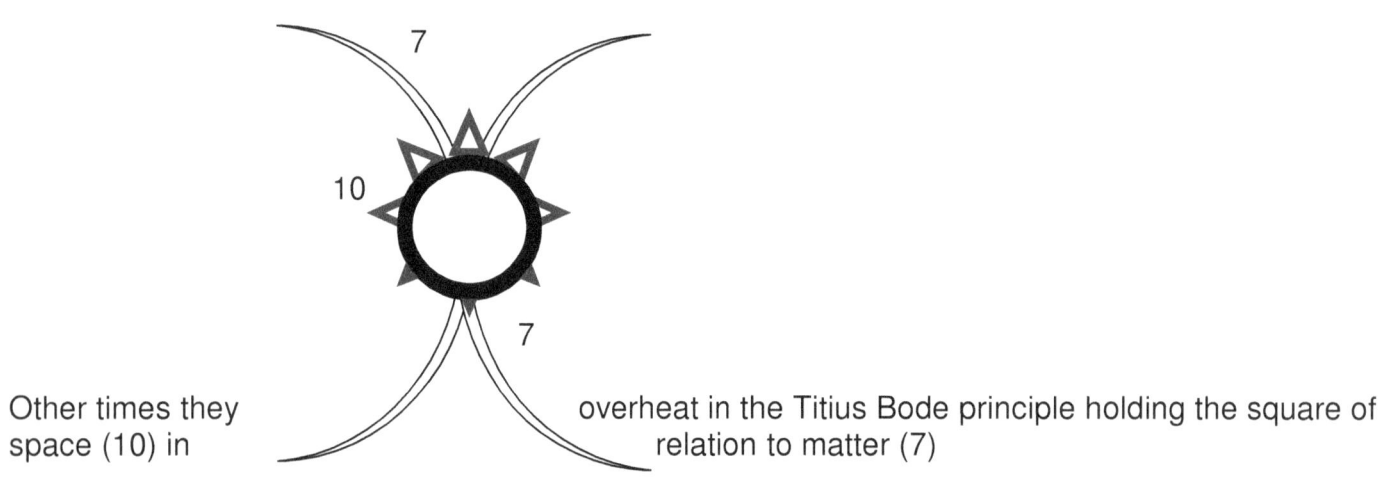

Other times they   overheat in the Titius Bode principle holding the square of
space (10) in   relation to matter (7)

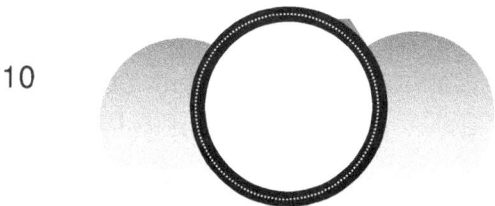

7

10

This comes about through the overheating of singularity (7+7)/10 (top) or layer overheating 10 / 7 (bottom)

It is no longer an issue that stars overheat but the issue shifted to the question of why stars overheat. Applying the Titius Bode laws (shown above), the Roche and the Lagrangian principles correctly, I can prove that :

1) **There is no gravity and therefore GRAVITY DOES NOT EXIST.**

2) **With no gravity it stands to reason that I also maintain that both NEWTON AND EINSTEIN ARE wrong about their views on cosmology.**

3) **With no gravity NEWTONIAN VIEWS ON THE WAY as how CREATION CAME ABOUT IS HALF CORRECT BUT HALF IS INCORRECT.**

4) **And shocking as it at first may seem but true enough is the fact that THE BIBLE IS CORRECT ABOUT CREATION.**

Could any one ever make claims as I do and at the same time claiming sanity as being one of my virtues. Well I shall try to explain how humans see the universe in total contradiction to how animals see the universe and to top the lot how atheist and animals see the universe alike.

Animals use eyes but we humans have more too see with when using what we have by using our minds to see with. We should see the universe with the light of understanding shining in our minds. When one look at the night sky one see darkness with little specks on light. Why would anybody see darkness because darkness has no light. Yet we see the darkness. The darkness should be invisible if we are seeing light because the one contradict the other. If the night sky was black then black is what we should see but then again black is the absence of colour and colour is the visibility of light.

We see the darkness because it has light it is withholding from us and while withholding the light from us we see the withholding as darkness and the darkness we do not see because we are not suppose to see darkness. That makes the darkness we see not darkness to be but it is in fact light we can see as darkness. On the other hand there is the brightly lit dots we can see because the light shining as dots are darkness as the darkness are stars giving us the light they are not withholding. As they are not withholding the light but pouring it into the vast container of light the stars then become the darkness we cannot see because they give us they're light and by giving us they're light they then have no light to have. That means by giving us they're light they withhold they're darkness from us and that makes the stars filled with darkness. That means what we see as light is light and what we see as darkness is also light instead of what we ought to see as darkness because that we cannot not see is the darkness.

Therefore when God gave the command "Let there be light" it was the command "Let the universe begin" because the universe we see is light we do not see and the stars we see is the darkness we do not see. When one is looking at the darkness as an animal you will be seeing the darkness because the mind you use is that of the animal. Then, yes you may be an atheist because all animals are atheists. I have never herd of one bleeding heart or philanthropist of whatever kind convert animals to any religion there is available. If you are an atheist and you see the night sky as darkness that would mean that is what you see as an animal; a darkness that if you had the sense of a man you ought to know that it is impossible to see darkness therefore it must be light. On the other side of the relevancy you also ought to know that by seeing the light the star is giving away the light and when giving away the light it has to hold all the darkness it is claiming for own use because only by claiming back the darkness can it give the universe the light as it does otherwise it would give its darkness by withholding the light. The darkness is singularity uncommitted to specifics, spinning at the speed of light never pointing in one direction long enough to shine as light but shining long enough not to be darkness. But being light we can see it but because it is in random spin, spinning at the same speed we use electrons to convey massages when translating information we cannot see it. The light then becomes darkness because it is extracting all the light through the one singularity line uncommitted not energising it because of the absence of a replacing source converting new light. In contrast to that is the light we see because of reasons forthcoming from not being able to see darkness we can see it as energised uncommitted singularity with the aid of a sustaining in singularity from a committed form replenishing the uncommitted singularity to maintain direction.

I do not see how one can be an atheist and put claim to being a human while observing what there is out there in the way animals observe by only and purely relying on the eyes without incorporating the mind. I cannot see how any human claiming to be human cannot see past the barriers restraining the animals from being human. With minds it is so clear what the Word of God says, but to be human and not see what humans should see is a dangerous reflection on the mind you use. God did not say "Let light be visible" or "See the light", He said "Let there be light" and that is what there is. If humans then see darkness where they know that one cannot see darkness the darkness are within their minds and therefore they are atheists not withstanding that they may or may not claim faith as part of their thinking. If you cannot read the Bible through human vision and as a consequence not understand the Bible don't blame the Bible for your inabilities but blame yourself and your inabilities. It is not the Bible you cannot read it is you that cannot read the Bible. Place the relevancy where it and as it belongs.

We are human therefore we have light in our minds and ought to make use of that! This very afternoon as was writing this part I took a break and lo and behold, one of my sons came to me with a problem of a religious nature. I shall not go into detail about his problem but I asked him to define religion and what life is. To strike some sense between his problem and the size he sees it in I asked him to tell me in his view about the contents of the Bible according to the Bible and the dominie (Afrikaans for preacher man) how would they define life because some parts in discussion about his problem was the discussion involving tackling the issue and thereby the issue turned to how far can you go in solving matters and leave the rest to preying and doing prayer. I am of the opinion and will die by that opinion that prayer only serves a purpose in thanks when you yourself completed the task without preying for some force to help you complete the task at hand. Life is the manipulation of your surrounding and that means you do things yourself if you want things done and you do not prey for things to be done on your behalf by God. That is the definition of life. It is the manipulation of space-time and involves neither magic nor divine prayer but you go about changing your surrounding to match your needs. What all preaching never advocate is that we are in the seventh day of creation where that specifically states that God went to rest and from that I draw the conclusion man can and man must do everything by himself because God clearly says He has gone to rest. We do things on merit by ourselves or not at all. That is the energy we think of as life. The fact that we have the ability to self-sustaining and not being fixed to the universal position space-time landed us in gives us life. With life in hand you manipulate what ever you can as you replace positions to suit the required changing of objects where changes are needed. Then your acquired needs changed them to be to your taste and there is no other way out. Life is about changing your surrounding for the better of yourself or others and to improve all around you. The ability to manipulate space-time is the energy I have and Old Neighbour lost. Still it is energy. It is neither food nor electricity but it is a more advanced form of energy than the energy mentioned. Another part of life is tacking the responsibility for change your manipulation may bring about and the effect such change may have for other beings sharing space-time with you. Never confuse the needs of others with needs of your own and project such needs about yourself as beneficial to others without consulting others. This is very typical human behaviour.

With the Newtonian confusion raging man has mixed matters bringing about a highly unsatisfactory climate where we try to pin cosmic value and pre-conditions on life and place very stringent condition suitable only for life onto the cosmos. That leaves science in disarray and confusion. Heat sustaining life as pre-condition Xepted science projects to stars and where stars fade we allow them to die as if blessed with life's changing and renewing. Stars certainly do not have emotions and when they erupt it is not in anger. The chemicals stars need to maintain singularity is very poisonous to man and the matter making life sustainable will have no chance of surviving even as a flash in the star. We think of a star being hot in the manner we translate life's pre-condition to what is hot. It is to the letter the same way that we take outer space as being unsustainably cold where it is quite the opposite applying.

While looking at the earth we think of the cosmos. We reflect what we conceive as conditions to match life being normal to the cosmos. Planets have to be plenty full because even we have one in hand and eight others in the back yard as spare should we make this one we have untenable to life. And should we run out of planets to ruin there then should be others nearby carrying life on one in nine, as is the case with us. We try to find life everywhere because life has such abundance on earth in everything we see. We even reflect our vision of time to mach time in the cosmos giving the start of creation an earth bound time range never thinking that the universe is growing and not dying. In the same manner we think of the universe as a living organism while the universe constitute every aspect we relate to death. In fact, the universe is the ultimate death. In the universe everything will only be once and never again whereas with life there was as much as there will be and even more will come than what was. That is the last thing one will find in the cosmos. If time ran out for whatever time will not replace or bring back what ever. Even the way we portray the earth's surface we wish to reflect to space using the same methods we use on earth. One mile will be one mile wherever you wish to take the mile. After all one mile is one thousand seven hundred and sixty yards (if my memory serves me correctly because this is still part of my culture when I was at school and South Africa used the British yard stick). Not once comes the thought that man cannot step one yard in space. Still one yard will be one yard wherever the yard may follow man. Man has acquired the inability to divorce life and the cosmos for some reason we can presume as cultural. Unfortunately we go in accordance to what we see and that is more cultural than culture it self. We see a shining light and presume it is a star in the same manner as we see a large dark antelope with horns on it head exactly in the same way as that of a buffalo and presume that what we see is a buffalo. In the case of the buffalo the past thought us such observations are correct and hence we grew accustomed to the culture of believing our eyes.

Never do science take charge of thought and divide flesh from energy in the manner I have done during this the writing of this article. Outside the view we have we can locate a something that is there but needs some vision in extelegence to locate. It is a small part of life that has an attachment to the physical but an overwhelming comes attach to something indescribable to define. In other books of mine I try my best to prove that our view of trailing outside the sphere of the sun is a myth and even travelling to another planet is not the same "as going abroad". There are so many dimensional barriers attached to what we can see without our locating or even knowing of such existing barriers because they remain unobservable barriers. There are so much more than what ever may meet the eye. In part 7 of the Theses I touch on the subject about the age of the earth and how short sighted (once again) the Newtonian view are on this matter. The earth is in truth not 4.5×10^9 years old but the core was part of the cosmos during the birth, the very first moment of the cosmic birth. Many processes came to change and shape the earth to what we enjoy today, but the inner-core-value came from the first parting of the singularity Alfa.

How life started as such I do not wish to speculate on, but logic tells that what ever was at day one of singularity Alfa, nothing since was added or removed and that puts the carbon carrying life at the very start as well. It would be reasonable to suspect that all cosmic structures holds the carbon but not all structures can present a satisfying environment to sustain and protect the singularity of the carbon in order to bring it to a point of holding divinity secured.

One opinion that I strongly hold is that Chandrasekhar is as misinformed about his carbon-a-plenty theory as he was about his crushing stars in weight. Carbon cannot come from the cosmos and go through the Π limit unscathed to infest the earth. That is as Newtonian as all other bullshit can be. Life in carbon was a part of the earth as it was part of the sun, but it had its being burnt to blisters and could on that account not develops on the sun.

What ever the earth went through was also a survival test for species on earth. What ever the sun threw at the earth the form of life that was dominant then, had to make do or die. The fact that life made do is testimony to life's survival skills. Life will last, no matter what man may throw at it. It is man that places man in jeopardy. Man is the prize of life's achievement that I do believe. Man is the accomplishment all other species carried the burden of. Life is built into man and all qualities of life manifested in man.

That makes man the youngest and the least protected. That makes man the weakest link in surviving. I have my sincere doubt about modern civilised man's ability to survive even the onslaught of a brake down in civilisation. One harsh winter and not one in a thousand would be able to see the next summer rains bring relief. Picture a big city without electricity for one month and think who would survive even such a limited test. One hundred years ago such a remark would have made me as silly looking as the claim I make about gravity. But man has gone down the tube, at the end of the ladder although to man's thinking he is at the top of where he ever was before.

We are launching a chemical war at all pests we do not seek. We kill and destroy them without thought. Bacteria, fungi and, viruses have been at tests far grater than man can produce and survived to tell the story to the next generation. It is written in their life code for the next generation to read and fight. When a species are at it greatest danger of not surviving an onslaught on its very existence a factor much dormant in normal conditions kick in. That factor rewrites the coded massage and the following generation find armoured protection. Man is weakening with all the chemical aid we see as medicine protecting us while we put the most dangerous forms of life on a survival course we cannot afford the luxury of. The day will come when there is no stopping these killing-surviving machines and we, man will stand defenceless while they go about killing and maiming on sight. Every little headache is a call for aspirin. Every cough is a call for anti biotic. One day we will find the disease and ourselves defenceless well and truly developed. Man will die and the count will become more than man can destroy human bodies. That will leave corpses for more viruses to grow and plan more attacks. Payday has to arrive we must see that coming and not be as arrogant in our self believe.

The fashion of the century is to place all, as equal and life holding space in a dog is equal to life holding space in an ant. That can never be for the single reason that all life in the body of a human cannot be equal. Any person can go without a limb notwithstanding the sacrifice they endure in whatever function. Losing an arm or a leg does not risk life at all on the condition that it is removed before it may infect disease to other organs in the body. Losing a liver is serious but machines may provide such an organ function replacement and life goes on, fairly difficult but without eminent danger of death to the rest of the body. The same argument can be said about the heart longs kidneys and such. The function organs play in maintaining the body is crucial but not vital. Losing such an organ does not mean death by necessity and can become even to some of minor significance. When losing the head or part of the brain things turn to a lot more serious nature.

I have witnessed friends of mine that were motorbike maniacs like I am, falling off their bikes and receiving head injury. After the recovery those persons changed in a manner where they became alien to themselves. They became another person no one new before and none can recognise. Such an injury is very serious and lethal, more to the persons that love him than lethal to himself. The persons that love him has lost a love one and gained a stranger they do not care for. Even in one body all life does not stand equal let alone from specie to specie. Losing my arm is not the same as losing my life because I can still live (more unpleasant but that is not the argument) with such a loss. The conclusion of logic is that the arm is not the "me" I lose, as did Old Neighbour when he went missing leaving his remains behind. Some of my body is life in issue for use to be discarded when no longer required for service but other part is much more closely connected to me as life.

This brought about the atheists campaign that life comes as part of a wholesale package wrapped in a carbon container and all philosophy centred around this argument went missing when some connection was proved between electricity and motorized motion of body muscle and fibre. This was the dawn of electricity and the wish-wash that went around with miraculous curing by only sending impulses of electric devises that could cure all and almost bring death back to life. Some devises remained proving through time their worth but in general it was a lot of quack and most disappeared where they came from.

Then came the theory that life was only electricity flowing from the brain to where ever body motion required the flow and all other philosophy went silent. It is not hard to imagine why because physics place electricity as a force with the same presumption (though they will die before admitting it) that a force has a control in similar fashion to a ghost or some unknown free spirit running around to every one's amazement. That mentality sticks like glue and much of that influenced scientific arguments to be in apathy to the philosophical and since 1945 when the physics got hold of the German nuclear bomb and let it loose on Japan it is mathematics ruling logic to the point of madness. No one since then had any inclination to touch this aspect again since all were satisfied that everything was flawless. Flawless indeed but at the heart of mathematics and in the very start of physics lured a flaw that became more apparent every year and the flaw eluded every one to date. It even diminishes all sensible argumentative possibilities to a stand still.

Losing a limb might not kill and it might not change any personally but it is loss to life. If some one acts promptly and in time doctors commonly have the ability to connect the lost limb and with some minor complication the limb may even restore to normal application. Would such prompt action work in the case of Old Neighbour being officially dead for say twenty minutes. The answer may be yes and more likely no because it depends on the brain damage that occurred in time laps where the brain fibre were starved of blood and more important oxygen. As was the case with some of my biker friends brain damage can and more likely will result in a mild to drastic personality change and in some cases dangerous insight attacks may occur.

Changes of such a nature are very serious and symptomatic of injury to the brain. In the brain damaged victim likes and dislikes behaviour pattern and mood swings will change the personality of the individual. The changes may result from a blocking of the flow of blood and it may result from a nerve area that lost function culpabilities but life still remains present. From physics point of view I am of the opinion that it is a natural phenomenon gone very bad and such changes in personality takes place with or without injury. The Romans believed that when a person breaks a mirror he is doomed for seven years because the broken mirror damaged his sole. This we modern people know is just another folk law tale but with some angle of truth. Of course the mirror part is the untruth but there is quite some truth behind the personality changes with an interval fluctuation of seven years. I would not go as far as putting a stop watch to the date in seven years but in a more or less manner we all show some changes in personality and a man of fifty will not find the company of a few teenagers to be friendship bonding and neither will the teenagers like a fifty year cold going gallivanting with girls very pleasant. Of course once again there are many exceptions to the rule and as with all else in the cosmos there are relevancies changing circumstances that may occur. What is without doubt is that the link between life playing a part and the fibre connection playing a part and it will be as silly to claim the carbon has no influence on the life energy as it would be to deny that there is another energy present above and beyond the fibre. With this I wish to introduce my Theory on the Seven Dimensions and I put it to you as I originally started with without changing some of it to fit my present day views.

1.4   THE SEVEN HEAVENS

Although from the name one may have the idea the article is exclusively attached to the spiritual as much as it is about religion and has nothing to do with physics. When a friend of mine saw my article in one of my scribbling pads (this was years ago before any idea of writing a book ever entered my head) he was astonished by my claim that it was pure and unadulterated physics. This was my advance from nowhere into physics. Justifiably you may say as my friend did so many years ago that the seven heavens have no bearing on physics but by saying that the biggest mistake comes into the open. I admit whole-heartedly I did not realise the importance it had back when I wrote down the loose ideas but in retrospect that was my initiating although not my first ideas. Every aspect of every aspect connects in some way leaving only nothing unconnected. It should be somewhat obvious by now that I see "nothing" having no claim in any form of nothing as part of mathematics or physics and to my view that is the main difference between arithmetic and mathematics. In arithmetic there are an allowance for a number or a marker such as zero or nil whereas in mathematics no such number can be found because no such pointer can claim any position from the origin.

Even when I wrote the thoughts down that many years ago I did not yet dispute zero as a number, but I have to admit I had some difficulty with the value of nothing. For instance what was more nothing and what was less nothing when there was two of nothing facing each other. In all of mathematics there has to be growth as much as there has to be decline from wherever any marker may be. In the article I show that the line the half circle and the triangle have on common factor in as much as all being 180^0

A straight line cannot start at zero and still be a straight line because zero extending to wherever brings about a full zero. A straight line starts at the point where the pen point meets paper. That point may be any distance from infinity to a measurable dot, but it cannot be zero.

$$180^0 \times 2 = 360^0$$

Any straight line is also half a square be cause the line forming the square cannot start at zero for the reasons I just mentioned. That is singularity pointing an eternal direction from a point of infinity and that is the basis of the cosmos as much as that is the basis of mathematics. To escape from nothing one has to become something and by doing that one could not have been in nothing in the first place. If one holds a point in nothing one cannot become something because of the nothing value.

To back this argument that no line can ever start at zero is to ask the simple question: what will the length of the shortest possible line be. It must be a line where the starting point is so close to the ending point the distance parting the two is incalculable yet there is the line therefore the end and the start is apart still sharing the same spot.

The difference in the circle and the square is the direction the indicator follows and a square cannot spin, as a circle cannot be motionless The factor of Π indicate eternal motion and NOT zero motion. There is a massive difference in that concept. If no line can have a zero point to start with where will the circle get the zero to indicate motion! This principle is the most basic mathematic rule The method applied when calculating a wave is by finding an average in the triangle continuing from the straight line to the pitch of the wave and then the decline will form a duplicate presenting the other side.

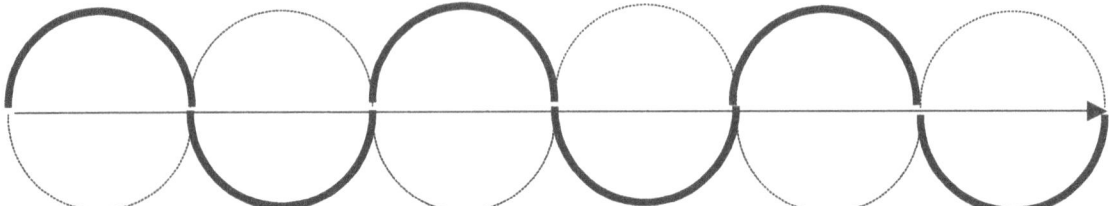

When the end of the rotation arrives the end rotation also announce the beginning of another rotation and not nullifying of the previous rotation because the rotation will have a line showing the effort it made and as it forms a wave, the wave will be there forever. The pitch may decline to a straight line, but the line remains. The wave confirms rotating directions followed by the circle as it spins. By stating that a wheel has a relevancy of zero by completion of a rotation such a claim denies the wave its rite of existing. The wave going flat, as it becomes a straight line also has an indication to singularity.

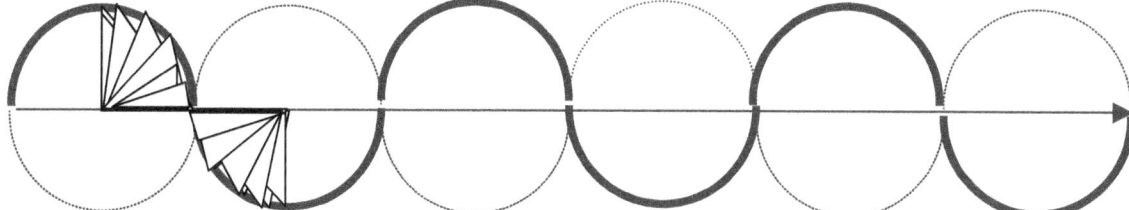

Being a circle means the thing must be round and spinning. In that case, let us take an example well known to all, the spinning top. The top spins on the thinnest of points, and still maintains a balance. By being a calculating value to match the work done in the rotating half circle the triangle depicts the flow of the straight line.

The straight-line holds a duplicate value of 180^0 to the half circle as well as the triangle all being part of singularity as much as being positions from singularity. That alone has to confirm the connection existing in the dimensional aspect. The dynamics behind the two principles is much, much more complicated than what the illustrations as shown above would suggest. However by using such basic of illustrations the simplicity might be tending somewhat to come across as misleadingly simple, but taken down to the core of factors behind the principles that forms the most basic of the principles, the illustrations prove rather effective in explaining the crude idea. However, please do not be fooled by such simplicity, in the very detail analysis it is as complex as can come. From the star holding a dominant point or most valued point in singularity it affirm all five other structure each holding singularity individually.

The universe link in so many ways we will not begin to realise the manner within the next thousand years. Electricity is one part of the link, but there are other links we may never come to know about because there is always another part of the cosmos above and below our perception and abilities that will elude us.

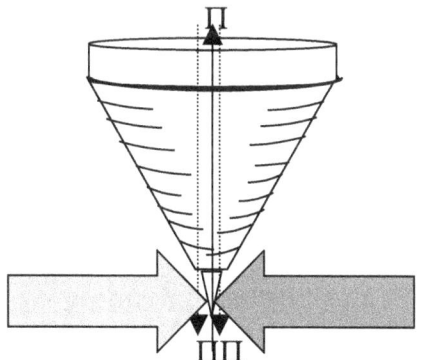

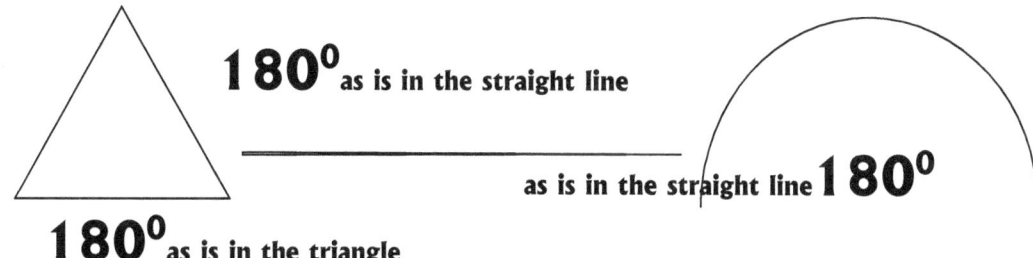

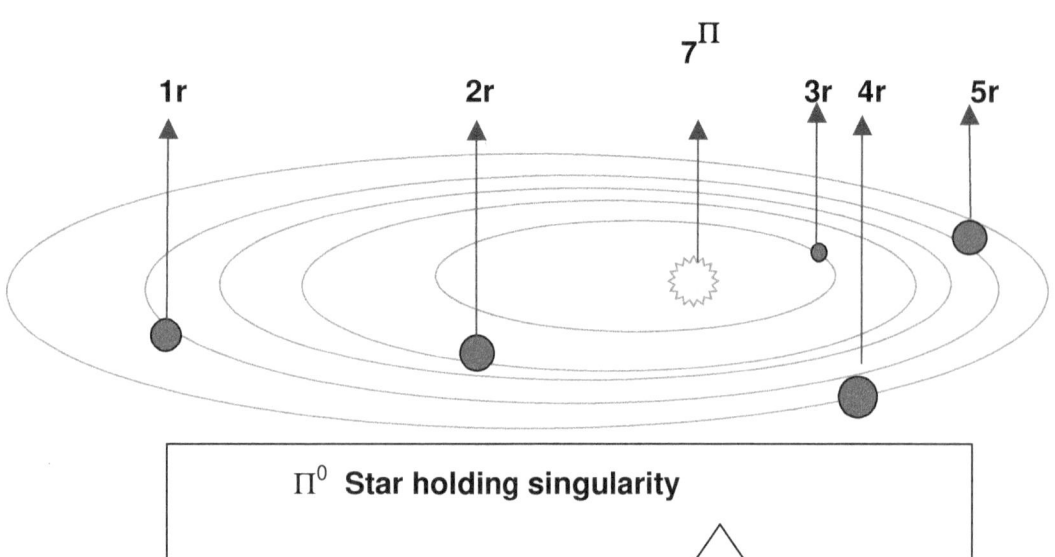

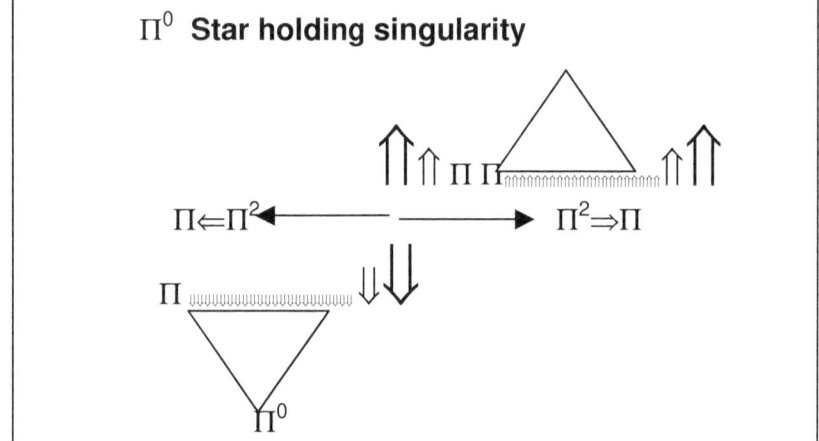

The network of individual singularity not only provide spinning through governing singularity in the sphere but also provide spinning in the geodesic through out the cosmos linking all matter to matter in a network no one will ever come to understand in full. In the sphere the foursquare triangle holds space in time maintaining singularity of different assortments. In view of the matter-to-matter Roche factor where the factor consists forming relation between particles occupying densified space-time of where ($\Pi / 2 \times \Pi / 2$) relating to the foursquare triangle the value of gravity Π^2 comes in position as $\Pi^2 / 4 \times 4 = \Pi^2$.

A STRAIGHT LINE , TRIANGLE AND HALF A CIRCLE WILL ALWAYS HAVE EQUALITY IN DIMENSIONAL CAPACITY PROVIDING EQUILBRIUM BEING 180^0 BECAUSE EACH ONE SHARES A COMMON DINOMINATOR IN SINGULARITY. As the straight line averts a zero it holds another straight line in place to set about such an averting where the two lines will always carry a relevancy in elation to progress (the triangle) and a common denominator in the start from singularity.

With the normal extending of singularity it will always form the triangle in a half circle whereby Π relates to the cube by 5 points to either side of the line singularity forms. Thus there are 10 standing related to seven and visa versa.

As singularity holds the straight line the triangle and the half circle as a base to form giving all and everything next to connected to and adjoining any form being of a straight line half a circle or a triangle forms space time. From singularity in the straight line (180^0) the half circle(180^0) and the and the (180^0) triangle matter form space in holding, claiming space by controlling space to influence space, but as maintaining singularity insist on space in spinning to the time singularity dictates time sets from such spin motion and by diverting from singularity time forms the law of Pythagoras in the square of space –time.

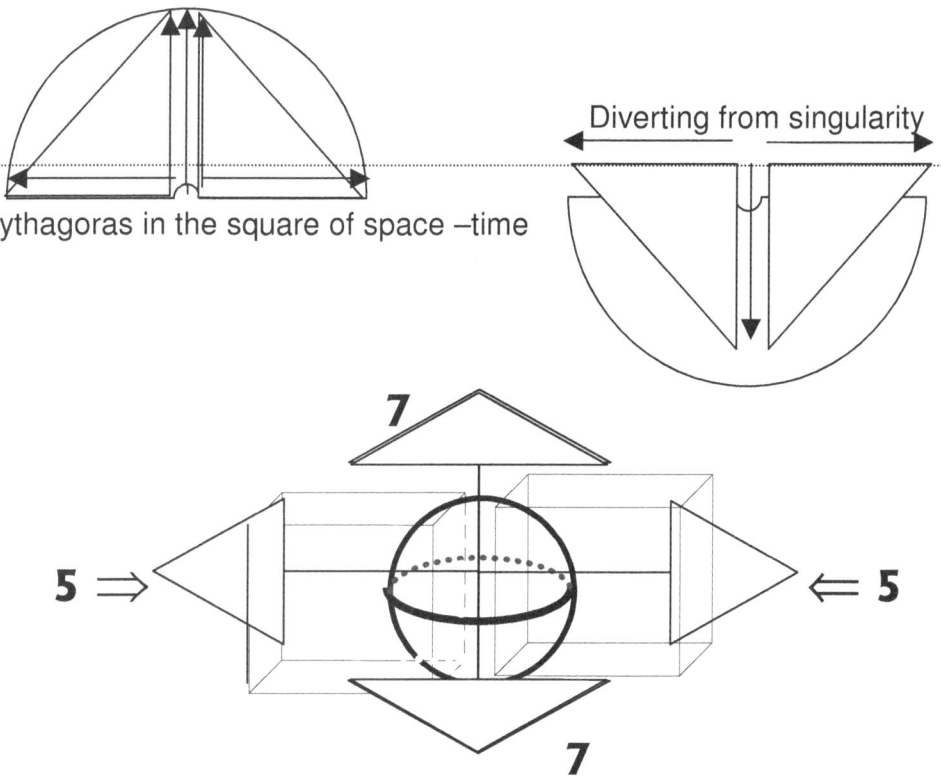

The normal flow will allow singularity extending to 10Π but when singularity blocks another sphere in singularity the two will form a joint value and by joining the larger will dominate the space as well as the time of the lesser taking control of the surface and the atmosphere. Through this the Roche lobe comes about with all its other dynamics I describe farther on in the theses.

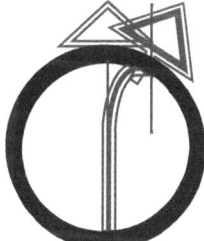

The result of the seven markers that matter diverts from singularity is present in the 7^0 inclinations the earth holds as a sphere as all spheres have. Matter is always moving seven points way from singularity as it progress in space through time.

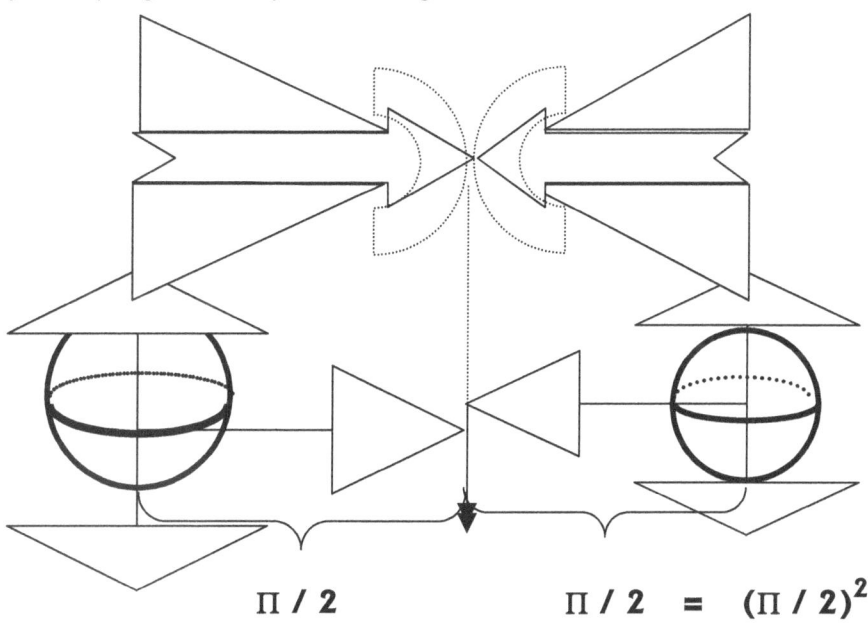

$$\Pi / 2 \qquad \Pi / 2 = (\Pi / 2)^2$$

SINGULARITY MEETS AND COMPLIMENTS EACH OTHER.

The diameter of the cosmic structure holds the value of r and singularity holds the dimensional value of Π meaning that the radius or diameter (r) extends to become the diameter multiplying the value of singularity. But since r already consists of the square of space holding a definite positional relation with the value of singularity being Π the diameter comes into effect.

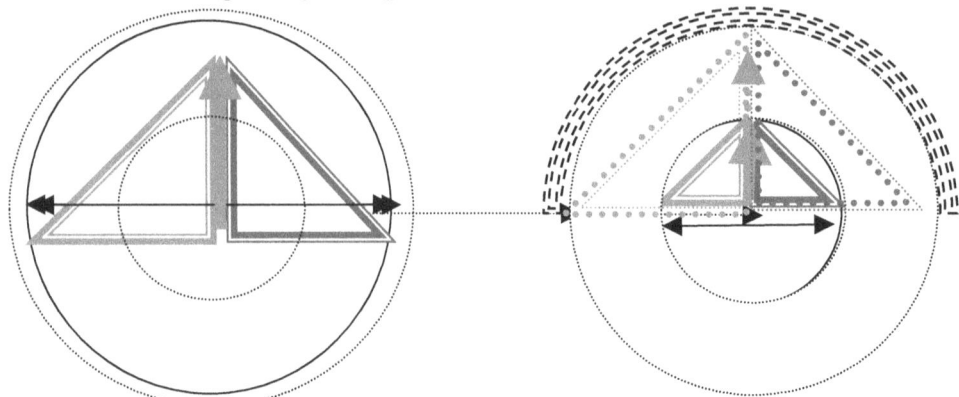

At this point the equality of the straight-line dimension to the triangle and the half circle holds prominence as a straight line, a half circle and a triangle is dimensionally equal. The common denominator will bolster all factors to an equivalent ratio,

When singularity by the straight line increases the singularity by the triangle will also bolster giving equal potency in singularity by the half circle. As the singularity of the major component revives the lesser singularity to equality, the triangle in singularity will match the performance and so would the half circle respond in precise ratio setting equilibrium in order.

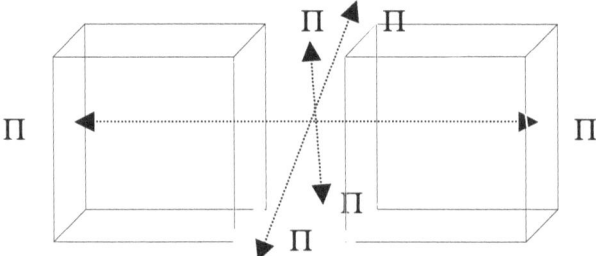

From this the lesser partner will fill by the extent of the larger partner and as soon as equilibrium sets in the growth will duplex in both accounts, normally to the fatality f the lesser partner as the lesser partner will not be capitulating under the straight of the duo. The Titius Bode configuration in accordance to orbiting formation holds a slightly different explanation to the explanation that applies to cosmic structure surrounded by space. It is moreover the individual singularity in maintaining the major singularity, which sustains the governing singularity providing equilibrium in space-time.

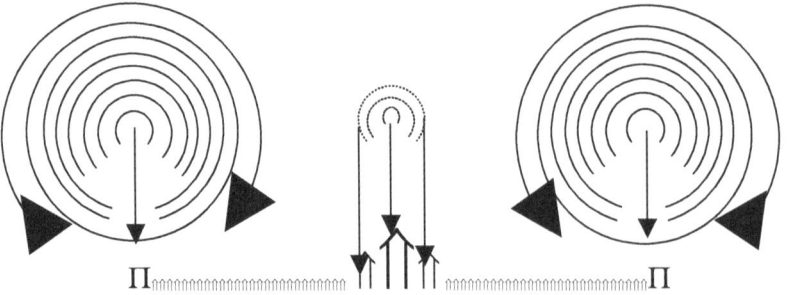

Not only does atomic individual singularity maintain self preservation, but in doing that it also sustain a governing singularity holding structural composition and form within a cluster of matter for example a star. Between stars there are a mutual or bonding singularity between atoms and stars.

The sectors provide individual singularity a means in sustaining governing singularity by which provision comes through maintaining governing singularity the required spin in maintaining cooling.

If this process did not apply, there would be no connecting individual singularity to major singularity. The sectors provide individual singularity a means in sustaining governing singularity by which provision comes through maintaining governing singularity the required spin in maintaining cooling. If this process did not apply, there would be no connecting individual singularity to major singularity In this maintaining of cross referencing of singularity providing spin to the governing singularity many factors of singularity all form a close knit network inseparable one unity but also strictly individual to a point of destructing.

Singularity has three part and five points with Π as matter being sixth and space (r) as light the seventh.

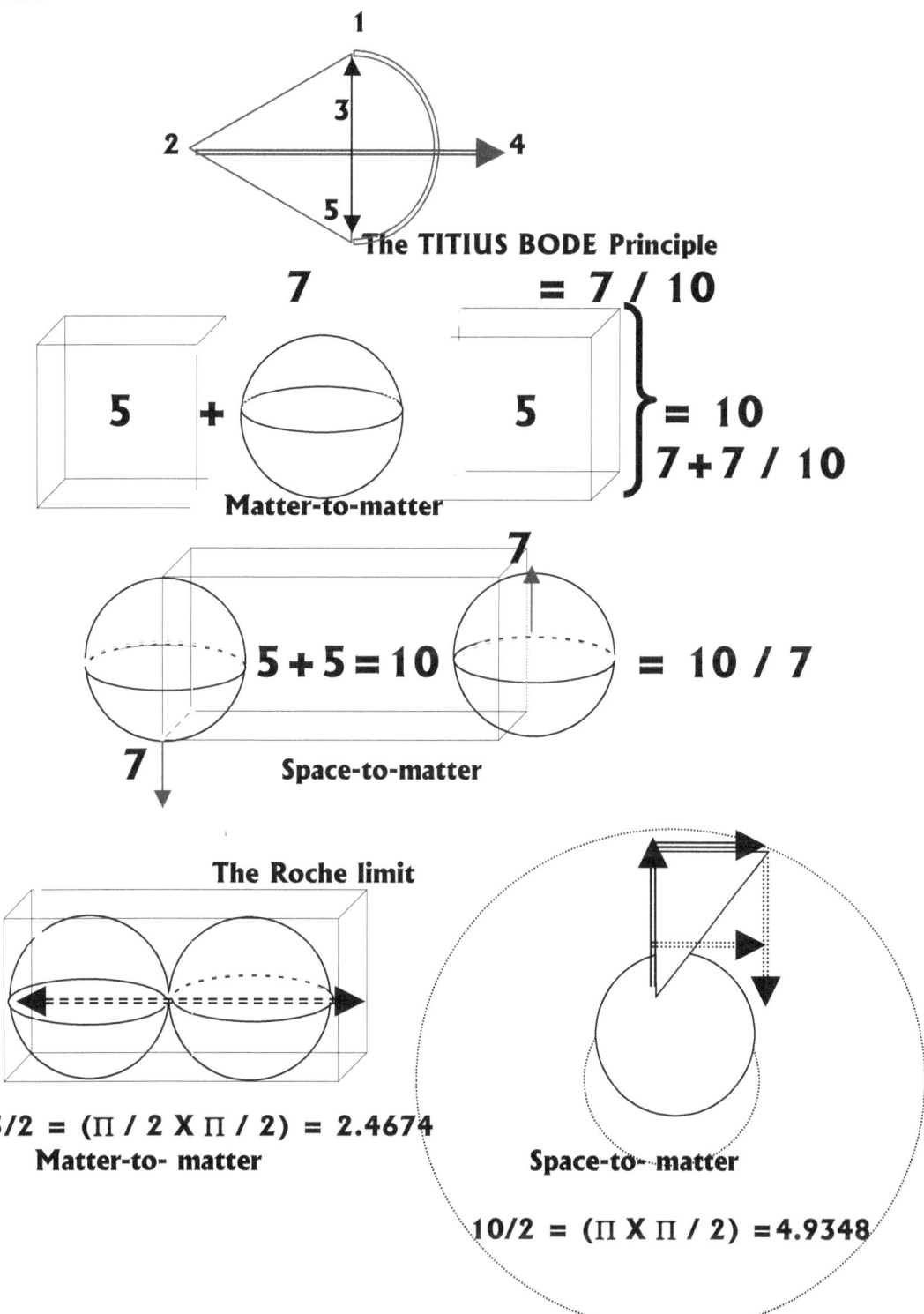

According to me, as a believer of the Holy Bible, there are seven heavens. The Holy Creator sits in the seventh heaven. His Word tells me that. That means if his Word says there are seven heaves that means there are seven heavens, that means there must be six undetected heavens. Before those atheists start shouting their lungs out, first go about answering the question put by me in a previous article. Answer the question where do life, being energy, go after it has left the body after death. Let us presume that the question is still without an answer. If the Creator lives in the seventh heaven, then He must have created the other six. By acknowledging the creation, I echo what millions of intellectual people believe. The so-called Christians, Judaism and Islamic faiths all accept the first five parts of the same religious Bible. Therefore, I feel free to speak on behalf of millions of people who consider themselves religious and these people come from all lifestyles and all over the world.

First let us consider the dimensions accepted by science and later argue those forms that are excluded. The so-called heavens have another name, which are dimensions. These dimensions could be regarded as planes or spheres. In looking for the dimensions that form the universe, one must look for sides, which has a combined value, but exists in total isolation from one another. Any object that can be visualized has to contain at least three sides with six obvious different spheres. These spheres do have single applications in the universe. Allow me to explain.

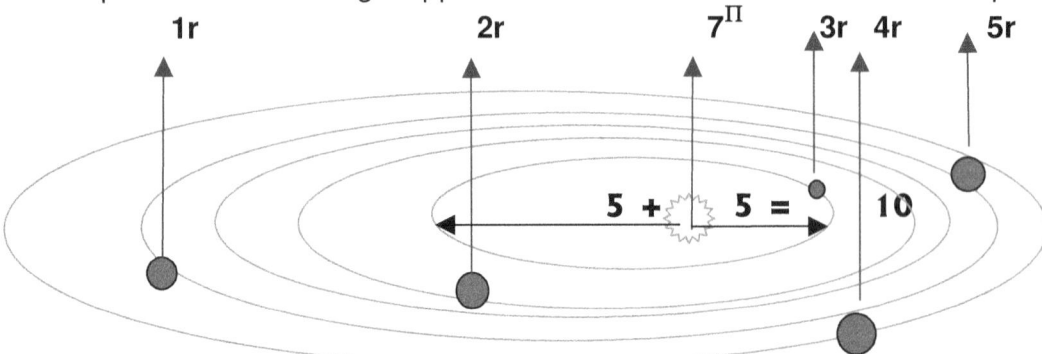

From the matter-to-matter relation in the Titius Bode configuration there are 7 / 10 + 7 / 10 = .7 + .7 = 1.4

From the space-to-matter relation in the Titius Bode configuration there is 10 / 7 = 1.42

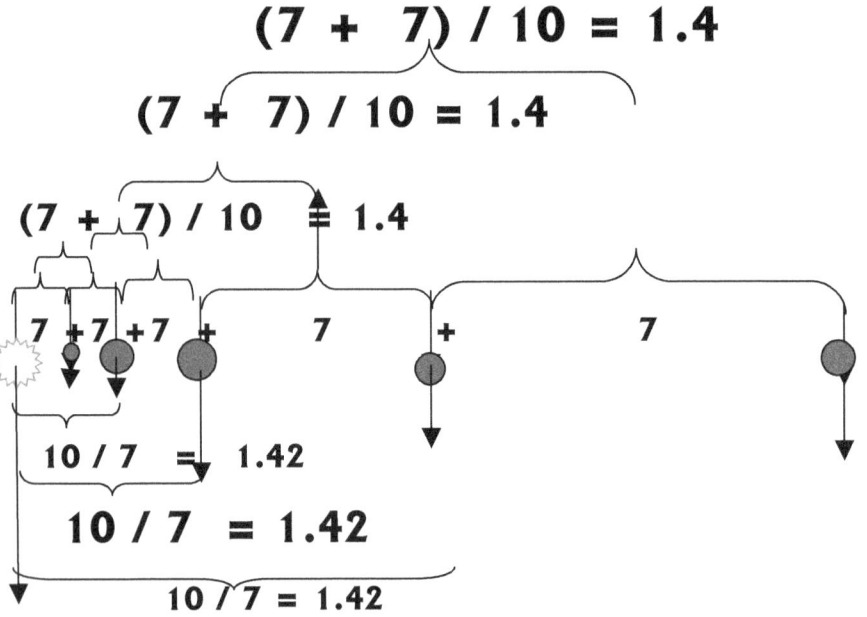

= .7 /Λ\ 1.42
= 1.4 /Λ\ 1.42 **Because the space-to-matter is in the square at 10 placing the matter-to-matter at a square of .7 + .7 = 1.4 the space-to-matter forces the matter-to-matter to double the distance by number as structures are place father from the mainΠ^0 maintaining singularity.**

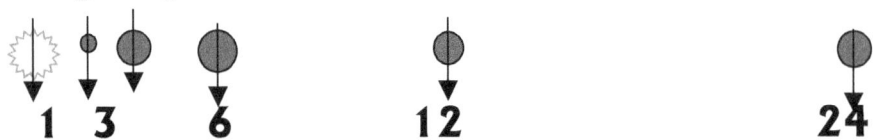

Reasons why this does not fully apply to the solar system I give in book # 7.

The space between the spheres divide in half, but because of the extending of Π and not applying r as ordinary mathematics will suggest where Π replaces r the singularity extending from $Π^0$ will be half of Π in the square of $Π = (Π/2)^2 = 2.4674$. In this lies the dynamics why planets have a positional (be it rather a dimensional) relation of 7 / 10 The second Roche limit is within the sphere as $(Π^2/2) = 4.9348$.

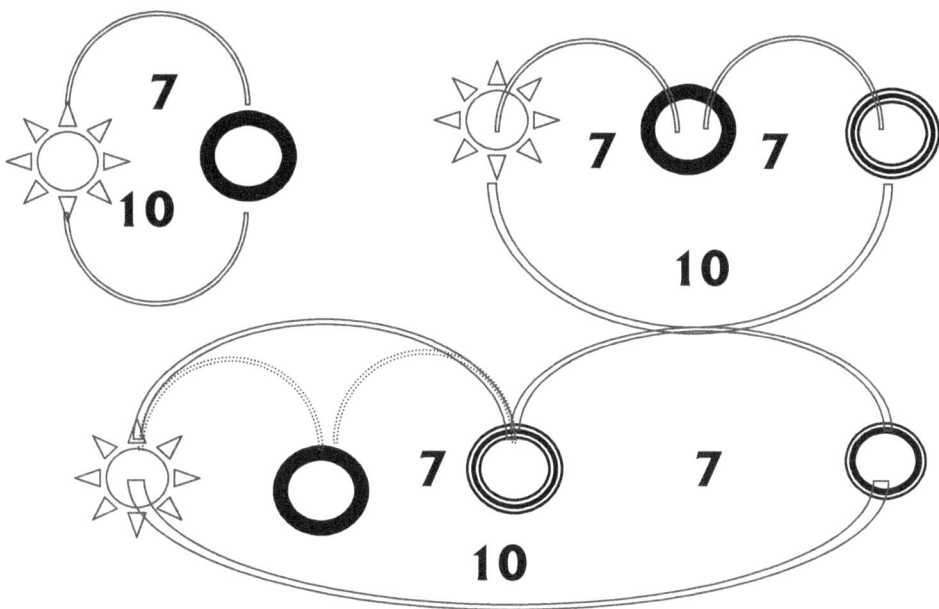

THE SINGLE OR FIRST DIMENSION (GRAVITY)

In the single dimension, one finds gravity or a pulling force that every cosmic body has.

If any person does not believe in gravity, then try to jump high of far. One would find that there is a force, which pulls your body back in the direction of the centre of the earth. This force is calculated to be 9,81 N m/s² and is almost precise the same value right over the world. This force can only be overcome if an object is hurled into space at a speed of 11,20 km/s² and an angle of 90° with the earth. That brings about that any object on earth is moving at a speed of 11,17 km/s² at any given moment.

As far as my knowledge goes, this force is perpendicular to the earth, with a distortion of 7% due to the inclination of the earth. A person in China will be pulled in an opposite direction to a person in America, because China and America are approximately in opposite directions to one another on the face of the globe. This means a person in America moves towards the earth at 11,17 km/s² which is directly opposite in direction to the 11,17 km/s² which the person in China is drawn to America. Taken these viewpoints into consideration, gravity is a single directional force moving only in one direction that depends on the body position to the earth. Where does this force stop? Nobody knows, because the crust of the earth is being drawn towards the centre of the earth. Even the see and the crust underneath it moves statically towards the centre of the earth at an even pace. If one looks at this force, one gets the impression that there is only one direction towards the centre of the earth without any given point where it will stop. On the other side there is no given any definite starting point. Without a star or an end, only a direction remains that envelops the whole idea.

I hope I was clear enough about the idea I tried to explain. The whole idea of gravity consists of only one single directional movement in one direction seen from all directions. There is no starting point, no point where it ends, only the definite direction that the body moves towards the centre of the earth relative to a point where all directions are measured. Another strong candidate of this dimension is magnetism. The polarity of the atom in the iron core is only towards one direction. But I shall explain a little further down the road, why magnetism do not comply with the single dimension force when I have brought some more facts and arguments in explaining what forces come into play.

THE SECOND DIMENSION : THE WAVE

In the context of the wave, for example water, the circle of the wave flows from one given point outwards. That implies that there is a definite direction. However, there is a second dimension in the wavelength. The moment that the breadth of the circle is determined, the wave has already altered it. That means there are only two values to be considered realistically without freezing time and that is the direction and the frequency. Sound and light are two such values. The moment the wave is frozen in time, which means it is standing still, it is no longer a wave, but an unnatural structure man made for his own benefit. It does not occur in nature because a wave can never stand still.

Light, sound and waves are two-dimensional values. The dimension consists of a point of origin and a frequency. There is no distance, because the wave beams out in all directions simultaneously. The Doppler effect in sound has an influence but that is because the point of origin keeps altering due to the movement of the source. The speed of sound has a definite ratio to the density of the medium it moves through and that applies to the time value of the space-time occupied by matter. In a later part, I shall explain this in detail.

Due to the changes in space time occupation of the matter the frequency alters and tarnish directly in ration to the space and the time the matter occupies. As the frequency of the space deteriorates in time, the wave becomes less in value but still remains. Therefore, all the sounds the dinosaurs made are still with us but the space-time occupation that the sound waves had, rendered it undetectable to us. Proof of my argument is found in the heat that still can be detected with the Big Bang event. Although the space-time value is only round about 3°K, the waves can still be measured, because these waves form part of space-time. The light wave is transmitted from its source in a sphere and will transfer itself through space-time until such time that a few particles hits a solid object. The wave consists of billions of light particles that move by wave density from the transmitter. The photon moves at a rate of ± 300 000 km/so. That means that the photon displaces space-time by negatively at the above-mentioned rate. This negative displacement of space-time by photons is called a light beam.

The light rays follow wave upon wave in endless motion through space in time. As soon as one wave of photons hit solid matter, only a very small number of the total wave is stopped. These meaningless stopping causes a shadow to form, but the large number of photons available would fill the gap formed by the loss of photons. That means that a shadow as large as the earth, will cast darkness for a very small distance / time span that means a tiny part in space-time.

The reason why the heaven does not light up is that the beam is scattered as it spreads out in a balloon formation and therefore the number of photons lessens in density as it becomes "duller". The more space-time it has to displace the thinner the layer of photons would become and the less intense the beam would seem. It is not so much the space (distance) that should be regarded but the time it spends in space has the ultimate segregation of the light beam. In space and time, the beam would become so minute that it would alternately become unobserved by the human eye.

All that we know, all the knowledge we have accumulated through time is based on the light that sends information to us. The size, the distance, the structure, the density and its position are all determined from the light that sends information to us. The light that reaches our planet determines all the size, the distance, the structure, the density and its position. All our facts are based on this small amount of evidence. Light is so widely connected with insight and knowledge that we assume that we have seen the light when a picture of a person is projected with a light bulb next to the person's head. Light is seen as the same as knowledge. Light is actually a very poor source of information. Magicians and con artists use the information of light reflected by mirrors, lenses and colours to mesmerize our wits.

Let us consider light as a source of information in daily use by humans on this planet. Light is a very poor medium of information, due to all kinds of illusions, which it causes. Take for instance how mirrors and lenses can disturb a person's image. Why then can we categorically and without doubt, be assured that the information that we collect are the truth. Still, those in power disregard any phenomenon they find as the truth. A very practical example to prove my point can be put as follows. Light shines on a green leaf. The leaf is reflecting a certain variation of light and the rest is absorbed. Let us picture the leaf olive green. The olive green leaf accepts all the colours in the spectrum, except that olive green. The olive green are not accepted by the leaf but are rejected. The part of the light spectrum that seems to be unacceptable to the leaf is the very colour we associate it with. If the leaf accepts all the colours except green, it must consist of all the colours except olive green. Another example we use every day, is "How bright the moon shines", when in fact, everybody knows that the moon does not shine !

This might seem trivial and irrelevant, but remember, because of such trivial terminology, people were burnt on stakes. When Georigiano Bruno tried to persuade the church it was not the sun rising and setting, this became his fate.

Man's attachment to his visual sense is actually a little funny and a lot deplorable. Science can calculate the density of a neutron star, they can determine the heat value in such a star, but when this star becomes more dens, to such an extent that the density causes the neutron star's atoms to disintegrate, these very same scientists declare that:

1. the star has vanished;
2. it has gone through to another universe
3. it does not exist any more
4. because it is dark, it must be cold
5. it is lost in the creation, never to be recovered again
6. with this, all other laws of science are disregarded that says an atom consists of frozen energy;
7. the gravity fields can still be mesmerized therefore the matter it consists of must still be there;
8. All that matter can still be placed at one certain, predetermined place, and therefore it is position is a fact as far as the universe is concerned.

When taking the above mentally into account, one cannot but wonder, how far did man actually progressed from those darken middle age mentalities.

With these very obvious facts in the fore ground of our minds, we still reassure all the doubtful thoughts of our ancestry, which regarded everything the unknown comprised of as magic and mysterious.

We still cannot appreciate anything that falls outside the boundaries of the visual senses that the instruments can detect and determine. How small we are and how inflated we regard us to be.

I have in the past been asked the question "Why are all these other stars and galaxies necessary , if we as humans can't use it?" The human arrogance has no limits and the common position man occupies in the micro as well as macro space, is completely distorted by our sense of self-appreciation and self-importance.

THE THIRD DIMENSION (THE ATOM AND MAGNETISM)

In the article that deals with the meaning about "nothing" the structure and lay out of the atom is explained to some extent. The atom is the smallest and the largest single unit that the universe comprises of. In the most basic form, the atom exists of one single electron that orbits to energy levels around a nucleus in a single electron lay out. This electron to nucleus balance is the precise force that keeps the universe in perpetual notion bound by time. In a later stage, I shall explain this in more detail. The energy levels that the electron positions itself to the nucleus are valued in quantum leaps. However, even this complies with a three dimensional length, width and depth. This means there are three definite measurements, although they seem to be microscopically.

Taken one step further, the atom is made up of frozen energy. Any matter that moves at the speed of light is pure energy ($E = MC^2$) according to Prof. A. Einstein. That means this atom cannot be changed in shape or size without enormous energy loss or gain and this would lead to extremely serious consequences. The Japanese at Hiroshima and Nagasaki can declare what extreme consequences are hidden underneath the structural disfiguration of the atom structures. Therefore, can the people in the Tunguska river valley give evidence to the outcome of atoms that gain in mass.

In this process, man has released the worst kind of destruction available. The energy released is of such vast dimensions those 50 years after the explosion the shadows of the victims are still edged out on the background in the cement and bricks. The intensity of the light that was released in a billionth of a second, has almost forever changed the face of the material it shone on. Spare a thought for the people who received that light onto their bodies as their shadows remain as testament to those in power who damned them forever.

This process was not caused by atoms that was demolished, but merely by changing some of the element value from one atom to another. This leads to a spontaneous thought: "Why has no American ever been brought to justice for horrendous acts of war crimes?" Needless to say this act was the mother of all war inflicted war crimes by killing and maiming hundreds of thousands woman and children and civilians. These bombs made those in power, who ordered the release of these bombs, the biggest sadists and mass murderers ever known to man and when compared to evil-minded monsters like Nero, Nero with all his menace, suddenly becomes a silly and naughty boy. Even decades after the release of these highly toxic energy sources, they still kill and maim the innocent. However, this only ties in with another admirable fact of the 20th century. Almost all nations on earth have produced war criminals and warmongers, except the English speaking nations on earth. This group is blessed with the innocence to such an extreme that not one has ever been charged with one single act of a war crime. If none has been charged, none could be found guilty and then not one can be guilty of war misconduct. Let us go back to the atom's structure. Because the atom moves about at the speed of light, the structure is timeless. All matter that is timeless, will last forever or eternally.

This brings about the atom's structure to be forever and timeless forming the third dimension. In a later part, the reader will find that I disagree totally with professor Einstein's assumption about the fact that the speed of light has the same value as time itself. In order not to start confusion this early in the book, we shall accept dr. Einstein's theory as for now. The message I am trying to bring across in this article is the comprehension of what the third dimension contains. It is made up of three dimensions without having time as a factor.

MAGNETISM: THE MISCONCEPTION OF THE ATOM

As I already pointed out, the atom is pure energy. Seeing that the electrons rotates about the nucleus at the speed of light, and the nucleus vibrates (?) at the speed of light, that brings about a confined cell made up of pure energy.

Seen in the whole picture, the fact that the atom is driven at the speed of light and is in total balance and harmony, its permanence lies in the fact that it exists confined to a structure, but not to time itself.

There are length, width and height. The tree dimensions it composes of will render the structure eternal life, or that is how it is regarded by science. The property of the atom might change from star to star, but the structure remains with its three dimensional qualities varying in size, but it remains the same. You, the reader may ask: "What is the common factor between the atom and magnetism?"

Magnetism on the other hand flows between two points without stop in a closed circuit, not influenced by time. There is a direction (length), a circuit (height) and a start / finish point. This forms a closed ring formation, which has the same qualities as the atom. Time plays no role in this energy displacement, because this movement is coupled to the speed of light.

To prove its existence, lies in the fact that the human civilization is mostly driven by electricity.

The energy determines the magnetization, but the circuit remains permanent although some times in a latent form. Because of this, it actually consists of all the ingredients to qualify as a third dimensional force.

The ferromagnetic field proves that space-time is being displaced as it is being done in the case of the atom structure, but the displacing of the space-time is in a continuous and constant closed circuit.

The poles that attract each other displace space-time in the same direction and those that repulse each other, displace space-time in opposite directions.
In bodies as insignificant as the earth, the difference between gravity and magnetism will be enormous in comparison with structures the size of the sun. The magnetism in the sun is comparatively much stronger than the earth, but the difference between gravity and magnetism would be less. In a structure, the size of a White Dwarf the electromagnetism might only be twice the force of gravity and accordingly as the star becomes bigger, the force would become equal.

In structures that compose of the mass of a neutron star, magnetism would be dominated by gravity to such an extent that the force of gravity would not allow any magnetism to exist. In the so-called black hole, there can be no such a thing as magnetism because the electron does not exist any more.

Electromagnetism is the "short circuit" in the flow of positive space-time displacement and can only exist in a structure that compiles of an iron-based core.

THE FOURTH DIMENSION: THE TIME FACTOR OF TIME IN SPACE

The previous three dimensions all dealt with the space factor in time in space, disregarding the time factor. Let us consider an everyday household item in normal use like a table to explain the value of time. The table is made of wood. The wood is comprised of timeless atoms. (That is, if the reader accepts the previous argument about the atom that lies outside the boundaries of time). The form and the shape that the wood is in, is not timeless. It started as a seed that was enlarged by cell multiplication as it germinated to become a tree. The process of germination took a certain time and the growing of the tree took another period in time. That means the tree occupied more and more space in a given time period in the space-time it shared with the earth in the form of a tree.

Afterwards the tree was chopped off. This felling of the tree comprised of a certain given space in a certain given time, as did the falling of the tree. Both these periods consisted of different space in different times that are coupled to the space-time the earth occupied. For a certain period, the tree remained in an upright position occupying a little more space as time moved on. Then at a predetermined point in time, the tree was felled. Every blow by die axe displaced a certain piece of wood to a different place in space and time. That meant that every splinter of wood that broke from the tree was given its own place to occupy space in time. The position of the wood has been altered and even if one try as hard as they may, every piece of wood has received its own space in time to occupy and could never be regarded as a tree again. It occupies a complete different position in the space it shares with the earth in time.

Afterwards this tree is stripped of its branches and leaves. The stripping takes a certain period and position in space and time. Each branch occupies a different position in space and time and is forever dispositional from its original position in space and time, because every part that is not part of the tree anymore, is in its own position in space and time.

Then this, the tree was taken from the plantation to the mill and moved through a different space in each fraction of time as it was transported. Every millisecond held a different position in space in relation to the next time fraction. At the mill, it was sawed into planks and the structure's position was altered even more in space and time. The planks were bought by a carpenter and were given a completely new form as a table in space and time.

Although this particular table I am writing on now, at this present moment, has occupied space in time since 1921 as a table, the wood occupied space in time much longer than the table has in its present state. This wood can never be made a tree again, because it is space and time has been altered indefinitely. So, how does this fit into the big picture of the universe as all things in the universe are connected and related?

I was born on a given second, minute and hour and I shall die on a certain second, minute and hour. From the day of my birth until the day of my death, I shall constantly alter my position in space and time and will never be able to occupy the very same spot of space-time because space and time will not allow it. According to my visual observation, I can presume to occupy the same space in time, but the geodesic outlay of the universe is altered as every millisecond goes by. That means I can never remain in the same space in time, because my body's position alters by the rotation of the earth, the sun and the universe.

When my time is up, my space and time is altered to such a position where as I am no longer in control of it. From then on, I will never be able to determine my own position in space and time again. I am in a state called death. In this state my body will be broken up by microbes into gas and heat, every second time moves on, my body's occupation of space, and time would diminish. This will carry on in space and time until only the elements my body comprises of, remains. The atoms I am made of, have previously been used to form plants, trees, animals and even humans. After my death, it will never ever form another combined unit to match my exact replica. Therefore, I have departed in more senses than one.

That means my position as Peet Schutte, in space, and in time, is suspended, unconditionally, forever.

This means that for certain duration of time a certain combination of atoms is forced together to form the elements that are dedicated to me. This dedication is temporary, which means that I was for a certain designated space in certain duration of time-sharing space-time with the earth.

Although this is a known fact to every person on earth, it is astonishing to see how every person yearns to maintain a youthful and vigorous maternal structure. This structure is condemned to destruct the minute a person is born, yet everybody guards his attachment to that structure with a jealous observation. The cosmetic industry cashes billions of dollars a year and all that income is based on this fear of ageing, which ultimately leads to death of the person and destruction of the body. The ongoing process in the envelope of the fourth dimension connects all the previous three dimensions, we regard as time. This means that time itself is not created by man, but is part of the physical universe and is only recognized by man, the very same way man has recognized the existence of the other three dimensions.

THE FIFTH DIMENSION: UNOBSERVABLE - THAT MEANS TIMELESS TO MAN'S SENSES

In the article "Life after death," I touched on the subject that life is a form of energy, and therefore is indestructible cannot be destroyed. If any reader cannot accept this argument, then please prove the opposite and let me know. Therefore, until the opposite is proven, I shall regard this argument to be correct. Because the generator, which I regard to be me, I of the electrons, is not the" me" that will be placed in a box and will never be able to share space time (because the worms are going to feast on me) I am above time and space and I shall not be in that coffin. Only the "me" that am made of flesh and bones and was considered "me" and which I had temporarily control over, will deteriorate in that coffin. If the generator of the electrons that is somewhere connected to the brain, and from where I control the me, which is the muscle bone and tissue held in place by the body I consider to be me, where do I , the generator of electrons go after I was disconnected from the me of flesh and bone me. I know where I, the energy-less dead body is going. I, the part I consider me, is to be thrown into a wooden box, dumped into a hole which is dug in the ground and where the worms are going to enjoy a feast of a meal.

This feast of a meal by the worms cannot be me, because I am not part of that decomposing structure. I was part of that decomposing structure, until the very second, I lost control of that decomposing structure. Thoughts, that travels faster than the speed of light is the actual part that is

I. I can elaborate on this line of argument, but those that do not wish to be convinced will remain so because they prefer to remain unconvinced, not because they know they are right, but just the opposite. The I which is regarded as me, has received the ability, for a short while at least, to manipulate space-time, whether it was in the form of my body, or other matter I came into contact with, or unoccupied space-time as I moved about, occupying unoccupied space-time in a random fashion at free will.

The part that is I, and which is not part of the corpse any longer, can think of one thing and then think of something completely else. The very following second I can change my surroundings as soon as I change my thoughts by creating new thoughts. These thoughts are not connected to time or space. It moves arbitrary and involuntary through time and space, from present to past to future. Sometimes these thoughts are so strong that a person loses track of reality.

The terminology we use to describe this, is daydreaming. When I was teaching as a pedagogue in class, this condition was my biggest enemy. While I was conducting my class, the students would sit there wide eyed, listening, but at the same time they were miles away from school living a daydream that had no connection to matters of schooling. When they were asked to reply what was just said in class, they were flabbergasted and completely unaware of their surroundings.

I think I may presume with some certainty that the reader would follow the two parts of the same person I meant. I agree that an argument can be made that these thoughts are part of brain cells that are stored by nerve tissue, but the information in those cells are created by emotion. These thoughts can be depressed by emotion or be prominent on the foreground because of emotion. Man is made up of a stream of emotion that flows continuously through these emotion fields and the flow of emotions is that that puts meaning into a person's life. One has to consider emotion to be one of the most pure sources of energy.

Energy creates thought, creates ideas of a spiritual as well as physical nature, it establishes a flow of electrons that drives the human mind and body, it controls all muscle groups in the body like heartbeat and when the body is in danger, it produces chemicals which enables the body to react far better than it normally does. That means these other organs are also under the control of emotion. The emotion drives the body to produce chemical substances, which enables the body to perform at levels far above its known abilities. This emotional control defines heroes from cowards, sportsman from the ordinary and even philosophers from the masses. That means the emotional part is the part of me that cannot be destroyed and that is the part that generates emotion. All this comes down to one value that life possesses, and that is the manipulation and control of space-time.

When this emotion driver, that generates electrons, leaves the body, a person is considered dead. When one takes the moment that life leaves the body, nothing physical leaves the body. There is no lightning like electric conduction, there is no immediate spontaneous combustion, but there is no emotion either. As I already pointed out, the only reason why a person is considered dead is that he is considered lifeless, meaning without energy.

A lot of energy has to be deplaned somewhere. It is no longer part of the physical world and being energy it cannot just vanish into nothing. What is factual, is that life as energy, is no longer attached or bound to the fourth dimension of time. At this point, I have to prevent my ego and self-importance not to get the better of me and put myself on a pedestal equal to our Creator.

It would be much better if I were grateful and thankful for the time I was allowed to use the atoms loaned to me for my own personal use. The loan period was of such a short duration in space and time, that the extent of the period of loan becomes oblivious in space and time and space time.

THE SIXTH DIMENSION: THE LIGHT AND THE TRUTH

I must confess that in this my faith and religion plays a large role, as one of my beliefs is that the Messiah has already come.

Maybe the Jews, Moslems, Hindus and other religious groupings would find their own explanation according to their faith. My Messiah said, "In the house of my Father there are many mansions.", which I interpret that these groups should be left to find their own salvation. I may not condemn them or denounce them or try and convert them, because each should have its own mansion. However, personally I accept my Messiah and He declared that: "I am the Light and the Truth and no one can enter the house of My Father but through "Me."

If my Messiah points out to be a light, it stands to reason that there must be darkness. That will be a dark fifth dimension and light sixth dimension. Seeing that the majority of English speaking persons do not share my beliefs and religion, and I do not believe in converting any person to my faith, I shall leave this matter at this point.

It is not that I am ashamed of the religious beliefs that I follow; to the contrary, I believe that only Israelites may be converted to my religion. I think I made enough argument to prove the existence of a fifth dimension to which life has to go after death. I presume, because I am not familiar with the contents of other religions, that their religion will allow them a passage by what means their religion chooses, out of the dark fifth dimension.

This dimension, the sixth can only is entered by human life. Let us then see what human life is all about. A human does not have to be a human because the creature is compiled of human D.N.A. D.N.A. cannot form a human. The gorilla is 97,8% human and the orang-utan is 98,2% human. However, there are obviously big differences between these species and a human being. No cultured person on earth will consider one of the ape species to be human. That means a human is more human by culture than by physical appearance. The additional 1,8% that a human have, is not even enough to explain the physical differences that exists between humans and apes. However, we know that the main difference between the species is the human's ability to reason, think, argue and control their emotions and instincts.

The more a human explore and scrutinize his feelings, himself, his surroundings and his universe, the more such a person would qualify to be human. After all, that is how other species evolved away from the animal and that should categorize us to be part of the sixth dimension.

THE SEVENTH DIMENSION: THE SUPREME ALMIGHTY

Because so few English-speaking persons consider themselves Israelites, they do not fall under the law of the Israelites. In such a case, I would consider myself blasphemous to share such knowledge with those that do not regard themselves to be law-abiding in all ways. Any person that does want to know more about this matter should read the book of Henog. I will say this much, that those that read Henog would find out why the book of Henog has been left out of the Holy Bible by the Roman Catholic Church and the other churches. The Supreme Being lives in the Seventh Heaven as Creator of all. Because I regard my fellow Boere as brothers, I did explain to a very small extent the seventh heaven in the original Afrikaans version. But all Christians and only Christians read this…all churches are part of the Anti-Christ being the Body –Of- The- Anti-Christ. Do not look for the coming of the Anti-Christ for he is among us. They crucified Christ for His throwing over the money tables and throwing out the money offerings (a lucrative business in any society) ridding the Temple of money some two thousand years back, and today all Christian religions fight one another to feed that which Christ threw out…the money tables…and best of all is they feel righteous in doing so! All denominations are a part of the Anti-Christ and BEING THE ANTI CHRIST. Christ threw out the money tables because He said you cannot serve two Masters…you cannot love God and Mammon for one you shall love and one you shall hate. You cannot serve Mammon and God. If Christ came back today and once again threw out the money tables all Churches and worshiping priests will once more shout for His crucifixion as they did two thousand years ago. As the Pharisees were the Anti –Christ back when… so is all Christian churches and denominators, Priests, Pastors, Reverends, Bishops, name them what you like, small and large…they're all taking part in the crucifixion every time they ask for money "In the Name Of The Lord" and if He came to destabilize they're Money machinery today, they will hang him tomorrow morning at they're earliest convenience and even on the same cross (if they can find it). Every preacher is more into collecting than pouring out the Word. It is a trade off that the preacher will bring the Gospel in exchange for collecting offerings. They heel, bless, pray, condemn and condone on behalf of the name Mammon. I challenge every purist of heart to show me one preacher of the Gospel that sends donations back with the message that such donations is condoning the Crucifixion and as preacher will not except having the blood of Christ on his conscience. If you cannot show me one, I can show the body of the Anti-Christ for they will kill again if some one should try to diminish the lucrative trading done in the Name of the Lord.

9.6 THE TRINITY... and you.

Singularity holds five dimensions inside and five outside singularity as matter and space forming space-time. The ten dimensions I named the atomic relevancy is also showing the double value of singularity as singularity extends into as well as beyond space. The atomic relevancy is $(\Pi^2+\Pi^2)(\Pi^2 \times \Pi \times 3) = 1836$ that is the mass relation between the electron (3) and the proton. Proton = $(\Pi^2+\Pi^2)$ Neutron = $\Pi^2 \Pi$. The atomic relevancy holds the dynamics of singularity control.

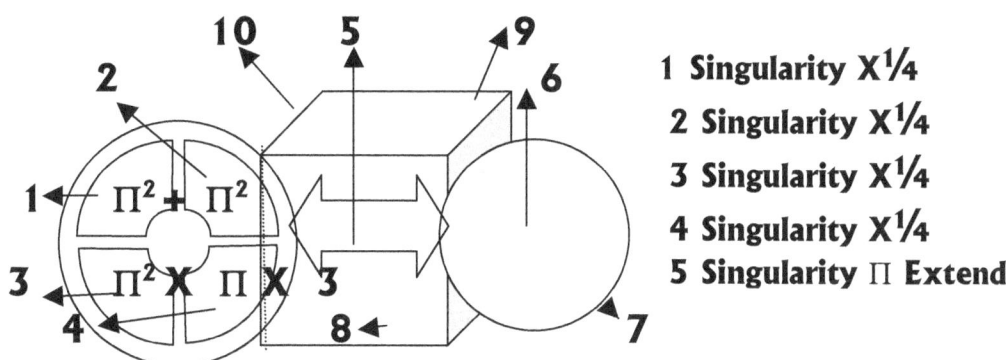

1 Singularity X¼
2 Singularity X¼
3 Singularity X¼
4 Singularity X¼
5 Singularity Π Extend

The challenge is to find a manner that life repeats this tendency to connect in four stages plus one extending. In singularity there are five positions extending to space-time with five positions and therefore life connecting to singularity should indicate five positions. As far as the aspects concerning the physics side goes I could prove in some way that my initial way of reasoning brought fruit to bear, but what about the dimensions past four and even past five. I think I may presume that in some way I did manage to show that the body has its place in the fourth dimension and without life it becomes just one more cosmic structure without any form of moving ability when all forms of life (including the bacteria that will decompose it to atoms) removes from the carbon and other elements where the elements then are exclusively and only cosmic particles. With good reason one may believe that the ability in conducting the manipulation of the space-time within the body has diminished to all extent and none is any longer present. To find a means of putting mathematical formulas to use in applying proof to indicate the fifth dimension is beyond me and that is where human extelegence plays the part. By the same token that atheists can say they wish for more proof about life and the fifth dimension I can demand the explanation to prove otherwise and ask an explanation about the energy presence in life-holding bodies and the energy absence in lifeless bodies and where as I doubt I may force atheists to sound understanding they too must admit that something does seem out of place in they're arguments about the energy being of a pure physical nature and only stubbornness will win by the days end.

So to them and those, as I might not find a way to prove beyond doubt them and those also must admit I have sown some doubt in their minds. Getting them to admit with some gallantry about the doubt factor, well that I must gallantly admit is a horse of another colour. With the fifth dimension seemingly impossible to prove mathematically the sixth will be much more difficult to prove and I shall not even attempt such an act. Fortunately man is not mathematics but much more complicated than mathematical equations can ever bring forth. Only good old fashion arguing with a dash of logic sprinkled when and where necessary will pave the way to understanding man and beast. To find out what man regard as good or bad and what beast evidently regard as good or bad must be different as everything else about man and beast are different and we must scrutinize beyond where mathematics and physics can prove. Even the most convinced Newtonian should see that there is a point such as that.

Being human every reader must have an opinion about good and bad and what is evil and what is not. I can scarcely imagine a lioness feeling bad about a kill while her cubs are filling their bellies with mouth-watering flesh of an antelope kill. Neither have I detected sorrow and anger as a new male kills the previous litter to establish his new domain. Neither the female nor the male bears sorrow after the deed of him destroying her litter sired by a previous dominant male although the lioness will protect her cubs while they are alive almost to the point where she may put her own life on the line. That how ever has nothing to do with rite or wrong good or evil and after the lion male did his killing of her cubs she shows no remorse or blame for that matter as she follows him back to the rest of the waiting pride. This is not exclusive to lions or even to predators but has a wide range of animals following the very same living style. Horses kill without thought because when a stallion wants to find a mare that he knows may follow him but for the foal by her side he will kill the unwanted foal and not be bothered by her reaction, being in the knowledge she will follow him after the foal is dead in any event. While the foal is still alive she may do some protecting of the foal but she will normally not go as far as the lioness in preventing the death of the foal. Baboons, monkeys and a variety of animals have this approach to life. During such an attack by the new dominant male baboon the females will fight off the onslaught either by grouping together or by fighting him off in ones and twos but the usual is that the new male is big young and strong and even a group of females are no match in a fight. However after all the noise and the shouting blood sweat and the rage of adrenalin has died down, the babies are dead and no female attacks the male after the fact in heart felt sorrow for the loss they feel such remorse over. No, it is clear that the deed was done and life goes on.

In humans such behaviour does take place and every time we read about such a deed committed against a harmless infant even the biggest humanist find a moment where he or she wishes the death penalty on the criminal for acting in such brutality. Why will humans shout for blood in punishment while other species take it in stride?

To find a solution I do what I always do. I turn to the Bible for an answer. (You atheists deny yourself a wonderful encyclopaedia of information and when reading the last book in this Theses you will come to understand my saying so.) In The Theory which is the seventh part of The Theses I explain to very detail the events of the first six days of creation as recorded by The Authentic Biblical Author but even after that some more explaining arrive when it is correctly translated. In the book called the Bible there are two trees described with distinction and I am of no opinion to whether they were in wooden fibre or just symbols to explain the complicated issues to persons with even lesser education back then than I have at present. It is an undeniable fact that man has an inclination about what is good and what is evil. Man would not kill an infant because he cannot find the infant's mother and the infant is crying of hunger. In the animal world any adult of the specie will walk past such an infant in distress with no feelings of care what so ever even if the specie does not show a normal tendency to destroy such an uncared suckling. Where one mail may kill another mail in a fight to establish dominance the group does not cry for justice as they loath such a deed. When a superior member of a pack relieve a lesser member of food or eating rank they do not hold congress in judgement to provide sanction for the lesser member with accompanying reprimanding about such incorrect behaviour by members of the group. Such is the caring of man that any human notwithstanding whatever urgency drives him at that particular moment that human with maternal instincts or not will stop to care when coming across a deserted human infant hungry and all alone. Why will man show such behaviour as normal through out all races on earth without any cultural distinction in any way and this may be the only distinction that races share because some eat their dead and some burry them with pity filled emotion and… Oh, I can go on writing a book on this topic alone but that will be useless because every one should know what I mean. With this shared by all where and what does mention this distinguishing behaviour of man's ability in judging between good and bad for the very first time as a landmark to man. Accept the Bible or not, it remains the oldest Book available on matters reported by man since no one knows when because Moses may have assembled the research on information but the information as such dates to times predating even what Moses' research may indicate. In addition it may be correctly presumed that many or most of the facts he recorded he was taught as a prince in the house of pharaoh and that then may explain some detail about matters he actually knew nothing of. His being adopted by pharaoh's sister must be a plan with some significance and must result in some meaning other than to give Moses a childhood of luxury alone.

One of the trees the Bible mention carries a name specifically as life and from that I draw a conclusion that may reflect that life chose to go the way of having a variety as sells with complexity in the evidence we now gather from DNA strands whereas a choice of life would indicate a sell of simplicity in structure as we find with lowly developed insects and other specimen of life. I have seen how a corn-crake of a specific kind only found in the desert and semi desert regions where I live can start a pest becoming so out of control that when run over on the road they form layers of millimetres thick trampled and squashed by cars to the extent the tar on the road is no longer visible. It truly is a pest of Biblical proportions but fortunately to unleash this pest it has to rain in November in one specific week. If it does not rain at that specific date, and I do not mean approximately but precisely they don't show at all. That makes their return very sporadic and it only occurred about five times where two of the five times became a pest like none can repeat in the more than twenty years I farmed on that farm. The pattern also follow a distinction where at first one or two may show very sporadic in places. Then mating starts and the pest develop where by it truly come to a climax at the end of February and dies down in May. From a few eggs they develop millions on millions and I do not exaggerate in the least when I refer to this phenomena as a plague of Biblical proportions. Poison does not kill them and when one gets hold of another one the bigger one just start feeding on the smaller one. At the end of the meal where the bigger one devoured the smaller one in totality (and I mean boots and all) the specimen will shed its skin immediately, eat it or not eat it and walk off. The one is a precise duplicate of the other with no distinction amongst them of whatever nature. Seeing one is seeing the lot. They share genes in precise replica with no differentiation of any kind what so ever whereby the one may have even the tiniest of difference in any form. They come from a line where life was still very basic and the mother specie of the very original has not changed in any way through millions and possibly billions of years. That I say on the grounds that to my judgement that specie is of a very basic nature and has developed with one single aim in life and that is to survive. They eat everything from the most poisonous plants to fruit to meat and bones of animals lying dead in the veld to dry hide and even one another. Fortunately too they only occur in the most severe droughts and when developing into the pest, which I describe, there is little to nothing for even grasshoppers to feed. Seeing the specie for what it is it made me realise that life somewhere after them made a choice to form complexity and variety or remain as they have and form a universal gene where the original mother is still present in all her offspring even after so many billions(?) of years having the opportunity to progress from where they were. They made the choice to remain the same where as the line that man developed by the original parents may have made the choice to evolve through complexity.

With all the explaining I do not wish to prove that man had a nibble or never had a nibble from that specific tree but I only wish to indicate that there are a variety of interpretations and clues around and when sanctioned they may deliver a vastness of possibilities. There is one other tree of distinction mentioned and also a mention of some eating by the female at first and later on by the male. This tree also was named and it was the tree of good and evil but according to the Afrikaans Bible the mentioning of the name says specifically and I quote: *Boom van kennis, kennis van goed en kennis van kwaad"*. Directly translated it reads as follows "tree of knowledge, knowledge about good and knowledge about evil" and that is where my argument starts in my attempt to indicate the possibility of the sixth dimension belonging exclusively to man.

In the detailed analyses the specifics concentrate on the "knowledge" and then distinguishing between "knowledge of a good nature" and "knowledge of a bad nature". Please note there are three mentions of knowledge one only about knowledge then about the knowledge of the good and thirdly about the knowledge of the bad, but most important separating the three by distinction.

After clearing this part we may return to the animal world of some being wise and not very wise. Animals by nature and by genes acquire a base for knowledge to carry the specie through dangers and more important even, to the survival of the future of the surviving gene pool. Surviving as far as the animals go is the good and the evil and there, at that point all other definitions stop. All intelligence the specie holds and all the intelligence the specie acquired contains the one underlining element being individual and specie surviving. In saying that I do certainly not say that rules amongst members of a group of animals is non-existent. There are certain criteria the individuals have to meet to establish rank in the tribe. At the same time such rules do not centre around emotions of ethics but they are practical well placed and directed to ensuring stability and it seems the higher evolved the specie became the more sophistication there are amongst rules constraining some members to the advantage of others.

The Matriarch in the Elephant herd is Boss and that is in capital letters! No bull will dare to push members and lesser infants around and she takes much less shit from young elephant males than the females. She is the rule and the law and every one abide by that. Should a male wish to afflict his attentions on some elephant cow the Matriarch will condition the visit to her satisfaction or the visiting male may even pay with his life. I say this as a result from knowledge I acquired as some game-farming friends of mine has elephants in captivity. Crossing electric fencing is hardly an issue for the matriarch because she takes a teenage male place the young male (and it is always a male) between her and the fence and let him walk unsuspectingly alongside her as she deliberately holds him in a position where he walks between her and the fence with the current. Then at a moment she decides on she thrushes the young male allowing him to plough through the fence and take the electric current shocks while he goes on his way braking the fence altogether. After that the fence is open and clear for the rest of the herd to cross. She will never act in this manner towards young virgins but the young males get the stick every so often. With the advances the African elephant show would the African elephant be a good pet. I would hate to find out because they may set the rules and not me.

I know for a fact that a Nile crocodile does not make a very obedient housebroken pet. Should any one have an idea to keep one in his swimming pool be warned the pet would not distinguish very well between his owner and his next meal. And his love for children might be somewhat different to that which a good pet should have as is the case with dogs. With dogs man had thousands of years in breading good pets but it is unlikely that the first relations were as timid between master and dog as that we grew accustomed to.

1. Through many generations of exclusive inclusive breading did we finally manage dogs to have become what we wish them to be. In this lies another fact to analyse. Many different breeds make many different dogs and the one race has characteristics setting that race apart from other races but in the race itself the variety of characteristics find more prominence in the race than in individualism. Characteristics of dogs connect more to the type of dog than to the individual dog and therefore some races have inborn hunting skills where others may have guarding skills. It is the breed that brings the selection and not the individualism in personalised characteristics. Therefore it cannot be said that a dog has a conscience but it is better said that a dog has a better breading line.

2. Some evidence suggests that when Cro-Magnon – man arrived agriculture replaced hunting as the feeding method and we are confident man exclusively kept that dog for it's hunting and sharing abilities. With the arriving of agriculture man then extended his space-time manipulation not only beyond his physical abilities in hunting but also his physical strength in working with tools. This must be the biggest leap of all even much bigger than the leap of the electronic age but such comparisons are extremely difficult to make.

What would be man's drive to not only manipulate his personal surrounding but also manipulate surroundings of other forms of life to their benefit but moreover to his benefit. What would give man such judgement as to select species beyond him and feed them to eventfully find more benefit from their feeding than they did benefit from their feeding. Genes it cannot be . The orang-utan has 98 . 2 % of the genes man has and the gorilla has 97.8 % of the same genes man has. The Gorilla still lives in woods and is destined to disappear while the orang-utan lives in trees and holds no better future prospects. Genes would at least give the species having such close relation in the gene pool with man an idea to follow the trend set by man and copy some of the abilities. Genes it cannot be and that just about excludes the last cosmic or natural physical explanation from the list of possibilities.

There seems to be a massive gap between what man became and what ape became. Science makes a great singsong about chimpanzees with the ability to use tools for their benefit but man has surpassed that so long ago science have no tracking record about the time and the way that came about. It seems as if man was not, then man was with agriculture and all other providing the manipulation of the other species under the control of man and by increasing all benefiting that the animal enjoy man could bring benefiting all around to benefit man. One day man was ape and the next day man became super-specie-of-the-world, the world champions in space-time manipulation or in other words of controlling life. Not only the life of man but also life of others to some mutual benefit slanting heavily in the favour of man. Still the benefit of the other species holds so much that only species that benefit man started to dominate world population with man.

Of course as usual and as with most of Newtonian scientists' findings, I question the accuracy of the gene pool percentages strongly and I am of an opinion that such percentages are in use for political issues more than scientific proof. I prefer leaving it at that. I am a very small fish in a very large pond sharing the pond with very powerful other fish that can destroy a small fish like me with one gulp. Going into the development journey as man followed the trail one has to look at not what man achieved by own ability but with own measure in manipulating others in life. When man started with chips and flint it was progress but it was also very limited. Only when man acquired the muscles of more powerful animals and took their ability in measure with mans manipulative power did success arrive at a level that brought progress in leaps and bounds. It is not mans hands or legs that brought man's domination and control but man's brain that brought response from other life to benefit man and find benefit in shared life styles where mutuality brought about safety and mutual prosperity. It is surviving more than anything else that means good or bad to animals and most of all surviving of the species and in that the animal only find man and man's company good because man holds its safety. When a lion brings down it's pray the rest of the flock will start grazing immediately without showing even slight remorse to the victim and for the loss the close relatives are faced with. By the death of one the rest find safety and to animals that is good. To the rest it is about surviving and if one pays the price, little concern goes to who paid the price. That is what annoyed me about our all-famous-pop-star-wife. She truly go beyond what nature puts down as rules applying her liking (and selling her book with cheap bluff) while other sheep walking on two legs like only humans ought to cheer her stupidity as if the stupidly were they're own. They have not even got that much brains to acquire that much stupidity but has to borrow to get their tally that High. The biggest annoyance is that with such stupidity those can vote and choose my future because of democracy. They prove almost all ways not to have the thinking power of a mouse but they have the rite to choose my future and I have no say in the matter but to follow what such morons may wish upon us. With the novel idea (novel as it is only man that uses mutuality single-minded and still provide beneficial good from all angles) of widening the use of abilities provided to different developed species, man gained extensively in progress and comfort. After all it is much lighter work riding a horse than walking all the way.

But gains in comfort goes both ways as the horse find protection against predator attacks while finding good nourishment in winter and the best hay to feed. Such a diet provides the horse with the strength to carry the rider and enjoy own comfort with the fact of much reduced fear and anxiety. By having male and female and promoting mating the good in the life of the horse becomes better. Did man rob horse from freedom? Did man take what was not his to take? Should man get some conscience attack leaving him with sleepless nights about his cruelty in robbing the horse of a natural life of freedom? Well the humanists will tell you with teary eyes and running noses that the horse should have its freedom as all are born to be free. But this emotional outcry comes with the comfort they, the humanist enjoy of secured sleeping a good all year round food supply and breading safety. I have not seen one humanist go running into the mountains never to return to civilisation, to enjoy the freedom they should enjoy as much as they wish that upon the animals in captivity. To the horse, after the initial fear of the subduing and the ultimate realising that the subduing is not life threatening as he can live with that, he finds comfort and even enjoyment in that. I see on a daily bases how horses get jealous when the owner takes one to ride when the rest wants to go first. They come and nestle with a desire to connect and in jealousy push each other away to be the one receiving the owner's attention. It is a bigger issue of the conscious to decide the likes of others in what they like and what you think they may like. No one of sane mind will hurt an animal in your care and when slaughtering that we do in the most humane way as described by law. No one cuts of a chunk of steak while the animal is alive. We humans have civil norms and values and by using our brains we can live and let live with more dignity going around sparing the animals huge cruelty than what the animals would have come to face if they're fate still was in freedom and being hunted down by wolves and hyenas Such is the difference between those having bleeding hearts and brainless skulls and others that can think. Now we arrive at an interesting question as how do we think and reason. I am sure all humanists will have as much to say about my way of being correct only as they will find many arguments as proof of the fact that by dislodging my logic they can prove me being beyond the norm of classifiably insane. There is ever a clear definition about rite or wrong and all principles we find appealing or appalling is within the brain

According to an article I read the brain holds more connecting lines than does the universe and I may even accept such a statement on the grounds that life has much more complexity than does the universe. After all life takes the dimensional barrier as far as the universe does and then beyond where the universe stops. This does not make the universe simple because I cannot see how any person may ever come to understand the flow of light as the light uses both the straight line of singularity, the half circle presented as the Roche limit in singularity and the Titius Bode triangle making light representative of every aspect which connects space-time away from singularity with singularity as light where Π meets r to become the value of C. But in the brain this is only one function as electricity holds an equivalent of light forming electricity as the messenger to whatever energy is above life and then in the human capacity above even what forms the barrier to life. The arm is not human life because a human can loose the arm send still be alive. Therefore what ever is in control of the arm is in control of life, which puts what we find as life at a higher dimension than that of life. You may argue that in case of animals such thought also control life because a dog may lose his legs but not his life. But even as complex as that may become there are relevancies between life and the physical because where the physical uses pain as a warning system the mind uses fear as a warning system.

By following such a line of argument one can freely deduct that an insect as our corncrake, which we discussed earlier on, is representative of life as the same life we find in the arms or legs of our body and the life we control but is not truly part of the energy "me". Clear to all it must be that life we find in mammals are advanced above and beyond the development the insect arrive at. If that is the case then I may claim that human life has more developed than what other mammals did because with my manipulation of space-time I may manipulate other mammals to harvest some of their manipulative abilities in benefit of our mutual relation inclining more to my benefit. The cow does not seem to mind when I milk her but can any person imagine experiencing a milking session involving a crocodile? Well, fortunately crocodiles have no milk but if they had I would never volunteer for the honours of being the first to train a crocodile how to behave in a steady manner when in a dairy session. You may have or may not have noticed but I am telling you that they have a sharp side and they have a blunt side and the sharp end holds rows of teeth they surly know how to use. Even the blunt side hits like any whip never can and I am sure a fully groan specimen may kill with that tail. Going down the order of evolution we come to bacteria and viruses, some of the lowest forms of life. Do bacteria and viruses count as animal and if not then surely they count for life because life they are. In that we find the equivalent of bacteria in higher developed species as we find that the insect may have the developed life mammals use in their bi-products included sustaining their superior life development. We can see evolution by applying a relevancy of devolution to siphon and separate life from life. I am trying to indicate that life becomes a compliment where the lesser developed formed a mutuality and aided the supreme form of developed who is controlling the master brain in that form of life wherever the master brain may be attached. Life is above and beyond the cosmos and surely even the most ardent supporter of atheism must grant me that much. By that grant must the atheist then add the fact that life cannot only fall onto a category of to be or not to be but there is a range in life forming a line of development and superiority. Life is more than life but has status of being lesser or more and that is the point I wish to address after all the talk. Within one body a range of life values combine in making whatever accomplishments the life form accumulated by extensive development that range in development. It is appreciable in concluding from a range of facts I mentioned but mostly from human common sense we all know that on top of the range being the model best manufactured and with all accessories all other models also having life envies and fear is man. What will make man that special?

In 1905 a case was reported for the first time of a woman that had a hand, which attacked her every night by trying to strangle her. In the manner the hand acted it was clearly out of her control as it was clearly out of control of whatever controls the brain have over body functions. She would wake at night and feel someone strangling her but the person strangling her was something she ought to have under her control. Imagine waking one night feeling someone squeeze all life out of you with a murderess motive. Even the thought of that will make most people get up and bolt their doors and windows just in case. It must be awful having a murderer wake you with such a horrific intention. Go one step further and think that person may be one of your house members you trust with your life. A thought of that becomes rather preposterous! Then for the ultimate in revulsion; think of the chances something acting in such a horrendous manner is something you know with every grain of your body as that thing acting is your body. There is nothing worse to be scared of than being scared of "you". How do you fight such an act. You cannot hide and you cannot go without sleep and as you go to sleep you know that there is some part of you yourself hat is after your life. If this is not enough to drive any person into hysteria I do not want to know of anything worse.

This flabbergasted the doters and no one seemed to make any sense of such phenomenon. I suppose if such an incident had occurred before it would have been denounced as an act of a demon of some kind but fortunately for medicine the art of healing had abandoned forces of nature as a scientific accepted fact unlike the likes of physics still clinging on to such madness. If my memory serves me correctly this case was in Germany. Please remember unlike our distinct academics I have no extended library to find all kinds of information but have to rely on a failing memory being destroyed by my diabetes. Then in France later on another case became known about a woman that had a hand also out of control where in this case that hand tried to forcibly turn the steering of the motor car she was driving to force an accident of a serious nature. This manipulation was seemingly as much out of control as was the previous recorded case. The common factor about the two cases was that in each case a person had an arm that was intent on destroying that person without the person aiming to do so in free will.

Later on in America two neurosurgeons planned an operation procedure where by they aimed to relieve patients having chronic and continuous convulsion attacks caused by epilepsies in the brain. These cases were dire and with the operation as a last resort all the serious after effects became a secondary factor to the superior motivation of saving life and improving the demented quality of life by the patients . The operations involved only the utter most serious cases that left no other option for improvement. It was this or death and not choosing death the patients chose the intended operation procedure instead, but still it was extremely serious and dire options in the choosing. They reasoned that the epilepsy was a result of the brain having vibration and with the vibration stimulating other vibrations through the brain in some cases the one caused the next vibration and it was more a reflex of the first causing the next as the symptoms was going on a prolonged non stop convulsion. To stop such reflex by the brain tissue they held the argument that when cutting the cortex the two lobes attached will not have the reflex and thus the continues convulsions will loose the continuous effect.

By separating the lobes the nerve attack coming about in the one side will not transfer to the other side thus it would not cure all elliptic attacks but the prolonging effect will be reduced. One vibrating lobe wills then being separated from the other part not cause the response in the next lobe and it was diagnosed that it was more a response to the reflex allowing a reflex to the respond and this brought about a never-ending cycle. The idea was that it would result in reducing the severity of such grand mal epilepsy

According to American law the doctors first had to show a high degree of success by operating on rats in order to prove that the consequences of such an operation is in acceptable levels before starting such a procedure on humans. To obtain the rite by law for the granting of the operation many rats underwent the procedure and the procedure then were extended to many other species.

Every aspect of recovery and side affects must be documented to an exact accuracy with no exception to the rule in the slightest. The behaviour of the animals before and after and the general physical data then goes to excessive detailed scrutiny by the finest the medical profession has to offer in America. Accuracy in the process of accumulating data and other relevant information is beyond question especially in the country with the highest standard in medical care. The after affects the procedure had on animals were indicating no serious side affects of any reason for concern. Many species went through the procedure eliminating defects if whatever possibilities there may be.

When monkeys went through the operation procedure our primate cousins had no side affects in any way. There was perfect hand eye co-ordination and the nerve system had no complications with the motorized operating functions in any way. This confirmed the surmise the medical profession then at the time had that this third lob was just fibre with no function of distinction. The fibre was in position to stabilise the two lobes and had no connection with the lobe in a functional manner at all. All the indicators brought about such positive results that the American government granted the licence for the first experimental operation conducted in a human in absolute confidence the procedure went about and with a very good outcome. But shock was looming to all medical experts. In every case the patients had one common disability. It showed a horrible disadvantage no one expected in the least.

All patients showed the science of a phenomenon later named after a movie Peter Sellers made famous. It was named the doctor Lovejoy syndrome because on of his characters in the movie was an eccentric half mad all crazy German general that had one arm always trying to strangle him. This was meant to be funny in the film but the patients suffering from the reaction of the aftermath are not so inclined to the humour. They all had one arm that went out of control and the arm showed serious signs of having a mind of its own by doing the most annoying things the patients had obviously no control over. The one arm had life apart from the person free will doing things that would embarrass or even threaten the so-to-be owner of the limb.

Well I am no brain- surgeon although I am inclined mostly to form an opinion of my own that may not always stroke with informed opinions by professionals. The test operations were conducted on a variety of animals including monkeys, the so-they-say close relative of man. Well as close as they can get but I am of the opinion there are other species being still closer to man, but that is somewhat off the point. From all the facts I mentioned the past pages I drew a conclusion of my own.

I showed that man has a higher evolved form of life than other animals.

In the Bible it reads that man was made the last but far from the least with more superior qualities than all life combined because man is all life combined and then added more than the fare share. Any one thinking of our Creator as a magician is mad and that I state without excluding even the Pope or preacher of whatever denomination. The Creator is a building architect applying mathematics and physics we can never come to appreciate. If there were any one that has an opinion that God spoke a word and magic was the word that person would have another opinion when thinking with some clarity about Creation as a whole. The Creator is Creating and by creating there is a building process involved. Every person starts an individual process of building a human just after birth. Looking at tribes living in regions far away from civilisation as one can still find in the Amazon River we find those individuals being adults by body still play games we find our children play with much amusement and childish enjoyment.

It is far from incomprehensible to make some sort of comparison between our children at play and those adults at play and the similarities are astonishing. Racist remark it may be but the grown ups are more child than the western child is child. They defiantly are backward in mind and mentality. The most logic deduction is that the child in us represents our development phases through a long journey. Humans at birth are animals. Babies can make noises to convey their needs and nothing more. As the little life grows it develop not only by growing but also by culture and what the parents put into the culture. It is more than likely that the developing pattern children follow is the same developing line humans evolved through as the generations brought more insight and better understanding to the following generation. It is only of late that there is some pattern of devolution taking place especially in the western world but that trend is set by a culture of greed. The parents are chasing a good life and instead of placing morals they push money in the hands of their children to get the children out of the way so that the parent will lead the life they choose and rid them of the burden of children while conveniently blame teachers for the children having unacceptable behaviour as much as they pay psychologists good money to correct they're mistakes. It is moreover a fact that parents do not actually pay the professionals to fix but pay the professionals to rid parents of responsibility, guilt and of course the children. The price society will pay for such luxury is far more expensive than affordable.

I so many times wished I could have my life over again. There is one study I shall conduct and that is follow the pattern the child indicates how the human developing process came to pass and draw the parallels from that to man's evolutionary path. It truly must be a study worth one life. To know man that well must be the ultimate there is to know and afterwards death can bring no regrets for wiser no one can ever be.

I think we should now return to what the Bible refer to as the tree of Knowledge and reflect once again on that verse. The quote was and I quote in firstly Afrikaans: Boom van kennis, kennis van goed en kennis van kwaad". Directly translated it reads as follows "tree of knowledge, knowledge about good and knowledge about evil" and that is where my argument starts in my attempt to indicate the possibility of the sixth dimension

From the stage of toddler I found eve dice of three characters in all persons. There could be more but not less. Let us call the three persons three entities of the good the bad and the ugly, but not in such a specific as being a saint a pleasant and a demon. The entities are rather more under cover that that straight foreword by definition. In every person's brain somewhere there are the trinity of entities ruling our lives.

The trinity are one person and not individual persons but the same although very integrated they are also very separated. It is not a question of schizophrenia with multiple personalities because I do not believe in that. I think that was made up to convince who ever needed convincing about something to do with nothing and has no base or then has the same base as physics hold gravity responsible for a variety of facts they otherwise have to admit they no nothing about. The three belongs to one person and in fact is the same person. In the Afrikaans book I named them Ek, my and myself translated as I, me and myself. To make matters more interesting and subdue confusion somewhat I wish to keep the Afrikaans as that would make explaining slightly less complicated. Ek is I. My is me and myself is will you believe it myself. The "My" one pronounce just as you pronounce the month of May in English and that would make the pronunciation of myself as you would say May-self in English with the pronouncing of self in English and Afrikaans exactly alike. Ek my and myself are the same and there are no distinguishing between the characters but at the same time they are as far apart as three that never met before. Every of the entities belonging to the same body, the same brain one has a character as unique and as far apart as another being on different sides of the universe.

Every personality has different likes and different dislikes and feels a different purpose in life as much as to life Once again I wish to press home the fact that this is (to my view) as normal as breathing and has nothing to do with the mental instability known as schizophrenia. Although only one personality claims occupation of the body at any given time, all three take responsibility for action all the time because all three are the compliment of one.

Any one such personality may claim occupation of the body at any single moment and normally do not relent occupation easily but of the three one is always in charge and the other two take position when the main personality loses concentration or relaxes guard of the situation. Every personally has own motives being apart as far as the north may be from the south. One may even think them in classes of being the person's personal god and personal devil but such a thought may place boundaries that are unfair. I would rather describe them as one is the charger being scared not even of the devil himself and the other will be the cautious the guard, the one always on the lookout for trouble coming. They form the one that is in charge and the one in charge cannot be excused because it was a totally foreign entity pushing in charge in the direction it never wished to go but was blackmailed in doing that wrong! The one in charge takes responsibility to the full for each dead the body did and every wrong committed. In the end they are only identifiable but not that clearly deniable and they always appose each other in complimenting one another. In the American cases where the neurosurgeons performed the operation in cutting the cortex to split the brain lobes of the operated patients a condition became a situation where the persons found the uncontrolled motorised motions of limbs under their control supposedly, but not under their control at all. Then for the first time the phenomenon became a syndrome with a name. It became "the alien hand syndrome" but also find referring by the use of the Doter Lovejoy syndrome named after the movie. There are cases known on record to result from a variety of brain damages and severe apoplexy. I can confirm as a witness that in the cases of my biker friends their faces changed with mood swing. Not bone structure or complex feature but the facial expression changed the way the muscle form the face. In cases they looked somewhat alien, but my referring to as normal is not about such extremes.

The alien hand syndrome observes cases where the one and the patient refer to as the naughty hand and the clever hand. The naughty hand seems as if it had a life and mind of its own sometimes acting to embarrass but some times even to endanger. The main connecting issue between these cases is the hands control not being within the owners authority and that the hands will obey as nothing ever change but then on occasion from the blue it will act on own impulse and with a motive clearly never matching the owners intension. I wish to underline with no exclusion to whatever intension that the alien hand syndrome does not prove the trinity that I refer to and the alien hand syndrome underlines two facets where as my view state clearly three identities. The reason for this mismatch holds a most intense connection to my personal religion, which I never share, with any person out side members of my immediate family. The alien hand syndrome only confirms (to me personally) a connection of some sorts in some way.

Many years ago after I had to overcome some personal problems at the time through which I was admitted on occasions to an institution. I spent time in nerve clinics where I had to recover from some brain disorder called endogenetic depression. It is a condition that the patient will get severe attacks of depression and the cause of this is totally inherited by nature. The disease comes from a bad gene carried from parent to child and as much as my father suffered from it so will my children and is not very scares or very serious. I would say it is as serious as you make it to be. My regrets about suffering from this condition is minimal because of the extent of learning I had the opportunity to come by that otherwise would never have come my way. But then from what I saw in the times I was admitted to the clinics I mentioned, my (unprofessional) conclusion is that in just about all cases the prognoses depend almost entirely on the patients willingness to recover and how serious the patients are about recovering as the recovering is al about the relevancy struck between obsessions and true problems. In that there rests a balance between fighting for the sympathy they wish to evoke in others and fighting a battle to achieve oppugn from the problem. In hindsight and after all the bad is forgotten it was worth wile because the learning of the human mind and the way others think and feel was enriching beyond my suffering. At the time the suffering was almost overwhelming but to escape ones own problems one can listen to others and by learning from them one get perception about their state of mind and your own state of mind. It was in this period I came to the conclusion about the trinity within us all. By increasing my personal learning curve I decreased my personal discomfort.

Inside all of us lurks to personalities above and beyond Ek and they go by the names of My and Myself. Most of the times they behave well but sometimes they can bee opprobrious without my consent in the matter. One of them and I leave the choice up to you are truly obnoxious and spiteful more to yourself than to others. The other one is normally timid and can be classified generally as your conscience. Between them you come and take the control because you are the boss. Being the boss and demanding control you are the balance as much as you take blame and shame when not being the boss and not in control where matters go out of control. You carry the consequences always as you should because whenever what ever goes wrong others will pass the blame on to you. And so they should because you are responsible even in such times that you are not responsible.

Ek finds himself in the middle of My and Myself and as Ek is in the middle Ek get advise from My and Myself. In accepting or rejecting comes regrets and jubilation but ultimately the final choice is with Ek. The characters are strong at times and are weak at times limbering and dominating whenever opportunity presents. In the Afrikaans I identified the character by names but found it somewhat complicating the issue in the English and so I did not name to identify by character.

Ek takes full responsibility for all actions for Ek is My and is Myself by only being Ek. We all have this in us and some find ways to fight it better and some of us are in a desperate fight for sanity and survival. The issue is not the guilt because every one is guilt ridden as we all have to cope with this behaviour in some degree. The stronger any one denies this the harder the one is in a fight for self-protection. Cases are sometimes most serious and other times just under the skin but it is there in every one. If you are human you are fighting. Being the one finally carrying the burden then becomes self.

It could be where he father feels he as person never accomplished anything and with that self detesting he puts all his blame, guilt and rejection for what he is into the passion he feels for his child. The passion he feels is carrying the burden of hate he has for his image and that drives him to expect from his child what he never could achieve. The child has to be many times better performing with many times more positive results, be a champion, be a scholastic genius, become the school president and make a mark as a pillar of the community although the child is only a child. The child has to outperform all others on all terrain in a fashion fitting the image of a champion, which the father never was. It could be as serious as I indicate or it could only be in little suggestions made to better his child.

These same feeling can bring about the very opposite where the father tries at every chance he gets to run his child into the mud because the father fears that the child will push him out of his role he has to fill but lack all abilities to fill the role. By destroying the child he is protecting he child because he is securing his position as the father figure. Knowing well the chid carries his genes and believing the genes the child is carrying is not worth much, he protects the child by showing the child what he (the child) is and allowing the child the realisation before the child will one day find out for himself in the cruel world. There is no good as much as there is no bad, and in the same breath everything is wrong.

The same behaviour may come from the father in fortune, the one with success, the pillar of the community. He is the top judge, the success fill attorney the town's top businessman the city mayor the admiration of others. He hates his child for that child will one day inherit all the goodness he now has and that makes him sick. He might feel the child will never become the doctor he now is but through his reputation that he worked so hard to achieve will become even greater than he now is. That makes him to push the child to live up to his personal greatness or destroys the child to show the child how much the child should be great fill for the admirable fortune the child has to have a father such as he.

It does not have to be about someone you love. I shall be very frank about my case and millions will recognise their fight.. My personal struggle involves money for one. I have not the slightest idea how to administrate my money affairs and sometimes I know I have this inward hate towards money. Whenever I have money I allow people to sucker me with a sob and crying story about their hard life and the bleeding heart dashes through, the knight within me with the glittering armour takes full control and helps me to give away sometimes even thousands and tens of thousands, knowing very well notwithstanding all the promises of repayment at the time, I shall never see the money again. It is not I being the bleeding heart and then yea it is I the bleeding heart but that is one character. I the bleeding heart am on the background where I the hating bastard am rite on the dot standing on all fours and then some shouting and urging me to help the others, or buy what ever. Forever it is ensuring me that there is thousands more coming my way in any way so what the hell, let some goodness flow. This will always come where for some reason some money fountain runs dry just afterwards and I drop my family into financial surviving periods allowing them, the ones I love most to suffer the hardships. Not once did this occur and was not the prelude to personal hard times! I am a middle-aged man. I know these characters. I recognise the precise feeling accompanying each one. I have scrutinised and analysed they're being part of me decades ago as I and the bastards still catch me again and again, sucker punching me over and over.

The first time I went gambling I recognised the one being very aggressive and on the forefront just under the skin filling me with anxious excitement and then and there I realised gambling was not meant for me. In that sense I have beaten him hands down by not starting the habit. Old positive me is about cars speed and going crazy in my mind and I know it is the strongest one because that one am the biggest I. A middle-aged man tearing down some street on a massive motorbike showing some youngster what the bike can do when there is someone on it that knows his onions are grossly irresponsible. How childish can you get, how in mature can one be. That is the I and that is the very me and that is Ek and I have as much control over that as a drunkard has control over his drinking. Saying that is also admitting I do not wish to fight him as I do not wish to beat him. I find his company very pleasant stimulating and destructive and when he comes to the foreground all other characters harmonise giving me all the different feeling each one of them should supply. I am then self-destroying giving the negative character his day, I am childish giving the neutral character his day while the positive one and I am the same person to let all power loos.

My characters bring me in conflict whenever I bay or sell or demand a price for my services rendered. I would bay the biggest shit at the highest price in the belief I am helping the poor slob. When I sell I get all guilt ridden when trying to make a profit because I get this idea I am cheating the poor fellow by insisting on the price I am aiming for. In all cases I feel so guilty to ask any person money that actually belongs to me I get a nerve attack or a running tummy. It is not out of fear because if any person gets violent or aggressive I know how to defend myself. You have to know that if you are a devoted biker because they always get drunk and strong at the same time and fortunately I was in motor racing during the age my mates learned to drink. But with racing cars always braking down and crashing there is never money or time to drink so I never got around to start the habit and by the time I stopped racing every one accepted me as a teetotaller so I never got around to develop the habit.. But asking for money or insisting on a fair price is more than I can achieve.

I shared with you this Ek, My and Myself part of the personal me-story so that you, every you will know what I mean because the objects may change but the objectivity and feeling and the motives never change. One example how the process works will be when a mother that hates her child for birth pains, unwanted pregnancy, feeling unfulfilled through a bad self image or carrying on where her parents left off with them treating her as a child in the very same manner she treats her child. The negative character prompts the mother to reject the child as she detest the child and loathe the child. The Positive character saddles her with a tremendous guilt punishing her as anxiety in blame riddles her.

In self-punishment aiming to destroy her because after all that is the purpose of the negative character the positive character punishes her as she fills her with self-loathe while neutral character tells her about the wickedness in her. She gets reminded at every opportunity about her love she must have for her child and her duties as a good mother to love her child. It is her responsibility to love and protect and because she strives to be the good in her the punishment is severe.

If she gave in to the negative character and start enjoying the hate and blame she feels towards the child the positive character let loose the pre-historic maternal instincts with a flow of torturous guilt hidden behind such strong emotions she deflects the hate onto her self. Then comes the neutral character reminding her that no one should ever know about her hate towards the child because as she detest herself so would the world detest her if ever someone became wise to her feelings and she will be driven from society with hammer and tongs. With the conflicting hate polarised and swinging between her and the child she knows that all of human kind will see her for what she is and with the hate she feels towards herself the negative character takes that hate and turn it into fear for others finding out about the truth. In realising that no one may ever know about her true feelings toward the child and know about the loathing she feels about herself she takes all precaution to hide it from the world.

In this mind game of rocking emotions the positive character supply her with advice in how to take charge of the situation never to the benefit of the child but to her benefit in protecting herself from the outside world. The advice will never have any concern about the child because she carries that burden by herself. The positive character continuously reprimands her of her evilness while the negative character fills her with hate running between the child and her feelings about herself. The neutral character reminds her constantly not to allow any one to find out about her feelings for her child and demands protection of the outside world finding out about her and her child. The neutral character pushes the fear to match the severity of the onslaught by the other characters in order to maintain equilibrium and equilibrium means almost insanity

As the insanity at times almost become intolerable she reflects the blame for her situation onto the child being there and making her life hell. This the negative character grabs by prompting her to punish the child as severe as she can and through this stop the child torturing her. The cycle leads to the next cycle and in this sanity guilt love and hate becomes one flowing emotion of disturbance. With this conflict within the mother the child's developing personality receives knocks the child cannot stand and less understand. The child starts behaving rebellious and unacceptable to the mother's neutral side and in the eyes of the community. This chance the positive character grabs and being positive only as far as the mother's well being goes the positive character advises the mother to leave the child and let be. This will show the world how much she loves her child by refusing to even punish the child when the need arrives.

In this advise the negative character joins by advising her to let the child become out of control establishing the fact that should her hate towards the child ever leak out, the world will not blame her for every one in the world that has contact with the child will hate the child in any case. When every one despises the child the negative character swings into action by filling the mother with more hatred towards the child and when opportunity comes and they find themselves alone the hate comes to the foreground and then she punishes the child with most cruelty. This can be as part of actual criminal prosecutable child brutality or it could be most cunning and devilish in conspiring but the brutality is all the same. With the presence of such severe child brutality all others in the community turn a blind eye not to get involved where each outside person will find some excuse not to become involved. As every one has a struggle of their own they too are in self-protecting not feeling the urge to come into the open and defend the child. Being in the open will unveil their personal fight for survival and they then will become the target of the community. All this is in the very distant back of our minds never in front where we can kill it but present in the way to be us and not to be us.

All three personalities agree on one thing and that is that the world must with all its people have a hate in the child as much as the mother. But as every one find the child unacceptable and revolting she can feel better about her feelings because now she is part of the crowd. Being part of the crowd will bring sympathy from others with their understanding about the hell and the torment this child inflicts on her every day. Such a feeling soothes the aguish of the wrong she feels she is committing as much as the wrong the child is committing to her. By allowing the child adverse behaviour and defending the punishment of such behaviour the child will become more unacceptable and the situation heads directly for the disaster she hopes to accomplish. She directs the child's personality in that direction while she feels all the torment others see her go through. She allows the child to go to nightclubs doing drugs and commit self-destruction while the mother is merely a spectator because after all that is the child she hates.

When the child is out at four in the morning she can feel good about not having the pest around. She can feel good about her hating the child. She can feel good about all the sympathy she receives about having such an unruly child. She can live the life she claimed, hating the child, feeling sorry for herself and good about others understanding what she is going through. Should the father, a teacher, a policeman or any other figure try to stop the madness and bring order to the child she will attack that person with all the hatred she feels towards herself and towards he child. She will destroy the prevention of her self-destruction because any body trying to discipline the child is fighting her aim to destroy the chid and after all then the world will see how she can fight for her child's protection by almost putting her life on the line.

This becomes the Sudan affair where the bleeding heart buys guilt relief and be god while the philanthropist pushes guilt as hard he can and be god to collect money on behalf of the Hoggenheimers that then can be god with such wealth distributing it to the Mammonites who can be god by baying from themselves as much as selling to themselves with unscrupulous profits making him god and allowing the Mammonists to be a slave driver and being god to the slaves. It is this sickness of society no one cares to see because every one gets what they want, even the luckless get what they want with the minor condition that when the luckless suffer most that becomes the region where most profits are for every one in the chain of gods. So the luckless must be in crises starving as they are dying to gain most profit for every one. The profit has little to do with money but with being god. Any attempt to stop the situation will never be tolerated by any party and therefore my remark that every one will press for my castration because of my suggestion to rectify and bring a solution.

The mother will fight any and all positive solutions with tooth and claw and no one should dare to lay a finger on the child because she know how successful she is in destroying the child and it is so easy to shout child abuse when someone wishes to correct the ways of the child to the benefit of the child in the interest of the child. Her devoting love will protect the child from such brutality as what a good hiding on the backside will bring if the hiding is done in love and the child knows it was on behalf of care. But that will stand in contrast to the mother's brutality and punishment and the child may recognise the difference and wise up to the difference.

The social worker will never accept such brutality after all it may cure the little brat and put our social worker out of a job. The lawyers and judges are on such a big job creation drive by minimising penalties and getting criminals back on the street for the next cycle in crime they will fight any interference that may reduce the crime and decrease their chances of money making. With so many to loose so much on the one side and only the child to gain on the other side all brutality in the name of positive punishment will become child molesting and will never be tolerated by those with influence in society.

The mothers behaviour becomes a reflection on the disease within society and it is in everybody's interest but the child's not to admit to any knowledge about the foundation behind the scenario, after all it is only a child going down the tube and to top it all it is a child no one cares for. Is there anyone out there that can see the parallels running here between the animals not caring for the unprotected infant crying in desperation for a mother while every one ells in the species cannot be bothered? May I now comment on the fact that we are going the way of the animal and devolution of our species is in progress?

Weather you care to admit it or not but greed and money is destroying man to the fullest while man is enjoying the destruction with all its lust. The drive in society after W.W.2 became progressively to feed the children to the hyenas of society, which are the crime bosses, the prostitution rings, the drug pushers because after all, they bank the money at the Hoggenheimers to the convenience of the Mammonites and Mammonists. The ones that are caught in police action are the ones not part of the official system and they become the offers the politicians demand in protection to show the public the system is doing what it can but unfortunately it can only do that much and if it is not good enough it is because we all are human. Hundreds of thousands of children disappear through out the world and no special task force has ever been set into action to get behind the problem This problem has no boundaries in as much as it is going on in every country there is world wide. I have an awful, awful feeling and please consider the next remark as a thought with no substance but that internationally oil is bought with children as payment because no politician through out the world shows much concern. But let three banks get robbed in one day, then a special task force comes into action and gets the culprits with extreme prejudice. This is the symptom while the mother's behaviour is the condition.

The positive character advises the mother to defend her child against disciplinary measure, the neutral character reminds her of the image problem and demands protection while the negative character sees the child slip into the ditch and everyone is happy. Every aspect is in line with almost one aim and that is the destroying of the child. What ever may bring positive results everyone shout down by making the connection where the punishment links directly to what may be extremely negative because from the onset it seems cruel and negative. Giving the child the spanking of his or her life driving the fear of god into the young person will have extreme negativity but that must stand in complete contrast to the love the child then must receive before and after the spanking. The child must know with one hundred percent certainty that the parent is and will always conduct the child's care with one aim and that is to ensure her or his well-being. But if the child knows that every aspect of the child care swings around the drive as far as getting the young person destroyed the yes, the child will find all connection to punishment intended on the destruction aspect but it will remind the child of the similarity and lack of contrast in the usual treatment because of the absence of the love and caring aspect is in harmony with destroying.

The whole aspect changes around when there is another person also punishing the child but with prejudice intended. When the punishment comes from despise and not from care the mother sits back and allow this to happen. If the father is sexually abusing the child the mother will suffer greatly all in silence all quiet not allowing outside intervention spoil the situation because then it is the father who is to blame. It is the father that is destroying the child and it is the father that is the devil. The other parent can then take responsibility and blame and the mother becomes the second blameless victim in the case where she does not participate in the abuse only because the father then plays the part. That is the only aspect that changes where as otherwise the scenario remains the same. All characters play their part as if she is doing the destroying because she is doing the destroying by helping to provide the perfect environment for the destroying and not allow any clue get outside the close knit intimate family circle. She takes a part in the abuse and takes as much enjoyment as if she was acting although the blame and the soothing shifts somewhat but all intensions still encircle the destruction of the child.

The conflict within the woman may drive her to protect the child she hates as much as the father does. Her neutral character tells her that now she is no longer to blame therefore what ever happens she can stand in the shadow of the male and he has to take the blame. But the negative character will not tolerate such idleness. The negative character wants the child's destruction but moreover the character wants her destruction. He advises her to action. When the father is at his most dangerous being overwhelm with cruelty she will jump in and save the child by physically protecting the child while knowing full well that she can do as little protection as the neighbours budgie can. This action will satisfy the positive character by her showing her unbound unlimited care and devotion as the epitome of true motherly love. With her actions she know she will unleash much more anger and the father will loose all control. The negative character finds stimulation in this action and supports more involvement to unleash more violence all the way. In such a rage she knows the father will then beat the daylights out of her before he turns onto the child with more rage than he had before. With such reaction all three characters are satisfied The positive finds a way where she can become good, the neutral character knows that society will condemn the father and the negative character will justify the cruelty as the correct way to go because the child the father and the mother is bent on destruction. Now the positive character can tell her she did what any good mother would have done, the neutral character knows the beating is the same as what she endured as child in any case and that did not kill her so this beating cannot be that bad while the negative character will enjoy the situation to its full as she and the child is being destroyed.

The balanced behaviour of a person with the trio not having violence to promote would run outside and call outside help from any source available at that moment. A woman beater and child abuser is always, always a coward and when real trouble arrives he will stop immediately. But if the trio is involved in violent provocation exemplifying hatred to the child, she would take charge in a different manner. Even if she does act in this manner she will not call the police or get the husband behind bars because she argues that the family will suffer with the father not providing at the time. No one receives any money while being locked up. The excuse she uses is that there will be will no one to provide for the needs of the family living expenses. The father's inability to provide is all but the truth as she wants the situation to continue because she is enjoying it as much as the father. Should she truly admit to the seriousness of the crime she would have the father locked up as if he died and never allow him close to any member of his family again. To her and her child the best will be if the father is dead because the father will destroy where ever he involves himself. The only sane thing to do would be to declare the father dead as far as the family concerns go because when he is released from prison she would have a new life to live that no longer will depend on him or his providing. She would recognise him for the monster he is and not see him as the senior partner in crime, which he truly is. Keeping this partner ship in place by using a lame excuse like ninety nine percent of woman finding themselves in such a situation uses would satisfy her three characters and once more the money matters more. Now the stage is set for the beginning of the next round because with the father as with her the creation of the next climax begins and develop until the next time hell brakes loos where the father then has even more hate and a lot more to prove and correct. His fury and outrage will cover his hate but also the negative character in him will demand revenge as compensation for lost pride. He will repeat his role and she will repeat her role and the child will have no role but suffer destruction and again and again the process goes on and on.

By throwing her body in between she knows no woman can stand against any man in a physical fight. Her excuse for aggravating the situation is that it is what any good mother will do and that also becomes the advise of the positive character. She knows she will outrage him blowing his week self esteem out of reality because now she takes him on as a man insulting him at the area he feels the weakest because he acts in the manner that he does because he knows from his weakness he can never manage to be a man in the company of men. The outrage she unleashes within him will satisfy the negative character. Reminding him of his coward ness and weakness will bring the monster in him to its full potential and that is exactly the plan. By her action he will be reminded of his weakness and that will be in response from advice given by the negative character as for the violence part and the shift in the blaming will be in responding to advice given by the neutral character. The child will now see how much she cares and that she does not hate but love the child to a point where she will sacrifice her life to protect the child. That will please the positive character. Then afterwards when the beating or molesting or whatever cruelty is completed she runs off with the male as the lioness did when the lion killed the cubs. The mere fact that she still remains with him runs parallel with the lioness accepting the animal behaviour because the human litter may not be dead but that is not her fault and in any event the fun can continue in full rage the next time around. Why call a halt to all the fun because everything will be alright after the husband pushes a few hundred dollars in the hand of the child or bay him a brand new whatever that will sooth all pain going around. He does not care for the chid as he can bay the child's silence. She does not care for the child because she allows the situation to continue regardless and the child does not care for the child because no one cares for the child in any case. After all the father did show remorse when he bought the child a new whatever. The theme is about money. If the father is caught he would most likely get a fine because the penal system does not encourage incarceration for such minor crime. He can bay his freedom even by penalty of payment and money wins the day.

In the event where the child stepped out of line and the father (or mother) comes down on the child as hard as he can, to shock the child with such force as to scare the child so much the child will fear any thought of repeating the incorrect dead ever again, the other parent holding the hatred will then come in and phone the police, contact the magistrate, get the executioner in preparedness and go on with all dignity going to madness. She will never allow such abuse. She will rather see him dead than punish her child. She will make such a fuss and such a scene that should anybody not get involved they will become assessors to crime and child brutality. This again is a charade where her characters enjoy every second of her instable behaviour urging her on to over react. Now the moment has arrived where she can show the child who is the devoted parent after all She can show the world just how much she cares for her poor little baby that only went about to destroy the child she was destroying in any event. She will never allow the destruction to stop because of an action a balanced parent sees fit. The law comes down so hard on this man as they wish to scare off any other parent that will ever try to correct the behaviour of a self destroying child. After all where will the next generation of criminals come from if there is a well balanced society and how many lawyers and judges may become jobless.

Every one in an influential position throws their weight behind the mother's instability destroying the caring parent for caring. The judge himself has three personalities to fight and no remorse but to uphold the law to the letter. The politician that helped creating the law realises there are many more unstable persons out there to vote than stable persons and with the majority being mad it is clear with whom he will side when writing the next law into the law books. And besides he also has this little fight going on inside himself. His negative character has an enormous advantage because to him was not only given one life to destroy but so many it can keep him busy as long as he can remain in office. His neutral side tells him to remain in office because that is such a lovely place to be and his positive character now with him in office can be the god he always new he was.

The madness runs deep as it runs wide leaving no pillar in any community in strength. The description I give may be mostly in exaggeration of the truth but the truth it is. Even if the slightest way of is applying or of finding such evidence that this is taking place it will show that with the least provocation the balances is shifting towards destruction and the way society is, is an indication pointing out more than just strongly that the human race is on the decline. The truth stands out as a sore thumb showing that there are massive problems waiting for man on his spiralling way down the devolution ladder.

This is part in all layers in society from the very rich to the very poor and every one puts the blame squarely on the others without accepting any blame. When this child becomes an adult the process not only continues but worsens. For the sake of argument let us make the child a grown up male. The child as an adult misses the mother that he had but also never had. From this he acquired the loss he felt with the loss he does not recognise. Now he is ready to find a mate. But being human we humans have the culture that mating is a life-time commitment.

The wife he is looking for must be someone like his mother but with a slight twitch. He hates his mother by now as much as she hated him all through his natural life. He confuses love and hate, compassion and punishment, caring and rejecting, in a way that allows him the freedom of becoming a most confused person. With all confusion running wild and seeing love and caring in the same light as not caring and running wild he goes on the prowl in search of a wife. He longs for stability, which he totally rejects. He wishes for companionship he finds to smother him. He hopes for security he does not care for. In everything there is a threat. What should bring devotion brings hatred because that is what he recognises but does not care for. Will this young man's three entities have a wonderful time. He came into the world unwanted because of free love, he was raised in anger because of a free and fair society, he was neglected because of democracy and now he wants to employ the culture that brought him destruction as a child. In loving his wife he hates his mother. In wishing for her companionship he fears for his life. In pleasing her he understands only rejection. He is even more bent on destruction than his parents were. Where there might have been someone that tried to show him rite from wrong he got that show of caring as massage that that is a person is out to get him. Any person alive including his wife that will make any effort to show him the wrongs in his ways he will reject to a point of committing violence. In that he will find the rejection he hates so much and he will kill to destroy that. He may or may not administrate violence when at home. That comes with the role of the dice.

The wife he loves he has to hate because to him that equates mother-love. To him loving someone is destroying that person. The one he cares for he has to destroy. That is the love he was taught to accept. This leads him on a journey of more self destruction than any attempt his mother ever made. His yearning for the mother he never had pushes him on to every woman he can find. His wife being at home does not please him because she is not the one he was looking for. His morals are mingled like a mixture of concrete. The good and bad are so intertwined he cannot see light from darkness. He starts a life of adulterous affairs partly to destroy his wife whom he confuses with his mother and partly on the hunt for the mother he has a desperate need to find.

With devilish cunning this sets his three characters in motion where they can join forces and destroy at will. His positive character allows him to bestow the love he feels but cannot share onto the woman he is with just because he does not love her. His neutral character finds her acceptable because it will last but a night and the negative character helps with the charm because he will once more destroy himself and the woman at home, which as a matter off fact he truly and dearly loves. Unfortunately he does not recognise the love as love because he does not know the feeling of love. The love he recognises as love is the feeling he feels for the woman he is having the adulterous relation with because he hates her as she reminds him at that moment as the female figure representing his mother. If he really hated his wife he would be at home destroying her but as he loves her he does not whish to destroy her so he destroys her by not being at home. The positive character puts all his attention into charm that he throws on the female he is with. But because it is the positive character it also helps him realise that he can have fun in destroying the adulterate as the adulterate is the one destroying the one at home which he loves. The neutral character tells him to carry on because he has to punish the one at home for not being the mother he is searching for and the negative character is in heaven as every one around goes to hurt.

He believes that his wife at home truly loves him and for that he does not wish to hurt her but in the connotations he has about love he also knows that if she truly loves him she will try to destroy him. After all that is what love ones do when they show love. But because for the simple reason that she does not try to destroy him the neutral character holds that against her and make him believe that she is acting in such a way simply because she is not caring about him. Such behaviour stands totally in contrast to what he thinks love is, while his positive character keeps reassuring him of his wife's devotion therefore he should set his mind at rest as he will not lose her. The negative character finds this unacceptable and reminds him to do onto her before she can do onto him. The outcome is a vote of three to none in favour of the affair and another round of cheating starts for another night.

While the cheating takes place the positive character will remind him to love the one he is with as the one he loves is not there and the neutral character will tell him that as long as no one at home finds out then no one gets heart while the negative character tells him should his wife ever find out she is in any case getting what she deserves for not being the mother he wishes to destroy back. He will not enjoy her company and may not even enjoy her sex. He is in search of something and that something she will no be able to provide. During the relation he will get bored and then dismiss her as a dirty rag.

The worst that could happen to her is to let him find out that she has true feeling for him. That would place her in his power a place no girl will wish to find herself. The mother hatred will come to the forefront and he will start destroying her as he then can inflict all the injury he does not wish to inflict on his wife. By chastising her he will find some accomplishment and relieve and seeing her anguish will fill some of the need he finds in repaying his mother. But that will only bring some satisfaction and it will only last for short periods. But he will still yearn for his mother and in that there will still be a need to run more woman down. Because he does not have a clear image of what his mother was and what love is the characters can play mind games he will not understand. We all get some notions when one with a clear image of a mother and love between mother and child come across a female. We all have thoughts about what may be but we discipline the thoughts because we know we love the one at home and do not wish to sacrifice what we have for what may be. Hell, there is some woman I have met that is as attractive as any creature can wish to be but the very last thing I ever wish is to spend even one night with her. It is not because she is unattractive but to the contrary. She knows what she is and she knows how she excites men and she uses that charm to get men to dance around her with pleasing delight. The worst fate that can ever come to any man is to get involved with such a woman or even worse than death will be to marry her. She is the female of the male I described and she is bent on having men flirting with them and then just throws the verbal cold water on them to enjoy the reaction they get. In a marriage the first signs of trouble will drive her into the first bar where she will pick up the first victim and destroy her partner and the one she is with one blow just because she did not get her way in the argument she had with her husband. The poor slob that someday lands her as his wife will have enough information to write a book about hell.

With the conflict another situation with another disturbed child may provoke the complete opposing figure of which he is in search of.

In the next scenario of the Don Juan now in discussion holds the neutral character in place that will not have a clear picture of the woman of his dreams. With the image of his perfect woman being very vague any of the other two character will come in and pour their versions of the perfect woman in his mind. The reference picture that he has about the woman he wants will be completely out of focus diluting his perception completely. He will want the woman every one desires. The disco queen or the brothel bitch or the bar tender out every night with another guy. He will wish to find the woman that treats him like dirt. Being treated this way will so kindly remind him of his mother and with his wish to please his mother as a child he will transfer that to his wife to be.

He will forever find some tart he wishes to please that has no wish to be pleased as she is in search of a man like her father. Her hopes are to find a man, and usually with success is one that beats the daylights out of her. Her first her second her third husbands will all have one thing in common. They will be woman beaters without exception. Or they will be drunkards, or womanisers, irresponsible persons but that is precisely what the woman has in mind although she will die before she admits it. If someone that truly loves her for what she can be to him turns up she will hate such a person because his devotion confirms her rejection. To her that man is the representation of her father and she in her twisted mind thinks her father was such a nice and devoted man because her characters will never allow her the opportunity to see her father for what he truly was. As her attraction and his attraction does not meet the requirement of their characters he will follow her like a lost puppy because her rejection of him is what tells him of her true love and devotion and that is precisely what she find so revolting about him. She wants a man that loathes her and here is a man that adores her. That throws her characters into disarray just as much as it throws his characters in disarray.

She hated her father as much as he loved his mother and because her father was the personification of brutality as was his mother they had to accept what they received as parents. But since neither had a real figure to relate to their characters turned the image they built around that parent around as to make them very acceptable. The characters they have will for the rest of their natural life bring to them the opposite of what they had as parent. That leads them to the very opposite of what they are in search of and what they find is what they wish for although it is exactly what they do not want. They both are condemned to one life of misery and if there ever is a hell, that place will be a merciful relief when they die compared to the life they have.

I do realise from the examples one must deduct that these cases are only the mental cases and anything more serious would find the person a patient in an asylum kept under lock and key by the President's special request ordered through the highest court in the land but it is not like that. The extreme I underlined because the extreme is the easiest to understand. But the destruction the parent has in mind for the child could be baying the child a very expensive pair of shoes only to let the child feel slightly important or giving money for a movie you would not wish your child to see but from the expectations your child has you cannot deny the child. It could be that you allow your child to go out with friends knowing the next day the child will write an important test. I do not wish to go into detail why that is part of the destruction or why it may be destroying the child because that is not the issue. The issue is the personalities lurking and being you. There is no slip of the tong with some wrong words slipping out. It is a deliberate intentional conveying of a massage to the other person about the true nature of your personal feeling and thoughts concerning the other person. What is the slip is your allowing one of the characters taking control for that split second while your guard was down. The turning of the cars steering wheel landing your mother-in-law in front of an oncoming vehicle while you were looking the other way and can swear under oath you never saw the oncoming car or had no inclination to go that way in any case. Why that happened is a total mystery because you will never do such a thing intentionally or other wise while the truth was that you were quite enjoying the nagging old witch's' company and her on going tormenting in her criticising you in the way you handling her precious daughter . That calling your employer by his first name when you were actually out to impress him. Calling the young girl in the office my lovely in the presence of the biggest gossiping bitch in town and realising this will lead to direct link involving a phone call by informing of your wife within seconds and knowing there will be hell to pay that evening while the truth is that you truly never even noticed how beautiful and smart and lovely and sexy and gorgeous this young girl was. Any thought of her appearance never crossed your mind for one tiny second. In another case your unintentional looking at a girls legs as she gets up and finds herself in a very embarrassing position for that split second while she is glaring at you for being such a dirty old man. Your looking down the blouse of a very breasted beautiful girl as she was bending over while your wife has caught you with your hand in the cookie jar. This is every day incidents with no intension of ill on your part but happened when you did not have full control of that situation for one instant. It all happened by accident but you believe me the deed was as deliberate and intentional as any of the cases I mentioned. It could even be as serious as your kicking a business competitor on the shin while you slipped and almost fell. It is the one character shouting your innocence as the other character is calling on your record always showing good manners and polite conduct while the third will never let an incident slip by.

This can and does even go as far as a nation. I am an Afrikaner Boer and being that I am not blind for my people's mistakes. When four Boere gets marooned on a deserted beach of a desolated island far from any other culture you can rest assured that within the hour of landing between the four there are, they would have started five different Christian denominations and six different political parties. It is a well established historical fact the during the Anglo Boer war at the battle of Ladysmith the Boere had twelve thousand generals with not one soldier amongst them. That is quite typical because we listen to God through His Word and no one ells. This last remark does not exclude our personal characters promoting twelve thousand times three different opinions.

....And then there are the Bible punchers...the ones that will convert you weather you need converting or not from ....well seeing that they never met you before it does not matter what faith you may hold, because only they can bring you absolution because they not only personify what ever they believe their god is but they see themselves as the direct extension of God, a finger or hand of God controlling life on earth.

They are hoity-toity, overbearing and haughtiness rolled in one god given container forming the BIG They. They can recite hundreds of Bible verses for minutes on end and that they believe is the key to their absolute presumptuous claim to God.

They walk with God and they talk with God and they discuss with God matters of mutual concern giving God advise where needed and as they see fit and where God is in their opinion straying from decisions taken at their previous meeting.

I am not referring to the normal God-fearing pious person that goes to Church and feels his thirst needs quenching on Sundays. I have no rite even to discuss any person's religious thoughts and belief. What I am referring to is religiosity to the extreme, a mental unstable drive of laying on the hands to heal, involve every one in religious debates every second of the day, starting to pray out loud as to draw attention of all persons around that should observe how their closeness to God has become as they are in constant prayer and will grant you some time between prayer because you should remember, they are keeping God on hold and the line is busy. Only they have the rite to prayer because after all when ever they get hold of you they wish to pray for you as if you have no connection to God, God has only given him a direct line and all others have to go via the switch board and wait their turn if they get a turn. Normally and in most cases, actually always I let them be but times arrive where they interfere so much you have to put some perspective in them. When I take them on issues their argument has the same logic as that of a pregnant pig and I have learned not to let them off easily. You press home the point and make them as big an idiot in front of as many as that wishes to listen, and destroy their mental thinking ability for the rest of the day. I have had situations where they tried to run away from me and I would run after them taking with me as big a crowd as I can possibly gather at that moment and destroy their image to shit. After such a session they are normally so annoyed with me they ignore me flat where I then leave them alone. I would never do that to other people for no reason can be important enough to humiliate your fellow man. But with them they leave you no other choice.

Not once and I repeat not once in my life did one person ever come to me with the introduction of: I come to you in the Name Of the Lord, and that bastard did not cheat, swindle, rob me, or steal from me. I have reached a point where I decided that should any one in the future introduce himself and refer to "I come to you in the name of the Lord" I'll chase him from my property like a bad dog. They are the biggest crooks and con persons walking on earth.

They too have a massive trio rage where their characters use religion to go out of control knowing very well all people will respect God and their referring to God always bring the other person in obedience. That is when they hit home with the most devious cheating and swindling you can imagine. If someone of their likeliness offers to pray for you don't close your eyes, grab your purse!

They're knowledge of the Bible is astounding but they know nothing about the Bible. They learnt a thousand or two thousand texts and recite them with speed, throwing the one after the other without making sense or allow the meaning having any connection. That is only an eye blinder, a way to astonish you so that you will lower your guard and that is when they hit home. There is no such a thing as a free ride and they always want from you ten times more than what you are prepared to give and when finished with you in the very last paragraph of the small print area it is only all about money but of course "in the Name of the Lord".

You wish to spread a gossip or any untruth, well be sure to use their channel. Normally those services they provide for free but then it must be juicy, unfounded and completely void from truth. Their trio works on the basis that the positive character takes them to personify God, therefore they can do no wrong in the eyes of the Lord. The neutral character advances the notion that they may convert you and that may be useful in some future schemes where you then can fit into more devious plans while the negative character is of the opinion that by your not believing the way they do, you are doomed in any way so robbing the convicted bears no shame. God put soles like you on earth to be useful to their likes where after you will go to hell anyway so what the hell, they might turn some profit before you meet your final demise. After all if you cannot see the light they give you, you may as well be blind and being blind you don't need more than what you can see. By them taking from you and giving onto them they are receiving with a self help scheme what the hand of God on earth deserves and where you are going to lose everything by your departing to hell you might as well start at a point where you are still useful to their blood sucking. Should any reader not believe me take some time and start discussing non religious issues with intent on your part to learn some angles their characters maintain as informed opinions. They are not hard to find. As with all criminal hoodlums hanging out at places of criminal conspiracy they normally hang around at the tents of the evangelistic preachers commonly referred to as the "Happy Clappies".

When I shared time in clinics with persons having some psychological problems this was the beginning of my theorising in this direction. I wish to state once more it is merely an observation of a layman fighting to analyse his own condition and took time to see where similarities were between different humans in the same boat that was sharing a mutual difficulty. Some were alcoholics where I am a teetotaller but still behind the condition I came upon similar causes giving one person one crutch and another person another crutch but it is the crutch one has to loose and behind the crutch you have to find the pain causing the person to grab for the crutch. What ever I share must be taken as not even an informed opinion but as merely another opinion where we all have opinions and it is worth the while to share opinions of an assortment and a variety.

In the last part I indicated persons holding the righteous views about their religion to advance and use as an excuse for their almost and sometimes definite criminal behaviour and malice intent. On the other hand I have seen suffering where these characters dish out what no one can bear. The agony and torment some people go through is of a much higher pain than I ever suffered when I came off my bike. The pain is more real than physical, the fear is stronger than death, and the confusion is louder than not understanding. It is horrible because some of them feel the anguish moreover than they would if a genuine murderer was chasing them. With a genuine murderer you can try and escape or hide but in their suffering there are no such luxuries.

The way they suffered and reality of their hallucinations had put the fear of God into me and made me more than willing to get over whatever small difficulty I had because my luck is worse than the Irish. (To my mind no one can have worse luck than the Irish because they got themselves in a spot on earth from all the places they could chose next to Brits, where the forever meddling and interfering bossy Brits is occupying the very next island) With such bad luck going my way I might just find my problems increasing. I had electro convulsion treatment on several occasions and to my opinion the treatment has a healing affect as the neutral character loses some dominance with the loss of memory through the electric flow. The patient then loses confusion by gaining perspective where the dominance of the entities reduces. As the generating of electric flow brought about by the life factor decreases electric tension in the brain the location of the entities become affected as well

Gravity, electricity, time is all the same thing and in a more or lesser manner influence life and moreover life in the brain. By reducing the electric tension the mind stimulates and as the convulsions allow electricity to escape from the sells stabilising the brain activity and helps sorting the influence the characters have on the person. I must admit that directly after such treatment I don't feel such an excessive urge to speed. Fortunately the condition normalises quickly and I can get to enjoy the exhilaration of my crutch once more. That proved that influences of such characters do vary and can be in dimensions of interpreting and it seems that influences from outside sources can be a major consideration

You may believe it or not considering my poor academic background but I have an inquisitive nature and a need for mental stimulation by acquiring facts and information and that was my academic downfall. My positive character always urges me to test another person's knowledge base and interpretations of facts. This was present even as a scholar as I did forever test my teachers. My neutral character will then classify him in filing order typifying and classifying ranging from brilliant to shit where my negative character will fore ever test the teacher in relation to my personal abilities and from that stance supply the necessary admiration or animosity. I never allowed my teachers at school to escape and when I became suspicious of their depth of knowledge hell would be upon us, moreover on me because they had the cane and always knew how to use the thing. However in cases where the teacher became a source to quench my thirst for knowledge I would eat from his hand. My positive character would take charge and push the others to a silence where no one knew they existed but in the other events of me growing suspicious about the teachers abilities the negative character would destroy any form of harmony that may develop between us and that was the normal in all but a few cases.

In primary school the teachers thought my behaviour was cute but in high school it became intolerable for all parties concerned. I make this remark to indicate that outside influences does play a part in younger minds and through positive stimulation the influences on the characters can be directed to a positive outcome for the child. No one lives in a tight cased cement container never to have an ability for change. It is the duty of the teacher to recognise and direct the children's interests to the benefit of the child and that could lead to the benefit of the class.

I established my theory with all relevant information based on my personal case. While this was going on I also realise there are nothing about me being exceptional or unique and if this applied to me so would it then apply in other cases. With a clear objective I started discussing other people's situation with them to draw similarities that would match my case. It was similarities I was after and not parallels so every time I got behind the whole issue by befriending the person and in that way I could establish a confidence that no professional could. As there was no malice intended on my part and I did not brief the person on my theory or tried to offer remunerable advice I could see no harm coming to anybody. The information was never brought to paper establishing personal files and since gossiping is not one of crutches no information of any private nature slipped past me. If ever any advice came from me it was certainly not on the grounds of my theory so I could not harm any person in any way.

But the more I came involved with other persons the lesser the importance of my personal issue became and the more I detected some golden thread running along lines undetected. In some people I could even detect which character reigned supreme that day by remarks they made or the moods they had and in limited cases there was differences in facial muscles as the mood swings occurred. But in the Afrikaans book which up to now a very limited number of readers had access to, this is the first time I went as far as mentioning my observations to any person. I can even remember the precise moment the light of understanding went up when a psychiatrist Dr. Steenkamp, which is still treating me explained about the mind and the free will of persons' personal thought, the way a person react on they're dissensions and the total absence there are of demons or other spirits that may influence the mind. I state this categorically I have never commented about this theory I have and least of all to dr. Steenkamp so I do not wish for any person to conclude he had any personal opinion about my conclusions. I merely said this because only a few incidents stand out in my life as very memorable and this was one such moment.

What I found was that it was as good as a human trademark apparent in every one, slightly more apparent in some than in others, they are in every one all the same. The entities are mostly absent but come in when a person has his or her guard down or when a person has emotions with an influence stronger than the person's ability to control. All actions man make is with intent. Some might be under the guidance of one or more of the characters but every one has full control over all their deeds, without having an excuse for conduct. They rule your life as much as they are you and will promote your true intensions when ever you do not wish to. Fighting the characters is fighting yourself but you can and you must find a way to recognise them because their intentions are to harm and never to uphold. The characters come with certain emotions bringing along certain feelings. The feeling may be an excitement that does not match the situation or an anger that does not fit the occasion but if one is vigilant you may catch the feeling before the feeling catches you. It may be very slight in irritating handling like for ever pressing the wrong "t" being the "y" on the keyboard or turning a cup of tee over on some important guest or having a dislike in someone you never met before and should not have any special opinion about.

One incident to try and prove my point I wish to raise is from my personal recollection and I wish to share it in order to avoid other peoples' affairs. I suffer severely from acrophobia. Putting me on a double-decker bus is about as high as I can go. In all sanity taken seriously one cannot have acrophobia. One cannot be afraid of heights when steel bars inches thick will prevent your falling. One cannot be scared to look down a glass window when it is closed and you cannot fall through. It does not make sense and yet I can assure you it is a fear greater than the mind itself.

The fear is irrational but should any one try to loosen my grip once I grab onto something he is not only endangering my life but he is seriously messing with his own life. I am aware of the problem and believe I do have a rational mind until it comes to heights. There is no thought, there is no reason and there is no arguing about matters. It is instinctive irrational animal-like behaviour where I go into a survival mode. I cannot fly and yet I know flying is the safest form of transport so much in fact it may be a thousand time safer than my cars or bike, but that is the rational and it disappear when I look down and see something small down there realising it should be big. Even just the realising that I am about to leave the earth is more than I can control.

The fear is almost if not an obsession and becomes uncontrollable. Then one day i stood on an exceptional tall building (well exceptional tall for me a person coming from Ellisras the true one horse town in the middle of a semi desert) of about eight to ten storeys high. I did it purposely to see what the emotions was accompanying the acrophobia because the attack must have some prelude. It can't just hit you like a brick that is nonsense. Something has to form a fore play, a sign of what is coming. Even if it takes one second it still is there. Nothing can just overpower the mind instantaneously but every thing must be about a collective of factors and facts coming together.

Coming out of the lift as I was walking towards the corridor where after entering it I could look down for the first time I intentionally was waiting for what ever to come first and announce the shock. Then as I came to the open I felt the feeling of fear but it was first another fear, one I was use to and knew. It was one of the characters coming to the foreground as if called. That made me realise that it was not the heights I feared but the negative character. I feared the character might take control and make me do what I did not wish to do. The fear of the heights is there, and that is no maybe but that is an extension of the problem and not the problem as such I may not fear the character and I may feel uneasy about the heights and it could even be that I become insecure but the problem was outright the negative character coming to the front.

As my dominant personality losses confidence the negative character comes in. It is not a case of him pushing me or my jumping but it is something going on in the realms of my mind where I do not understand all things all the time. It was a fear of what I may do to myself and not of the heights. Under all of this was the presence of the negative character lingering almost like a shadow feeling not present and not absent but just there. Then came the shock of the actual height and all logic flew away like a little bird. I was clinging and grasping for dear life.

If I can take my mind back as far as I can go back my very first recollection I can recall is a scene where I was on this huge tractor and it was far down. I was definitely under two years of age because I know which farm it was and my Grand Father sold the farm in the Free State that had the tractors before my second birthday. When we moved to Tzaneen he farmed without tractors so it was definitely before my second birthday. I was on this enormous tractor (enormous because of my youth) shouting desperately as I was crying hysterically for help because I remember the thought that I had no chance of getting off that tractor all by myself. I do not know how I got off or who helped me off and being where I am now I must have gotten of because I am not on the tractor any more but that day my negative character and my acrophobia met and got mates. Of that I am sure and if I am correct, then parents should take care not to scare their children in an innocent prank of unintentional fun with their child's fear. It could have lasting consequences to the child. It is not the heights I fear but it is the character taking control even if I know there is no chance of that happening still there is no logic as far as the phobia holds ground. If that is the case with me it should be the norm. People are not scared of objects because objects hold no threat and every one knows that. My humble opinion is that one of the characters is dominant and the dominance is so much that the person feels threatened by that character. The character is in control so often through a depression or an anxiety or a mania of sorts that when situations arrive where the scene should be normal fear becomes the norm.

The negative character brings the threat the neutral character bring the warning about dangers and the positive character joins in by bringing the fear. The positive character brings the fear in to dislodge any attach the negative character may launch. The positive character on the advise of the neutral character disables the body and disables the ferocity and fierceness of the negative's dominance. By pumping adrenalin the body goes numb, the legs go weak, the arms shiver and the body has no strength to function while the person who is in the middle of the junior civil war see object as the reason for the attach but the object is only the trigger and not the show.

I too, am of the opinion that these characters and the way they perform their balance on the day and in the situation makes the hero or makes the coward, depending on the balance at that precise moment and occasion. Phobias connect to the negative character and mania to the positive character and by allowing un- protection through some situation triggering the mini civil war inside the mind; the person becomes a bystander where the person should be the controller.

A kleptomaniac may put something small in a bag and swear by the fact the kleptomaniac did not know about the action. That may be as much the truth as it is a lying because the actions was not deliberate, but the actions were intentional. It is easy to ignore the compulsive behaviour and claim non-participation when participation may have been semi unintentional but still enjoyable. The one character may distract the attention of the person but it is done with full participation because in the end responsibility is with the person. One would often hear the remark: "I knew it I knew it was going to happen" when something was going wrong. All humans can read situations and your mind told you something in the situation were desperately wrong. While the positive character does warn you of events coming it is a deliberate action to allow the neutral character to distract you while the negative character can play for time for whatever occurrence to take place. The whole scene was a deliberate action by the person to gain a negative outcome to produce some suffering or hard ship to some degree because that is why we are on earth. But I shall get to this last remark later on.

This takes us back to the bleeding heart baying off guilt by paying the philanthropist to collect on behalf of the Hoggenheimers dishing out to the Mammonites paying the Mammonists for some slave driving. The actions are deliberate but the true intentions are deliberately unintentional. We are bullshitting our conscience for gaining our mistrust. It is the lye of culture and all participate but some participate to a degree that does not please others. The degree might be to some extend not serious enough to bring commitment and the persons would stand on the side line and criticize without direct involvement because of fear of own guilt uncovering or even of a want to participate while others would come in and rescue but not to save but out of spite because of personal yearning for participation that the person knows would not be permissible.

Another scenario is where a person is drowning. The rescuer comes to save the drowning victim. The victim is exhausted beyond normal mind control. The lifeguard reaches the victim whereupon the victim tries to drown the lifeguard. The victim has lost rational thinking and is then in a mode of action versus reaction. The positive character grabs and clutches at the lifeguard in anticipation while the neutral character tries to survive the conscious and the negative character wants to save the situation by taking the lifeguard with the drowning effort because after all it is the responsibility of the negative character to destroy and destruct as much as possible.

Some gave these characters names. The negative entity goes by the name of a death wish. The neutral character goes by the name of don't care. The positive character goes by the name of optimism. By naming them we found once more a way of avoiding our duty to recognise. It is much easier to dismiss than to admit because admitting has to lead to prevention and prevention is no favourable option.

We say the drowning person got panicky but that is another word we use to escape from reality as much as an excuse in avoiding responsibility. By being panicky we deliberately excuse behaviour and responsibility about actions we may commit to explain irrational behaviour. The negative character wants to punish the lifeguard and even make him pay for his life in interfering with a situation the negative character is enjoying thorough rely while we others only recognise the efforts of the positive character because that will be the nice thing to do. The next time we behave in the manner as to kill our saviour we too can be acquitted on grounds of incompetence. The action of trying to kill the lifeguard is as intentional as the trying to grab onto him to be saved and as intentional as loosing control through the neutral character.

When I first read about the Lovejoy syndrome it brought to my attention that not all mortised control of the body is in the domain of the person all the time and sometimes there are some part of your life that can take control of your actions when you are not in absolute control. It is all a mind game you play with yourself in diverting responsibility with the compensation of enjoyment but the avoiding of dismay about yourself. There are no excuses because the final responsibility is in your power. It is in your power and much more even it is your birth duty to fight the characters but the moments you lose the fight you take the responsibility for the actions because you momentarily lost the fight. It is a win lose situation where only you walk away with the prize as much as the punishment. The muscles are under your control and you are in charge but the entities are little pests being you and are thorough testing you by grabbing control whenever they can. The entities are not only and absolutely negative but are positive as well. We have all been through situations where we admit afterwards we came through by the grace of God. We always hear some one remark that he or she does not know how they did "it" but "it" came through far better than "it" should under "normal" circumstances.

All our phobia, all our desires and all our hopes in achievements we pin on luck or the role of the dice but luck and the role of the dice has nothing to do with it because or achievement good or bad as our accomplishments wrong or correct, and our thoughts being acceptable or not is within us, in our control as much as it is us.

A paedophile should be hanged from the nearest tree because he gave in to the want of the characters and not because "he is not in control of his actions". He wishes for the characters to take charge and even deliberately set up situations where they may take charge, because he enjoys the dead as much as they do and more because they are he. They are in all of us but for some certain behaviour are unacceptable and for others it is not. The judge bringing judgement knows in his grain that the molester will commit again yet he sentences the criminal to a few years of incarceration and is fully aware that their is no chance of rehabilitation because the culprit does not wish to be rehabilitated. He will find his rehabilitation as a death sentence because molesting children is keeping him alive. When the judge do not bring the death penalty he, as much as the culprit participates in the next cycle and therefore must take responsibility and participation in the next round of child abuse. But the judge sits there with the idea " there am I but for the grace of God", which is true in a way but also is not true in a far bigger way. With my fast driving I do not wish for a cure, but I know something is going to go wrong somewhere some day. That is my chance I have to take. He upholds the same argument and goes to the molesting because the sentence is the chance he has to take, and is the chance he takes. You can bet your bottom dollar that should I know before hand the next road race would kill me I would not participate, and the same goes for the molester. With all certainty that the hangman's loose is waiting he will have second thoughts about his next molesting session. Murders always fight for their life by fighting the death penalty. The underlining is that there is no black white and grey. There are no clear-cut defining borders and sides. The neutral character can be as destructive as the negative character can bring a positive out come. When the mob comes out to lynch the actions of the mob are negative in lynching but as they do not wish the continuing of the criminal's behaviour it becomes positive when doing the demonstration. That it will bring conflict to the child's guilty feeling and in that sense their actions are neutral to the child, the participation is positive in preventing the repeat and there by positive in the negativity of lynching where the police prevention is negative by protecting the paedophile and that is positive by upholding justice as it is neutral by delaying another criminal's relapse in crime because relapse he will.

The atheist does not go on a disgraceful child molesting campaign because he thinks there is no God and if he does not get caught there is no punishment. He avoids indecency because he is human. The paedophile may be the biggest Christian around but argues that since he is only human and humans are sinners and sins are alike he can maintain his behaviour until judgement day. Such a line of argument is very typical of the trio being in charge. The responsibility is always with someone or something else and the person never pin it to specifics but shifts the blame to wherever is convenient. And so does the Judge! Hang the bastard and judgement day to him the sinner will come a little sooner. If he does not wish to control his characters help him by eliminating him with his characters. By molesting the child he starts another cycle producing one more child having little control over the characters the child will fight when he is an adult. Stop the violence by stopping the cycle by stopping the one not controlling his characters. If death waits as a surety he will mend his ways or seek help to accomplish change. It is easy not to change and difficult to change but we all can change.

One may have be opinion what I refer too is about the good and the bad, about the saint on the one shoulder and the devil on the other and think well…yea I've heard that before… But it is much more than that. People stop at serious motor crash sites not with the intension to help. If any one admits to that that person is untruthful. They stop to feed the urge. People watch blood sport, not for entertainment but to feed urges hidden deep within the mind. When an armed robbery takes place with possible killing spectators run to the scene. When there is a fight on the schoolyard the word spreads like fire and little else can generate more enthusiasm. The most brutal serial killer always receives the most male. Woman would throw themselves at the criminal misfits with marriage proposals; coming up with the excuse they have enough love to concur the beast's evils, but that is a hideous lie and they more than any one else know that.

They wish to share in the darkness of evil, find someone that could lift the veil covering the beast within. Stopping at the scene of a bloody accident evokes prime animal senses covered by social upgrading but is still very much lingering within. It serves as a reminder about the days when a feast came from such human blood. At some time all of man was a cannibal, man-eating monsters that feasted on the flesh of the enemy after a conquering battle brought victory to some and victimisation to the others. When faced with starvation it is a normal sense that kicks in where people will start eating human flesh. Shocking, as it is... the shock is about realising the urges more than revolting.

We all have the darkest desires of committing unspeakable atrocities and most barbaric acts that will shock the normal mind into panic and frenzy. Some people dislike blood sport, not for what they're purist of heart tells them to reject but the rejection comes from the craving they fear. They fear the need for such beastly acts may linger and run out of control. A story about a mass murderer is best- seller weather it is fiction or truth it is popular. It sells and there is a valid reason for that although our minds reject the reason.

When thinking about these crimes we put it in the basket of the negative because it is where it belongs. No one can ever be positive about such behaviour and with that we create fabrication of truths we wish. It is as negative as it is positive…it is neutral. The good / bad character is not the danger; it is the neutral that is dangerous. It is the neutral taking our minds on fantasy journeys. It is the neutral character we most easily identify with and mingle with. It is the neutral character setting our morals. When Ted Bundy killed the many woman that he did with the brutality that he applied he was not negative, he was super positive. To the woman "forgetting " to lock her door, walk alone down the dark ally at night, stopping to give a strange man a lift or hitching a ride the deed is not negative. They are super positive in their expectations of what may come to them. Sure, they do not ask to be raped, or beaten, they do not beg to fall victim to crime, but lower down in their minds they know of such a possibility and very, very deep, deep down they stand neutral to such an event because they do not mind the excitement or the sympathy afterwards. The bank robber, the pickpocket, the shoplifter, they are not negative they are supper positive. It is being neutral that is the negativity.

The animal cannot see the universe from any other position than the one the animal holds. Singularity placed divinity in the centre of his universe and the animal has no means or brainpower to translate his mind to another position other than his centre where his needs and desires are. From there comes the neutral holding positive or negative as well as the he. Negative means he should fear and positive means the other should fear. In this religion can be one form of atheism and atheism another form of religion. Atheism is the inability to see anything except from the centre of that person's universe and the inability to transform to another position seeing it from another viewpoint other than that persons universe centre. When Albert Fish sent the parents of the child he cannibalised a letter informing them how he devoured their child he was positive and the deliberate pain he caused was positive. Peter Kurten the vampire of Dusseldorf only became neutral with his last victim, but he always remained positive to his wife so much that in the end he forced her to betray him so that she could claim the reward. By her collecting the bounty he realised his final act of being positive as he always felt about her…but only to her and so his final offer he gave her was that the bounty that went her way. To him that deed was as positive as his killing and eating of three children in one night was positive because he was unable to translate his centre point to their position. That is being animal in every sense and animal on two legs. The Boston strangler was always neutral working his way to becoming positive for days although that meant the utmost negative to his victims. Being human and religious is the understanding of other concepts forming a centre way beyond your centre of the universe and moving away from the animal neutral to a human stance Not eating the flesh of the sheep does not in any way make you more positive but it makes you too stupid to see the universal picture of the totality of Creation and its wider meaning. That is trying to prove you are not the animal you know you are but cannot seem to separate from. In the neutral character is the one that no one propagates because that character is much to close to us to be comfortable with, that is the animal, the beast or civil the morally accepted or rejected but still nursed by all because of pre man mentality. From that the other characters stands positive or negative but the neutral character holds the centre and the key to light and darkness in the mind of man becoming man. Claiming the position all other characters including the self occupies a part of the mind. The neutral character sets the tone and the rest will follow. In the neutral character lurks the animal and depending on the person, the darkness of the animal comes to the foreground or stays in the back ground but is forever present in the mind. Identifying with or identifying the neutral character places us in the realms of man or animal. Setting us apart from the neutral character is what produces man and not animal. In the same manner as not eating sheep is the saving of one criminal life where every one involved knows that criminal is destroying dozens of future lives and expanding the problem by creating many future criminals where they then would create some more criminals at a ratio of a dozen to one. In three generations the one criminal is then the cause of hundreds paedophiles walking the earth. That is degenerating civilisation and it is all because the law enforcement from politicians and judges through the lawyers and civil servants down to the bleeding heart and the cop on the beat that wish to prove them not being the animal that the criminal is. By creating the environment and breeding ground for the animal and forcing their positive ness in neutrality they destroy the future of the following generations of man. Law not taking blame for their actions in the neutral stance is as much being a criminal as the criminal's neutrality creating his positive to negative relation. The law enforcement' officials excuse for not wanting to be as bad as the criminal is more destructive and a much bigger misdeed to society than what they do to the criminal because after all it was the criminals free choice to commit or not commit and as much criminal not to wish to pay for his deeds. Either we teach the criminal there are billions of centres to the universe or remove him from our centre but allowing him to become forcefully our centre proves as little as it is destructive to every body. But by throwing him in prison where he shares time with others the same as he accomplishes only that law enforcement may promote crime to establish more cycles for their and all the other criminal's benefit but to the disadvantage of man in general.

The paedophile or any other criminal or anti social behaviour is quite rectifiable where psychologists must teach the criminal in recognising the incorrectness of behaviour, the recognising of the characters and the control of the characters by thought control. Criminality starts with a thought and that is the point I started in my discovery of the characters. Being suicidal as I am and that being my problem starts with a thought starting with a feeling leading to a depression. I am never suicidal on my bike or my car doing high speeds. Then there is only the positive character with the neutral character keeping guard for traffic control. Never do I at any stage exceed my personal limits or endanger lives through recklessness by outsmarting my personal ability where my mind works at the speed matching my vehicle and that is the secret to success. It starts with a thought. It starts with a feeling. That is the gate to progress the characters follow. The remedy is striking a link where one will recognise the thought sparking the feeling and recognising the felling sparking the thought. There is countering the feeling by thought as much as suppressing the thought by feeling but the BIG issue is that you HAVE to know yourself. The alcoholic feels the urge for booze as much as the thought for liquor but when recognising it he counter acts and that is where alcoholic anonymous has such a great success. If he did not wish to recognise the urge as much as he did not wish to suppress the thought alcoholic anonymous can do little. It is the difference between cure and tramping. But it is fighting day and night and fighting yourself knowing you are in the fight for your life for the rest of your life. It is fight you can never win and must never lose. It is continuous round after round with no victory and no defeat, no prizes and no glory. Only shame to follow defeat and the winning part has no recognition for effort or accomplishment but to you. That alcoholic fighting himself and finding his determination is the human victor. To me staying neutral is staying alive. That is what prison should teach the criminal in recognising thought and control thereof. The big courtroom confessions these murderers make are about showing the world how big their universe are, its about how they are the centre of other's universe and the control they had in destroying the others with their universe but it is never about apologizing to show they can understand what others have in the universe they destroyed or that in fact are others with an own universe to have. Should he think he can outsmart be unable to rehabilitate he is animal and should become destroyed like all other raging animals being out of control. He should therefore fear death or find death. In the modern penal system rehabilitation is a word used in the courtroom with no other place to have. Life should be about improving and not sustaining.

Alcoholics and all drug addicts lose the neutral character when intoxicated. The person becomes super positive loving the world for the world loves him right back or becomes super negative by getting aggressive or drunk with remorse hating his parents for what they did or never did. In this the addiction of the parents always plays a part when the child of the alcoholic also becomes an alcoholic. When the addicted becomes sober the neutral character takes control with vengeance, as the addict needs the next round of substance. To the neutral character surviving means finding the next round of intoxication. There is never a balance and when the positive becomes positive, drunk or sober, it shares a spot with the negative and of course the other way around also applies. The drunkard is prone to mood swings that is apparently out of his control, but that is a fairy tale. When sober the same applies, as his moodiness seems to follow his every move. The drunkard uses the substance to escape from himself and his own adequacies he feel about him in his self-portrait and therefore allows the characters a free hand when intoxicated. This we in South Africa call Dutch courage and will have different names in different regions but all names apply to the same behaviour.

Things are normally not as serious as in the case of the child molester or the mass murderer turned cannibal. On average it does not have such deep and intense underlying emotions and fights to the bitter end. It could be a case of the young man is meeting his in laws to be for the first time. The dinner is in great preparation as it is in great anticipation for all involved. Every one but most of all the lover-boy and soon to be in-law is under pressure to impress. He is the new face in the family having all eyes on him because every one knows all the other faces and he only knows his face, which he cannot see in any case. He is out to impress and is tuned for this all out effort of do or die. In this effort he relies without relying on some help from the characters because he can use all the help he finds. He lowers his guard completely to allow the positive character unhindered passage in the situation arriving. In the background lurks another character that is far less dominant in the whole affair.

It will take but a flash in a moment for the negative character to prove a point but as he opened all the doors so wide for the positive character he now is even positive in being neutral. With all the tail wagging he has to go through his negative character takes little appreciation in any possible discomfort on his part. Lover-boy is yearning for acceptance to such a degree this making of his all out effort is rather new and strange for the characters as they show an all out help line helping by the full range of their individual abilities and to use the chance in such an unhindered open channel participation. But one is waiting his chance in quiet anticipation. At the moment of climax where father in law to be wishes to make a toast will be the moment all three characters join the fun by creating the incident all three have been waiting for all night. It will be to the determent of boy-impressing of course. The negative character is very aware that the neutral character may prevent his planned action there fore the strike is lightning quickly. The neutral character sits lapping up all the attention as the positive character is all out helping with the impressing of all around. With every one well occupied the chance comes for the negative character to make his point for the night since everybody was having fun and he had to sit idle and unwanted.

As the tray carrying the wine glasses, which are filled to the top, passes by lover boy, the negative character motorizes the arm closest to the tray and hits it with force. The action holds the speed of a boxers punch and with that lightning speed lover-boy never thought he had such quick muscle movement. The action reaction reflex action reflex reactions is beyond the abilities our boy to be married ever dreamed he had. He is totally surprised at his quick ability in movement and that to his knowledge is miles ahead of his ability to move. In all his life his arm never moved that fast as he turns the tray over on the lily-white table linen.

With him being totally out of place and out of sorts he cannot recognise his ability and that is what his positive character ensures him. His neutral character will be very embarrassed and that embarrassment he places in lover-boy's private embarrassment about what happened. Shouting and proclaiming accident by him as well as the other two characters will bring all members of the in-laws-to-be under the impression that this was indeed an unfortunate incident completely convinced by his very genuine embarrassment. The embarrassment he proclaims are partly his but mostly the embarrassment comes from the other characters proclaiming innocence to him about their involvement and misuse of trust. This he uses to further his embarrassment to exclude all blame of deliberate action on his part in order to impress his in-laws-to-be.

Every one present will feel deeply sorry for him except his negative character that made the point that all the licking may be for tonight but he is still his own man and will have is own way in the future to come. The question is was his innocence really that innocent and was his embarrassment truly that hearty? On the first count no, he was just as deliberate in the overturning of the tray as all his other actions was though out the night and innocent is the last thing he can be guilty of. On the second count, well yes, in a way but not only for all the reasons he proclaims. No one can ever proclaim innocence and non-participation in deeds the person commits. All your actions are all your responsibility. No excuses can be maid without lying through your teeth. Being born means the fight is on and the fight will continue till your last breath is wind. Having trinity around is your birthright and fighting is the option you made before birth and not after birth.

Life of man is about fighting yourself with all the vigilance you can ever muster. That is why you are here having the time of your life for all your natural life. It was your inheritance the day you were born a human. It is not all about negativity but it is all about achievement. It could be quit within your self and it could be in the presence of thousand of spectators. Every sportsman has "on" and "off" days and there is such strong emphasis on the physical that by training the physical no one notices it is the spiritual in training. The spiritual always is the physiological but all preparations are psychological. Practising is about training life to manipulate space-time to the best affect. In all it is only about the physiological. Training is about telling the muscles to obey command and telling command not to obey the muscles. When the muscles shout in agony to stop, command must turn a deaf ear and when command shout to the muscles to go on the muscles must be like a dog and react without questioning or arguing. To describe this we use the name fitness. It is the conditioning of the flesh but it is much more about the mind having control over the body and all fitness is about life commanding the structure in occupation to do what life wishes to be done. Fitness is moreover about controlling the characters than the body. You have to control the negative character's destructiveness to be about the opponent and not about you. You have to control the neutral character to dismiss outside interfering with your efforts. You have to control the positive character to bring subduing confidence. You have to use the fear to your advantage in believing you are fighting for your life and not merely a trophy.

Every sportsman has a story of "absolute brilliance" and always the remark is about the sportsman not truly believing he had the ability in accomplishing what he did. When saying that the sportsman only considers the physical aspect and never the mental drive that pushed him beyond the limits he accepted. On another occasions it is the very opposite when the sportsman declares the day as a disaster because notwithstanding an all-out physical effort nothing went according to plan. His muscles did not respond, his legs were stiff; his arms were not in synchronisation or what ever the excuse for the disaster is. When shove comes to push it is the three characters we find again behind the success or disasters.

The apparentness comes through by him not admitting his efforts links directly to his state of mind and the blame goes to external factors. Everything went just rite or just wrong. It is everything that holds the responsibility for his success or his failure. It is external forces at work and in charge of his luck or bad luck. With that remark he indicates his absence in his actions. His lack of admitting participation and his unwillingness to claim success or admit failure proves that he is relying on something he feels he has little control over.

We all admit that when we are tired we make mistakes. When we lose concentration things go wrong. When being absent-minded we make accidents. That is admitting to the role the characters play. Being tired means relinquishing control, letting go and then the character take control. But also when in fear of one's life you find yourself in super control where you are miles better than ever. That too is the characters at work. It depends on which character takes charge and to what degree does the character take charge. I always teach my sons never to fight a man in front of his girl because you face a man much better than normal. If you are winning a fight leave the other person a way out to escape and never allow the impression the person has no way out. As soon as the opponent gets the idea he is fighting for survival, or he has to fight for position in the tribe in example for the favour of a female you have a monster on your hands and in that case be prepared to fight for your own life. Never allow the idea to enter the opponent's mind he has to win or ells... that will be to your determent. In the instance where you do not allow an escape route or a man thinks he is fighting for a female you have a raging bull and three characters to fight and you do not wish a fight him with his characters on his side fighting against you. That will be your death you wish upon yourself.

It is not schizophrenia I am referring too. I am no psychiatrist but as a complete novice I do not believe in schizophrenia. I do not believe the person is hearing voices from beyond because there can be no voice of beyond. It is his imagination and his characters playing mind games and if a psychiatrist or a psychologist come in and join the fun admitting to such a scenario. With such outside help and sympathy to help the dreaming along the characters then can and will come out to play. The "split personality" in the fight is within every body and can go into rage whenever allowed to do so by deliberate actions, provocation by other, tiredness, fear but also delusions, that is true.

From the article so far one may tend to get the impression it is about big issues like meeting you're in laws the first time but it is not. It is every day all the time things. Sitting in traffic waiting for the light to turn. The positive character slowly shifts to neutral and the neutral slowly joins force with the negative that was in the background all day. Without the person noticing the whole situation within him shifted in frame but he is unaware of it. Suddenly an explosion fitting a war burst to life. Another motorist sits daydreaming with his neutral taking him on long trips because his positive has nothing to do and his negative is in the background. The light turns green and the daydreamer is just that tad slow in responding because he was in thought. He was in thought as much as he was on Jupiter. His negative character helped the neutral create a situation the negative character can participate and not be bored. So his slow acting is as deliberate as his breathing is and the other person with the shifting emotion sees this, recognises the stunt that the daydreaming motorist is pulling and with the positive character standing on neutral ground and neutral fully in the negative territory, the negative character comes in with a punch like none. Suddenly two very timed persons go into a rage because of the smallest incident. They do not recognise their own behaving and if they do not get their characters on a leach quickly the characters of both men will take charge and blood might flow. This we call traffic rage. It is a very convenient name.

The man sees some one he may regard as rather attractive and smiles at her. The positive character becomes embroiled in the proceeding while the neutral character joins by taking control and urging the man to step just a tad closer, just to see… while the negative character is anticipating rejection in any event and deliberately steps on the girls' toes. Embarrassment is all around even with surrounding crowd because somewhere in the back they in the crowd all no what happened and that brings the embarrassment to their door. Every action on all accounts are very anticipated and pre-arranged. We use the name of an unfortunate incident to file this under. It's about calling your wife but using the neighbours' wife's name. It is knowing you have to cut the lawn but the drowsiness just will not let go so you sit down and close your eyes for a second, just to relax for a second. The one character shifts one position while you were not attending procedure and you miss the opportunity to mow the lawn for one more week. This goes by the name of slipping the tong and nodding off.

Very typical of this is the driving absentmindedness. How many times did we sit back after arriving at our destination and thought how on earth did I pass this or that town? While driving your car it becomes as routinely as breathing so there is little to be positive with and less to be negative about. That is the chance the neutral character has been waiting for all week and he takes charge by letting your mind wonder to many destinations but the one you are heading to. Little harm can come from this under the normal, but when danger suddenly strikes you are gone, your positive is gone, you negative with all the adrenalin is completely absent and your neutral being neutral does not bother in any case. By the time all the absentees arrive at the scene just that second later, a horrible accident is in progress. This we call driver fatigue. The process lingers on while we are absent minded, not totally in control of the moment and shit happens. But shit happens because we want it to happen for that bit of excitement we do not need, but the characters do because they make misery, To them it is a case to change the situation to something more exciting, more emotional, to press our social standing or just to ease boredom. This goes by the name of the mind is wondering.

Someone may say something out of the order and a rage follows. Why would a rage follow when you know very well that what the other person said was unfounded and if that person does believe what he said he is so misinformed you should not even listen to such nonsense? But confrontation is about to blow like a rocket on a launch pad. You will show him what he said was untrue and he will take it back or pay with his life. Once again we allow the characters to shift and completely obstruct our normality. The best is that with you allowing the shifting you will press old issues long ago forgotten but the characters suddenly helped you remembering this or that and this is the last straw! Under the normal the previous incident hold such minor importance you would never remember about it but at that moment it does not even strike you as odd remembering such trivially while not surprising at all it is clear in your memory that moment. You feel you can murder while never in your life did you ever have a thought about how it may feel taking another person's life. Everything you experience is out of the ordinary and out of order yet it seems to you at that moment as being as normal as discussing the papers. Does your characters enjoy the excitement and taking you along for the ride. They take control and you take the mess afterwards not knowing how you will ever show your face again after such an incident. This we call losing your temper.

You walk down the street with your best suit on feeling as chirpy as a robin in mating season. Then suddenly your foot misses the curb as you were walking and you land face down in a crowd of people you never saw previously or know any one around you. There is not the slightest chance of you meeting any one of the spectators ever again. Yet your world plummets down the deepest mine shaft. You cannot see the light of day ever rising again. You hang your head in shame while trying to conquer the incredible urge to run from the scene as fast as you can. Not for a moment do you stop to think that it is not that shameful and happens to everybody many time throughout they're lives. That the people you hold so important can never be important because you will never see any of them ever again. Again you characters turned a situation around as they turned on you. This we call embarrassment.

All of us meet the challenge in battle without ever realising. A young man newly wed has a friend of his wife staying with the couple for a month or two, just until she can find some other accommodation. This young man is going ballistic with the fight. He is in rage about this beauty sharing a roof with him and with him working shifts he finds himself alone in the company of this woman because she is still in the market for employment. She page thorough many papers per day to find a suitable position but that does not take all day and with his wife being at her work there are many hours pleasuring about unchecked. Now he goes in spin. His positive character tells him of what he has in his wife and that he should not endanger his fragile and young marriage, his neutral tells him his wife is at work and need not to know while his negative went positive by telling him how beautiful this young female is. This we use the name temptation to identify. It is the clearest way we know of identifying the threesome.

We were all at one or other stage young in our lives and young at heart and know the feeling when you are running the hundred meters or you are on the Rugby field and there is that special person looking from the spectators end. You feel her eyes burn in your back and the excitement blows your senses. It is like someone ells takes control and you become that much better to the degree you cannot believe yourself, or you have gone pinching fruit (a favourite pass time amongst the Afrikaner up and until I was a teenager. That was before money and greed came in and made kids criminals for helping themselves to fruit at night). The other side of the coin was that you knew who ever gets hold of you, will tear of the skin from your behind, but that was part of the fun being the challenge to the danger. Never was police involved one way or the other and even coming home safely did not mean security because if your parents catch you, you have to go back to whom fruit you pinched belongs and fetch your licking. I was a hundred meter athlete in my time and was quite quick, but when caught in the act I saw these real slow guise and can-not-run mates in crime pass me and if Carl Lewis was there they will pass him as well. They beat me by a hundred meters on the three hundred meters and cross one and a half meter fences as if it was hurdles on the track. The neutral character was on guard all-night and got the other two pumped but on stand by. The moment surprise takes over the positive gets negative but charging past the negative to get the body in motion. The negative sees that being negative is a splendid way of getting away from the danger and re-passes the positive while the neutral is pushing both to get out of the way so that the negative can take command of the muscles. With in less than a heartbeat there are four characters (including yourself) that take control of muscles and there can be no fitter and more potent athlete occupying the body than the fear factor presenting the next few minutes. This we call motivation.

Then there is the young person that is the one-week in the dump because he cannot see what the world is all about. His girlfriend dropped him, he is in the middle if an exam he cannot see how he will ever pass and to top it all is the fact that it has been raining for three days while he has this camping trip planned. The positive character finds a few days rest while laying low in anticipation of the coming holiday forcing the negative character to take charge of his outlook on life while his neutral character takes charge of the schoolwork making him take his vacation early while exams are pressing. The next week he finds himself being over positive bout life, about his happiness and about all the opportunities in life that are waiting on him. The exams are history, he is vacating and met the girl of his dreams and have a song in the heart. To tell the world, he is the man of the moment and to prove that he has long uncombed hair, walks with a swing and doing his thing just to annoy his square parents that are living they're life in history. That is the neutral character in charge and being all-positive for one week/ month and then negative for one week/month does not surprise any one. The name we attach to this is growing up. It is all about finding the characters, meeting the characters and blending with the characters in order to fine tune for preparedness for the fight ahead.

We name the events. We know the events. We suffer through such events but never stop to think why it is taking place. Why would your mind start to wonder? If you are in control and you are in that position you should not find yourself out of control or somewhere ells while you were being there. If you were your body why would the mind go absent? If you were your body and mind why would you stray or over react in anger, pity, shame or stupidity. If only being in the body you are in as the atheists believe, then you should be in the body, because where ells will you be but in the body you are. The fight is on and you have to recognise your opponent because your opponent is you in person. Your opponent knows your weaknesses as you know them because you are your weaknesses. Your life is your fight and it is on for the rest of you natural life.

It is as common in every one as miss-placing keys, spilling milk, forgetting some one's name and such minor incidents bringing great embarrassment at the time but is as normal as breathing. It is no big psychological problem forming in the dark side of the sole where the brave does not dare. Yet it can be there. When ever a person comes up with a brilliant remark astonishing the person that made the remark much more that the person to whom the remark was directed is an example of the trio. When trying to repeat the brilliance the very next time we become the stuttering idiot that wishes the floor would dissolve the human body. We find an inability to repeat such brilliance. We all have astonished ourselves from time to time as much as we embarrass our selves from time to time, so there is no exclusion and only inclusion.

It is the person being unwilling to do a task and finding himself repeating an error that he knows he should not repeat but something is driving the person to repast his actions. The more he repeats the error the more he gets annoyed with himself but getting annoyed with yourself must be the most unaccomplished task you can accomplish. His normal reaction would be then to become annoyed with any object or person he may find displeasing and although the object or person has little to nothing to do with his actions it proves the best way to blow off steam. This I refer to as the boss syndrome because that is one of the perks a boss seems to have. Being negative about the task makes your mind go wondering to more pleasant places to be. By your negativity you are suppressing the positive character and that puts the neutral character in charge. But the neutral character has as little interest in you're being there and takes you away on more pleasant day dreaming trips to where you would rather be and that leaves the negative character to be in charge. Who is doing the task with every one gone…it is the negative character, and an unpleasant one at that being all alone and hell is on its way!

The person starts making deliberate errors that can be avoided but with his lack of enthusiasm he becomes absent-minded and that is when the accidents stars cropping up. It could be small annoying things but also it can be very serious injury coming from such little absent mindedness. The negative character wishes to draw all three characters attention to the fact that the negative character is as displeased with the situation and wants to be relived of the duty. The LAPSE in CONCENTRATION is no lapse in concentration and the disaster following (big or small) is as deliberate as the person slipping away on his daydreaming trip. The whole affair is one big disaster waiting to happen and always does happen. All parties involved claims innocence and protest to any involvement accept the negative character that now is relieved of his duties. The net result is that the negative character is the only winner in the end.

Where does it come from? It comes from being human. What encourages the characters to become more dominant in some than in others are the better choice of question? There are at least a thousand possibilities I presume but it is more than just likely that it may present itself as a manifestation of a chemical imbalance in the brain. That does not explain the fact that it is present but may support the fact of more dominance in some individuals than in others. Being alive is about chemicals but the chemicals allow conducting electricity and are not life itself. The chemicals are a conductor but we all know the conductor is not electricity. But when the conductor goes hay-why the electricity goes hay-why. In that sense the chemicals will play a part but not play the part. When there are cross over of wire connections some life will flow in a direction where it actually is intended to be at other outlets. From what I have seen with some of my brain injured friends the control becomes absent but that does not mean the injury cause the characters because I have witnessed the characters in every person I have met. It is as much part of our personality s it is our personality.

The chemicals could be one aspect but underlying fears are the predominant issues that renders the characters the possibility of going out of control. An unhappy childhood brings an unhappy life. When the child is growing into a skew adult the adult will bring about another skew child. The main drive behind man is fear. Where fear is absent the man is a danger to him and to society. Fear brings about a conscience. The fear I refer to is not anxiety or being scared but having respect. Having respect for one's parents or teachers or the community and respect for the law. Above all the fear of God brings respect for God and that forms the basis for being man. Persons with personality problems and character flaws may have too much fear or none at all. It is a balance and it is the balance that keeps us upright. Underlying in the balance is respect for yourself and that you have with the respect you have for your parents. When being a child your parents are a substitute for God because only they can bring values and norms that will one day bring about a balanced member of society. Money cannot buy that.

I know that in itself having a conscious and adhering to the conscious does not say much because the alcoholic hides his booze from himself because his characters are in conflict. He knows he is on the road of self-destruction but normally that does not bother an alcoholic that much as they love drink more than life. It normally is the fact that the alcoholic knows about the destruction in his children and his love for his dear ones. He wishes to protect them from him but he has this massive problem that is stronger than his urge for life. To find a way to solve the problem he starts to hide the liquor from himself and in that way he stars to lie to himself. The characters are accusing him as they are destroying him and the escape is the destruction. When he is drunk he has no control and when he is sober he has no control. The characters are telling him he has no problem as much as they are accusing him about his problem and denial is also admitting where admitting then becomes blame and the blame he carries are more than he can carry therefore he drinks to escape the blame of his guilt.

It is not only liquor but also it is sex, drugs, pornography and gambling. Those are the weaknesses that puts people in the same situation as that of madness. It is living the life of lust knowing that that life of choice is what you choose and choosing such a life is the equivalent of destruction but modern society makes fighting thereof much harder than ever before and capitulating as easy as breathing. The guilt, fear, anger and despair comes in waves and in conflict where the alcoholic is in the middle with his problem out of control and out of his hands. His characters took charge and they destroy him as well as those surrounding him... all that loves and care for him but the biggest destruction is his destroying of the ones he love. Seeing the suffering in the lives of those poor, poor soles make me great fill for the small load I received and the ease with which I have to carry my small burden. In that way we're in a fight but the fight is all in private and we may acknowledge failure where there are great success as we may judge success where there is great failure. We on the outside judge what we see on the outside while the fight is on the inside where we can never see. A rehabilitated alcoholic must be a far bigger success than a successful achiever born with the golden spoon in the mouth but skins the cat in the dark by indulging in cravings of the night, all in the quiet his money can bay. It is about what you fight and not the way in which you fight or how you fight.

I am no philosopher of any sorts but thought about success and failure and my position in life where I as a person will never reach great achievements therefore the question I asked myself is will I die a failure because of that? My success is my children I leave behind and the success they may be. Not in great achievement because in the end even King Solomon declared it was all about chasing the wind. Leaving behind riches in money can bring as much despair as leaving behind poverty. The only way I may achieve success is leaving behind children that are of a better fabric than I was. The next generation must be better equipped, better evolved and better humans than was the previous generation to ensure evolution. What is taking place to my mind at present is devolution through out the western world. I do not have to bring proof about that because reading the paper or looking at nightlife will bring proof to any one wishing to see proof. How do we western man raise our future and how do we equip our future? Western society removed discipline from schools with admirable success. We removed the authority from the teachers as best we could. The teachers are through out the western world the lowest paid professionals in society. Being the lowest paid brings about the under achievers of society and any successful teacher leaves the profession for better pay in the private sector. This is the cancer of the new age we are facing. We all know my last remark is the truth and by that the parents wishes to compensate, but to modern man compensation is about money and paying to get rid of the guilt. Every nation culture community and individual admits that money is no measure for success and yet that is the norm we live by. Every person living a life is but a building block to establish the next generation. Man is an endangered species and animals are overpopulating the world and walking on two legs is not the criteria for human classification. The position you hold distancing yourself in thought and behaviour from the norms that animals uphold makes man or beast.

Animals are not in a struggle with their improving of the mental but in a struggle with the survival of the fittest, the one that can kill the best, run the fastest outsmart all others. That is not man. Man is about his fight to better his life and not his body or position. Man is about fighting the battle of the best in man. That is not modern man's ambitions. Modern man has ambitions making him the best animal on the planet because atheism propagates the fact that we are all members of the animal Kingdom. If man is that blind it may be best if man returns to the animal Kingdom where he thinks he belong. At least the loss will be small but smaller will be the gain of the animal world.

I wish to take you back to the indicial verse in that there were three things man would obtain when rating from the tree, and not only two…it is the three that every preacher misses …There was the tree…the tree of knowledge…knowledge of good, and knowledge of bad. You determine your knowledge…of good…of bad but above all it is your knowledge. The Bible does not name what is good or bad…that is your choice you make and that is your price you pay in the end…because the knowledge you accumulated will stand you either good or bad, as animal or man.

It is not what you may accomplish or accumulate that has importance but what you learn through being man and standing apart from beast, that is what you take with.

Accept the following or reject the following but consider the following.

Because there's mathematically no nothing there must be a God.

By excluding nothing you have to include God because if there was nothing there was some scope for God being nothing and therefore non existing but since nothing is the only excluded number in mathematics you have to include God as a factor. What ever you dare to prove or disprove you can only prove or disprove through the human thought and understanding of concepts only human concept can understand.

I do realise with my statement about nothing and God it is stretching matters beyond the argument. When taking the argument that far the argument can include the existing of fairies and other fantasies, that is quite true, but by my placing a fairy and other fables in the realms of insignificance brings only harm to what remained of the child in me and I can assure you there is not much left in that sense. What ever is there is also in the infinity of my childhood memory. Not believing much in the fabric of the fantasy does not affect my position relating to the animal as the animal has little regard in such matters. Fantasy may have significant when I connect ethics to it as the Roman Catholics do with saints, demons, angels and such. When one can disregard such fantasies without affecting norms and values of the civilised, not harming the moral fibre of society there is little harm done by such removal. But when removing such norms for the enjoyment of being superior while the correctness of the argument in its core is invalid and, the structure of society goes to hell, a lot of question marks appear about the morals behind the motivation as to the motive behind the act. Man is moreover morals than life.

All life will destroy other life to its own benefit. A virus will kill his host for the benefit of one life cycle and at that a virus life cycle. What waste we may think. Killing a host only to spurn seems a lot of waste. Not so, because ticks will devour a cow alive without feeling any remorse. The same apply to a lion killing its pray. If there were no chance of the antelope finding means to escape the lion would start eating without killing saving itself the effort. It does not kill quickly through pity like humans do. It kills because of self-interest alone. When one of the herd falls victim to a lion attack the rest will start grazing thanking in that manner the victim for securing their position for one more day. There is no compassion, just self-centred egoistic drive to self-protection. When a person jumps into flames in an effort to save another person he is a hero. He receives praise from all concerned and may even land a medal for his effort. Why would we humans consider that as brave, being exceptional and above average?

From our cosmic position we are in the centre of the universe. Where you may sit or stand is the very centre of the universe (your universe) because you can only relate to the universe having your individual singularity and where all other aspects in the cosmos are pointing away from your singularity to all other positions. We shall always be 56 in a universe of 112. The earth will always be Π^2 in relation to the sun with the speed of light being $3\Pi^2$ from our position. The universe will grow in all directions pointing away from our direction, just because we are the centre of our universe. It is not surprising we see ourselves on the edge of the Milky Way. From where we are we will be on the outer edge of our galactica. To the inside everything will be brighter and to the outside everything will be darker. That is a fact, but not a reality. The fact we may appreciate but the reality we can never understand. We are egocentric maniacs coming from the position of our singularity. That is the animal, the trio breathing our air. That is what the fight is about and what one may take with after death changes dimensions occupied. That is wisdom standing apart from knowledge and the animal has knowledge in intelligence but man has wisdom in extelegence. That parts man from beast. Man has to part from singularity's approach.

Us humans must fight to find our place outside singularity, outside self-interest and strive to better of our position not including finance. We have to relate from a position in divinity and not singularity to understand the cosmos and to understand life. If not we may well die as animals, and such possibility is there. If one person spent a lifetime killing his human in him, he may end as the animal he always strived to be. Being artist's engineers or preachers have nothing to do with it. Knowing the Bible or not has nothing to do with it. Being a believer or not has nothing to do with it. It is your approach to the cosmos that makes you part of the cosmos or that dimension above the cosmos. Accepting that one is not part of the cosmos but part of life in a cosmos puts one above the cosmos, but still in the cosmos. Never lose reality but accept responsibility. Persons believing in fairies and fantasies have the problem of over acclimating the positive while the neutral protects sanity with much of it going in the direction of a lost cause leaving the negative happy for destruction can only follow such obscurity and the arts and artistic are typical in this. This is the line the drug addict follows to the last letter. By pumping the acid the positive hallucinates, the neutral exclude the incorrectness and the negative character are in seventh heaven as destruction is deliberate and decisive. The very same apply to a soldier in war where the positive character enjoys the killing as much as the negative character enjoys the being killed and the neutral character wishes the cruelty on them before they are upon us. When the soldier arrives back in society, society understands him as little as he understands himself because the fabric of human principles has gone skew. His characters, including him is as mixed as a milkshake fruit salad. You may well ask why is the controlling of the characters of any importance, as they are obviously part of us? There is the middle, a precise middle where I as a person hold my personal relevancy as I see myself. What ever I attach or detach puts me in relevancy to others in life. That is the purpose of my fight with my characters. My precise middle must be very straight and when the middle leans toward any side in particular my middle is out of alignment. To secure a centre there must always be room for others and their opinion. When I say there is no God I go eccentric and when I say there is only God I go eccentric.

As it is in mathematics in the matter of the line and the dot, it is a question of you deciding the relevancy. It is your choice and only your choice to what relevancy you wish to place God as a presence or a factor. By declaring the absence of God your relevancy might reduce God to infinity but in your denial you place relevancy therefore relevancy remain be it in the infinity. It is to the peril of the denier that such a person excludes God as a factor because by placing God at a point of infinity next to zero the denier places himself next to the animal that excludes God because God excluded the animal from extelegence. Such a thought proves the fool. God allowed the animal excluding the admitting of a God because God exclude the animal from the dimension of being human therefore liability to questionability. In the perception lies the norm.

On the other side of the spectrum is the religiosity maniac and such is his idiocy he gives God the role of the animal being on call by prayer to serve the master's call. In all such cases the maniac places himself in eternity by creating a spot in the centre of eternity for himself providing him endless power having God as his slave or animal on a leach in acclimation to his eternal status he then places God just outside eternal big for that spot he reserved for himself. Through his effort by prayer he can heal, bless and doom…and God will obey as instructed. Such a fool proves the thought.

When a person buys a car worth millions (in South African Mickey Mouse Money), because his business is going grate and he can afford to, but has his sister and his brother in law is working for him at minimum wage, not making ends meet in providing for their school going children, battling every day to put food on the table while he is making money like water flowing every one admires him for he is rich. He is treated very softly and no one dares to stand up to such a person because he has money. Such a person as a human is wasting breath because he may walk on two, but he may as well walk on four for the humanity he portraits. Still society regards that animal as a pillar of the community because he has money making a "sharp and intellectual business man". He should be shot at dawn for impersonating man whilst being beast. His positive has only one aim and that is to better his financial position securing a better admiration in society and establishing a front with him being god to the rest he sees as lesser mortals. His neutral sees his money drive as security and that satisfy while his negative is bent on destroying others because his positive is telling him how great full his sister and her "useless" husband should be for his generosity of providing food on their plates. Not once for the shortest and briefest instant will he ever give a thought that it is precisely the other way around. By their self-denying they are enriching him, but through his ego madness that thought never comes to mind. Then the philosophers will say that is man.
But such a remark is bullshit, only protecting the beast in himself or herself as they protect him, our unkind and selfish millionaire.

While a human (as is the case with all other living things), holds life being intertwined with singularity there is a distinction we all know of between animal life and human life. Animals are lesser in life than are humans because they cannot sense a cosmic position other than eat and mate, and in that sense we are born with a natural same view animals hold as just merely being senseless objects in space-time, without having the ability to distinguish between what is permanent and what is of a passing nature and our position as cosmic objects occupying space-time. We maintain our position as if we do not realise we truly are not positioned in the centre of the universe as we seem to be, because animals have not the sense to distinguish such realisation. Everything else in the cosmos is not there for our satisfaction including other humans. Having the ability to distinguish makes us intelligent, and moreover, extellegent as humans might be. As long as the atheist, the criminal, the ego-centric maniac and all self-appointed gods connected to singularity and space-time consider them to be the star with all other objects in the universe as heat in space-time, there for one purpose and that is to be used, he will be an animal. Our purpose is to move away from the animal, because through a lack of better knowledge, that is the attitude of the animal. The animal holds the outlook that all objects are there for its pleasure or displeasure pivoting around the animal's view of eats or be eaten. The universe focus around the animal as he sees the universe, and has no ability to see other forms of life with equal importance.

As human life we have to see that there are billions of singularity that included with ours, always encircle a more important singularity as lesser singularity revolves around us, and in the epitome, there is the ultimate singularity connecting all. We are all planets, only planets, with some being bigger and some smaller, but we maintain a queue from the planets and objects in our solar system, not counting the number or size in the personal solar system. As successful humans we must see our position in relevance to the universe we occupy and our goal to serve by improving such position with evolutionary time laps. When we get the urge to grab what we do not own, we must realise the animal in us urging us to return to animal form. The fortune of man is his ability to recognise the stars above, but much more: to recognise time in space and his role in that. Animals have not received the ability to recognise stars, and therefore has no ability recognising the possibility of singularity and the ultimate Creator of singularity being God as such. Without that they become space-time for man to use and dismiss. In that man and mans superiority becomes the complex issue we do not understand. Above al, when we recognise singularity, our aim should be distinguishing life from singularity and parting their different purpose to the dimensions controlling creation.

Even when a planet finds itself to be just another planet without finding other planets to guide, and holds no relation to other planets as is the case with Mercury, its orbital pattern is very obscure. Since we humans are part of the cosmos with life intertwined with singularity (I delve into this aspect in another book) we must find similarities in life and the cosmos, because the cosmos takes after life as life would take after the Creator. When the Cosmos is gone to singularity, returned to the dot once again, life as energy would still be, because the universe is heat, and heat goes in three forms, but life is another energy, apart from heat in the cosmos very indestructible. By recognising the lay out of the cosmos, we can determine with an intellectual eye, what our purpose should be and we should place our objective to the future accordingly. What is the use of money-mania and the accumulating thereof, if one would die in any case and lose control over it afterwards. I could never understand that part of the gaining game.

Through your relevancy with God you become the God that makes you your own God as much as you're own devil and your own forgiver as much as your own accuser. You make the relevancy and the relevancy proves you. Being positive is loathing and being negative is damning and being neutral is obstinate and in everything you confirm the applying of norms be they right or wrong, good or bad as it is the free choice you make proving your distance you have as a factor claiming life and location as a being from the animal. In the case of the animal the ride is free for the animal knows no better but man does and is liable through conscience placing relevancy to deeds. On the physical aspect in the being named man, the human body is an accumulation of singularity positioning divinity as one comprising of three identities without ever separating the trinity. The one holds the straight line the other forms the triangle and the third is the dome, the inclusive sphere, and the container having seven sides to the entire outside world. You cannot be in two sides of the universe simultaneously but you may observe both sides by being centre.

CLAIMING A POSITION IN SINGULARITY'S INFINITY IS LIFE'S DIVINITY. YOU LEARN TO LIVE AS MUCH AS YOU LIVE TO LEARN

Singularity has three in parts as does divinity have three in places. On the other side of the divide is the divine and as things are in the one side of the divide in total relation to the other side because all relevancies align as much as match in equilibrium and in as much as it is the cosmos forming one half of the divide it has to be the divine duplicating to establish equilibrium.

The purpose of man in life is to know your other entities, to learn to live in recognising the other entities, to take from them strength they can give but also to detach from their weaknesses, and above everything else recognise yourself in what you do as pure or spoilt. Gain knowledge the knowledge you have in the good you can acquire and the knowledge in the bad you have to detest. Then only you may serve a life of gain. Did I prove anything, well you are my judge! As I am the one promoting all things connect, so I am the man believing all things connect.

A Road Leading to A ROAD LEADING TO THE CENTRE OF THE UIVERSE.

Without a zero there cannot be death. Without a nothing there cannot be no God. Accept one and you cannot except one.

If it is that simple why is it complicated.

BEST WISHES,

PETRUS. (PEET) S. J. SCHUTTE

COMPRISES OF

#	Title	Pages
1	AN OPEN LETTER "TO SELECTED ACADEMICS" ISBN 0-9584410-9-X	1 - 217
2	AN OPEN LETTER ABOUT "XEPTED MISTAKES" ISBN 0-9584410-1-4	218 - 611
3	AN OPEN LETTER ABOUT "INTER GALACTICA SPACE TRAVEL" ISBN 0-9584410-2-2	612-1059
4	AN OPEN LETTER ON "CORRECTING COSMOLOGY" ISBN 0-9584410-5-7	1060 - 1592
5.	MATTER'S TIME IN SPACE: THE HYOPOTHESIS ABOUT STARS. ISBN 0-9584410-6-5	1593- 2061
6	AN OPEN LETTER ON "STARSSTUFFN'" ISBN 0-9584410-3-0	2062 - 2612
7	"SEVEN DAYS OF CREATION". ISBN 0-9584410-4-9	2613 - 3326

FOR other related information, PLEASE VISIT THE WEB SITE, FOR YOUR CONVENIENCE

www.spacetime.co.za

The INDEX of MATTER'S TIME IN SPACE: THE THESIS ISBN 0-9584410-8-1

SELECTED ACADEMICS" A summery specially for academics convenience 1–217=	217
AN OPEN LETTER ON	
XEPTED MISTAKES An abbreviated Version of book four	393
AN OPEN LETTER ON	
INTER GALACTICA SPACE TRAVEL (Explaining the sound barrier in relation to singularity)	447
AN OPEN LETTER ON	
CORRECTING COSMOLOGY Forming the basic allowing a better understanding the basics	532
MATTER'S TIME IN SPACE:	
THE HYOPOTHESIS (concerning the way the universe started and how galactica develops)	468
AN OPEN LETTER ON	
"STARSSTUFFN' (concerning the way the universe started in as much as stars developing)	550
"SEVEN DAYS OF CREATION" ISBN 0-9584410-4-9	
Is about the forming of the planets and gas structures in the solar system)	713
MATTER'S TIME IN SPACE: The Thesis (As a combination of the above)	3320

You may also contact me by land mail at:

Po Box 1093,
Ellisras
0555
R. South Africa.